Environmental Toxicology

WITPRESS

WIT Press publishes leading books in Science and Technology.
Visit our website for the current list of titles.
www.witpress.com

WITeLibrary

Home of the Transactions of the Wessex Institute.
Papers presented at Environmental Toxicology 2006 are archived in the WIT eLibrary in
volume 10 of WIT Transactions on Biomedicine and Health (ISSN 1743-3525).
The WIT eLibrary provides the international scientific community with immediate and
permanent access to individual papers presented at WIT conferences.
Visit the WIT eLibrary at www.witpress.com.

FIRST INTERNATIONAL CONFERENCE ON
ENVIRONMENTAL TOXICOLOGY

Environmental Toxicology

CONFERENCE CHAIRMEN

A. Kungolos
University of Thessaly, Greece

C. A. Brebbia
Wessex Institute of Technology, UK

C. P. Samaras
TEI of West Macedonia, Greece

V. Popov
Wessex Institute of Technology, UK

INTERNATIONAL SCIENTIFIC ADVISORY COMMITTEE

D.S. Achilias	E. Gidarakos	M. Latif
M. Altstein	S. Grigson	N. Marmiroli
C. Blaise	P-D. Hansen	Z. Romanowska-Duda
T-S. Chon	J. Iliopoulou	A. Soupilas
D. Fayzieva	D. Kaliampakos	H. Takanashi
	A. Karagiannidis	
	D. Koronakis	

Organised by
*Wessex Institute of Technology, UK and
The Department of Planning and Regional Development,
University of Thessaly, Greece*

Sponsored by
WIT Transactions on Biomedicine and Health

Environmental Toxicology

Editors

A. Kungolos
University of Thessaly, Greece

C. A. Brebbia
Wessex Institute of Technology, UK

C. P. Samaras
TEI of West Macedonia, Greece

V. Popov
Wessex Institute of Technology, UK

WITPRESS Southampton, Boston

A. Kungolos
University of Thessaly, Greece

C. P. Samaras
TEI of West Macedonia, Greece

C. A. Brebbia
Wessex Institute of Technology, UK

V. Popov
Wessex Institute of Technology, UK

Published by

WIT Press
Ashurst Lodge, Ashurst, Southampton, SO40 7AA, UK
Tel: 44 (0) 238 029 3223; Fax: 44 (0) 238 029 2853
E-Mail: witpress@witpress.com
http://www.witpress.com

For USA, Canada and Mexico

Computational Mechanics Inc
25 Bridge Street, Billerica, MA 01821, USA
Tel: 978 667 5841; Fax: 978 667 7582
E-Mail: infousa@witpress.com
http://www.witpress.com

British Library Cataloguing-in-Publication Data

A Catalogue record for this book is available
from the British Library

ISBN: 1-84564-045-4
ISSN: 1747-4485 (print)
ISSN: 1743-3525 (online)

*The texts of the papers in this volume were set
individually by the authors or under their supervision.
Only minor corrections to the text may have been carried
out by the publisher.*

Preface

This volume contains the contributions presented at the 1st International Conference on Environmental Toxicology, which was held on the island of Mykonos, Greece, from September 11th – 13th 2006. The conference was organized by the Wessex Institute of Technology in collaboration with the Department of Planning and Regional Development at the University of Thessaly, Greece. It was sponsored by WIT Transactions on Biomedicine and Health.

Environmental Toxicology is one of the most interdisciplinary sciences. Biologists, microbiologists, chemists, engineers, environmentalists, ecologists and other scientists work hand in hand in this new discipline. Assessment of the environmental effects of chemicals is complicated as it depends on the organisms tested and involves not only the toxicity of individual chemicals, but also their interactive effects (including synergistic ones), and genotoxicity, mutagenicity and immunotoxicity testing. Hazardous waste management is closely related to environmental toxicology and there is a growing need for techniques and practices to minimize the environmental effects of chemicals, and for the implementation of the corresponding principles in policy planning and decision-making.

The editors would like to thank all the authors for their papers and, in particular, the members of the International Scientific Advisory Committee for their help during the review process.

The Editors,
Mykonos, 2006

Contents

Section 8: Hazardous waste environmental effects. Monitoring and remediation

Section 9: Biodegradation and bioremediation

Section 10: Biotests

Section 1
Risk assessment

How to predict the potential effect of chemicals on human health: an extended question of environmental toxicology

A. Mueller[1], C. Graebsch[1], G. Wichmann[2], M. Bauer[1] & O. Herbarth[1,3]

[1]Department of Human Exposure Research & Epidemiology, UFZ Leipzig-Halle, Germany
[2]Department of Environmental Immunology, UFZ Leipzig-Halle, Germany
[3]Environmental Medicine & Environmental Hygiene, Faculty of Medicine, University of Leipzig, Germany

Abstract

Often the question is how you can decide whether a chemical affects human health. The problem of human cells is that these cells loose some properties if they have been isolated from the tissue. Considering this the single cell organism *Tetrahymena pyriformis* was chosen. It is comparable in sensitiveness and responsiveness to human tissue cells, is an organism and independent from tissue and is a model for human respiratory epithelium-cell functionalities. The effects of extracts bioaerosols (mould spores) and associated mycotoxins (gliotoxin and penicilic acid) are of special interest. The following end points have been investigated: viability (cell count), energy levels (adenosine-5'-triphosphate content), and cell respiration (oxygen consumption). Effects on cell proliferation and on physiological processes are suitable indicators for the different impact of chemicals. *T. pyriformis* is a suitable organism to study effects of ubiquitous chemicals.
Keywords: toxicology, model system, Tetrahymena pyriformis, mycotoxin

1 Introduction

A lot of chemicals have been detected in the environment. The problem is to decide whether a chemical affects human health and human well being or not. To do a human oriented risk assessment a lot of test systems are applied for *in vitro* (and/or *ex vivo*) experiments (starting with single cell systems, animal cells up to human cells) and for *in vivo* investigations (using animals or epidemiologic studies). Although human cells may reflect the human properties the main problem of human cells is that these cells lost some properties if they have been isolated from the tissue. Otherwise the transferability of the results coming from animal cells or animals is limited. Preconditions for the selection of test systems are that the used test system should be simple to handle and should be deliver a result within a short time span.

Considering the pro and cons the single cell organism *Tetrahymena pyriformis (T. pyriformis)* was chosen. The advantages are that it is comparable in sensitiveness and responsiveness to human tissue cells, it is an organism and independent from tissue and it is a model for human respiratory epithelium-cell functionalities. The use of undifferentiated cells and moreover of single-cell organisms in toxicological investigations carries advantages in regard of making general statements about the reactivity and the influence of mycotoxins on cell metabolism. Furthermore, this avoids the difficulties of systems restricted to the analysis of only one cell type or tissue-specific effects. However, such cell models for assessing environmental risk and impact have to meet some important requirements: they should be eukaryotic, their biology and general responses should be well known, the laboratory handling should be relatively easy, and a short generation time is desirable whenever studies of long-term effects are necessary (Nilsson [15], Nicolau *et al.*, [14]). *Tetrahymena pyriformis* fulfills these requirements and was already used in several studies (Sauvant *et al.*, [19], Nicolau *et al.*, [13], Massolo *et al.*, [10]). Another advantage of this cell model is the organization of the cilia that is comparable to those of the human respiratory epithelium.

Two pathways are of interest for human health - the inhalative and the ingestive one. The investigated substances have been selected be relevant for both pathways and to give some examples of prediction of health effects using *T. pyriformis*.

Mycotoxins are of general interest. They are associated with bioaerosols (mould) and food. Regarding to the inhalative pathway exposure to indoor mould may be a risk factor for health effects (Rylander [17]; Husman [9]). Indoor exposure to mould is predominantly associated with *Aspergillus* and *Penicillium* species (Horak [8]; Flannigan *et al.*, [2]; Gravesen *et al.*, [5]; Müller *et al.*, [12]). Among other metabolites mycotoxins may be the responsible agents for the adverse health effects, like recurrent infections of the respiratory tract, asthma and fatigue (Flannigan *et al.*, [2]; Husman, [9]; Garrett *et al.*, [3]; Herbarth *et al.*, [7]; Meklin *et al.*, [11]; Müller *et al.*, [12]; Rylander [18]).

To receive an overview of the mode of action of mycotoxins and their influences on the metabolism, the vitality (cell proliferation) and parameters of energy providing processes were investigated which reflect the physiological responses of *T. pyriformis*.

The main question is whether it is possible to predict concentrations and kind of mycotoxins which are relevant for human health.

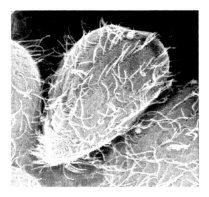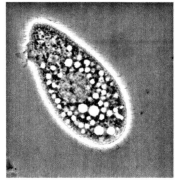

Figure 1:   *T. pyriformis* (left: laser scanning microscopy, right: electron microscopy).

2   Material and methods

The cell culture of *Tetrahymena pyriformis*, the exposure conditions (mycotoxin exposure), the cytotoxicity assay, the ATP extraction and ATP measurements, the measurements of oxygen consumption are described in detail in Gräbsch et al. 2006.

T. pyriformis was exposed against the mycotoxins gliotoxin and penicillin acid. Methanol and methanol-DMSO are used for solution and serial dilution. These substances and mixtures did not affect the development of the *T. pyriformis* populations within the range of used concentration.

The Kolmogoroff-Smirnow-test was used for testing the normal distribution of the data. The Welch t-test was used for statistical analysis using STATISTICA 6.1 (StatSoft Inc.). A *p* value below 0.05% was regarded as statistically significant.

3   Results

The cell concentrations were strongly affected after 24 hours in the same way as after 72 hours by the gliotoxin treatment (fig. 2). In contrast, the cultures which were treated with penicillic acid showed a decrease of cell concentration only after 24 hours (fig. 2). After 72 hours only those cultures treated with the highest concentration of penicillic acid show a statistically significant inhibition of cell concentration.

The EC_{50} values for the cell proliferation after 24 hours incubation was estimated to be a concentration of 0.38 µM gliotoxin (CI 95% 0.31-0.47) and 343.19 µM (CI 95% 306.18–383.92) penicillic acid.

Compared with human cells (PBMC – peripheral blood mononuclear cells) (Wichmann *et al.*, [21]) *T. pyriformis* is in the same order but in detail more sensitive for the indication of effects.

Table 1: Concentration with 50% decrease in proliferation (EC_{50}) after 24 hours.

	PBMC	*T. pyriformis*
gliotoxin [µM]	0.77	0.38

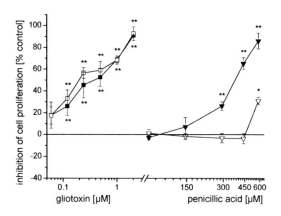

Figure 2: Effects of gliotoxin (■ 24 h of exposure; □ 72 h of exposure) and penicillic acid (▼ 24 h of exposure; ▽ 72 h of exposure) on: cell concentration of *T. pyriformis* (*p < 0.05, statistically significant; **p < 0.001, statistically significant)

Further investigations of the influence of mycotoxins on physiological processes by the measurement of the intracellular ATP content and the oxygen consumption were done.

The cell physiological parameters were influenced by gliotoxin only weakly in contrast to intensity of the effects on cell concentration (fig. 3).

The physiological parameters of *T. pyriformis* were influenced by penicillic acid in a totally different way. The oxygen consumption as well as the ATP content was stimulated by penicillic acid for $c \geq 294$ µM (fig. 3).

4 Discussion

The *in vivo / in vitro* results show that mycotoxins indicate their toxic effects not only in a different concentration range but also effect the exposed *T. pyriformis* cell's physiology in a different way. Until now, in some published studies various mycotoxins were investigated using the cell model *Tetrahymena* (Hayes *et al.*, [6]; Nishie *et al.*, [16]; Sparagano [20]; Dayeh *et al.*, [1]). For the determination of end points in toxicological studies mainly the proliferation of the cells or the assessment of vitality was used. However, this paper represents a study of the association between the influence of the mycotoxins gliotoxin and penicillic acid on proliferation and the impact of cell physiological processes more comprehensive.

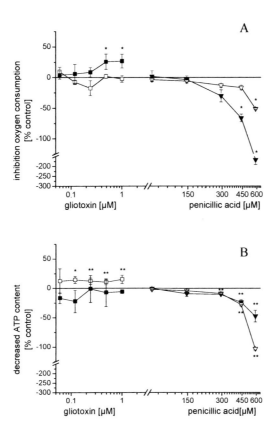

Figure 3:   Effects of gliotoxin (■ 24 h of exposure; □ 72 h of exposure) and penicillic acid (▼24 h of exposure; ▽ 72 h of exposure) on: (A) oxygen consumption and (B) ATP content of *T. pyriformis* (*p < 0.05, statistically significant; **p < 0.001, statistically significant)

   Contrary to the influence of gliotoxin on vitality and proliferation of *T. pyriformis*, the assessed concentrations of penicillic acid only incipiently showed cytotoxicity.

   In conclusion, the different cell physiological parameters in addition to the cell proliferation were suitable methods for detecting different mechanisms of mode of actions of chemicals like the tested mycotoxins gliotoxin and penicillic acid. Gliotoxin cause rather permanent effects on organisms due to its stronger and sustainable toxic effects, whereas penicillic acid might be metabolized and thus short time toxic effects are easily to compensate. Especially the results of penicillic acid demonstrate the requirement of experiments with long time exposure. For future studies it will be interesting to analyze the cellular behaviour in experiments within a long time permanent mycotoxin exposure in environmental relevant low dose ranges.

Summarizing it can be concluded that
(1) the effects observed using *T. pyriformis* can be used for the explanation of effects in humans observed within epidemiological studies
(2) *T. pyriformis* is a sensitive organism to indicate chemical stress
(3) *T. pyriformis* allows also to investigate chronic exposure which is more difficult to apply on human cells or cell lines
(4) the investigation of different vitality and physiology parameters in *T. pyriformis* indicate the different mode of action of the toxicant

Acknowledgements

The authors thank Karin Lange, Nicole Göbel and Jacqueline Kobelt for their technical support.

Parts of this paper has been published in Env Toxicology 21/2 (2006) 111-117. The authors thank the publisher John Wiley & Sons, Inc. for the possibility to republish parts of the published article.

References

[1]   Dayeh VR, Grominsky S, DeWitte-Ori SJ, Sotornik D, Yeung CR, Lee LEJ, Lynn DH, Bols NC. 2005. Comparing a ciliate and a fish cell line for their sensitivity to several classes of toxicants by the novel application of multiwell filter plates to *Tetrahymena*. Res Microbiol 156:93-103.
[2]   Flannigan B, McCabe EM, McGarry F. 1991. Allergenic and toxigenic micro-organisms in houses. Soc Appl Bacteriol Symp Ser 20: 61S-73S.
[3]   Garett MH, Rayment PR, Hooper MA, Abramson MJ, Hooper BM. 1998. Indoor airborne fungal spores, house dampness and associations with environmental factors and respiratory health in children. Clin Exp Allergy 28:459-467.
[4]   Gräbsch C, Wichmann G, Loffhagen N, Herbarth O, Müller A. 2006. Cytotoxicity Assessment of Gliotoxin and Penicillic Acid in Tetrahymena pyroformis. Environ Toxicol 21: 111-117
[5]   Gravesen S, Nielsen PA, Iversen R, Nielson KF. 1999. Microfungal contamination of damp buildings – examples of risk constructions and risk materials. Environ Health Persp 107:505-508.
[6]   Hayes AW. 1977. Effect of the mycotoxins, penicillic acid, on *Tetrahymena pyriformis*. Toxicon 15:497-504.
[7]   Herbarth O, Diez U, Lehmann I, Müller A, Rolle-Kampczyk U, Wichmann G. 2002. Umweltmedizinische Untersuchungen beim Vorliegen einer Schimmelpilzbelastung und Erfassung immunmodulatorischer Effekte. In Keller R, Senkpiel K, Samson RA, Hoekstra ES, editors. Umgebungsanalyse bei gesundheitlichen Beschwerden durch mikrobielle Belastung im Innenraum. Lübeck: p 35-90.
[8]   Horak, B. 1987. Preliminary study on the concentration and species composition of bacteria, fungi and mites in samples of house dust from Silesia (Poland). Allergol Immunopath (Madr) 15:161-166.

[9] Husman, T. 1996. Health effects of indoor-air microorganisms. Scand J Work Environ Health 22:5-13.

[10] Massolo L, Müller A, Tueros M, Rehwagen M, Franck U, Ronco A, Herbarth O. 2002. Assessment of mutagenic and toxicity in different size fraction of air particulates from La Plata (Argentine) and Leipzig (Germany). Environ Toxicol 17:219-231.

[11] Meklin T, Husman T, Vepsäläinen A, Vahtereisto M, Koivisto J, Halla-Aho J, Hyvärinen A, Moschandreas D, Nevalainen A. 2002. Indoor air microbes and respiration symptoms of children in moisture damaged and reference schools. Indoor Air 12:175-183.

[12] Müller A, Lehmann I, Seiffert A, Diez U, Wetzig H, Borte M, Herbarth O. 2002. Increased incidence of allergic sensitisation and respiratory diseases due to mould exposure: Results of the Leipzig Allergy Risk children Study (LARS). Int J Hyg Envir Heal 204:363-365.

[13] Nicolau A, Dias N, Mota M, Lima N. 2001. Trends in the use of protozoa in the assessment of waste water treatment. Res Microbiol 152:621-630.

[14] Nicolau A, Mota M, Lima N.2004. Effects of different toxic compounds on ATP content and acid phosphatase activity in axenic cultures of Tetrahymena pyriformis. EES 57 (2):129-135

[15] Nilsson JR. 1989. *Tetrahymena* in Cytotoxicology: with special reference to effects of heavy metals and selected drugs. Europ J Protistol 25:2-25.

[16] Nishie K, Cutler HG, Cole RJ.1989. Toxicity of trichothecenes, moniliformin, zearalenone/ol, griseofulvin, patulin, PR-toxin and rubratoxin B on protozoan Tetrahymena pyriformis. Res Commun Mol Path 65:197-210.

[17] Rylander R. 1995. Respiration disease caused by bioaerosols – exposure and diagnosis. In Johanning E, Yang CS, editors. Fungi and bacteria in indoor air environments – Proceedings of the international conference Saratoga Springs, New York p 45-55.

[18] Rylander R.2003. Environmental risk factors for respiratory infections. Arch Environ Health 55:300-303.

[19] Sauvant MP, Pepin D, Piccini E.1999. *Tetrahymena pyriformis* – A tool for toxicological studies – A review. Chemosphere 38:1631-1669.

[20] Sparagano OA. 1995. Griseofulvin: Generation time and ATP changes in the ciliate *Tetrahymena pyriformis*. Life Sci 57:897-901.

[21] Wichmann G, Herbarth O, Lehmann I. 2002. The mycotoxins citrinin, gliotoxin and patulin affect Th1 rather than Th2 cytokine production in vitro. Environ Toxicol 17:211-218.

Estimation of vapour pressure, solubility in water, Henry's law function, and Log Kow as a function of temperature for prediction of the environmental fate of chemicals

J. Paasivirta
Department of Chemistry, University of Jyväskylä, Finland

Abstract

Environmental risk R estimation of a chemical is based on predicting exposure of biota (PEC) in media and harmful effect potency as PNEC from (eco) toxicology. For default emission Eo, numerical relative risk is expressed as Ro = PEC/PNEC. Using the risk limit Ro ≥1, the risk emission is RE = Eo/Ro. PEC can be estimated by the fate modeling. Our FATEMOD model computes the fate of chemicals in an environment defined as a catchment area. In this model, the substance properties (liquid state) of vapour pressure Pl, water solubility S, Henry's law coefficient H (Pl/S), lipophility Kow, and the reaction half-life times are automatically computed as a function of temperature. This feature is unique, thus far, in fate models. The physical properties in the environmental temperature range are obtained from the equation: Log Prop(i) = Ai – Bi /T (T in degrees Kelvin). We have learnt to determine the coefficients Apl,Bpl (for Pl), As,Bs (for S), Ah,Bh (for H) and Aow,Bow (for Kow) from thermodynamic equations of these properties as functions of molecular parameters and temperature T by dividing the equation into two parts: 1) T absent and 2) T present. The results compare well with expensive direct measurement results and also with results from retention time comparisons by temperature-controlled GC (for Pl) and HPLC (for S). In reasonable risk predictions for variable climates such as in Nordic countries, the ambient temperature cannot be ignored. Examples of temperature-adjusting FATEMOD runs in chemical risk estimations are given.
Keywords: degradation rates, physical properties, regional model, relative risk, temperature corrections, thermodynamics.

WIT Transactions on Biomedicine and Health, Vol 10, © 2006 WIT Press
www.witpress.com, ISSN 1743-3525 (on-line)
doi:10.2495/ETOX060021

1 Introduction

Environmental fate estimation of anthropogenic chemicals is increasingly important for management of their risk of causing harmful effects to man and ecosystems. The risk can be presented as a product of exposure and the effect of potency of the chemical on the target species. Harmful effects of potency are evaluated by methods of (eco)toxicology e.g. as numerical value PNEC (predicted no-effect concentration, inverse to the potency). Exposure, as value PEC (predicted environmental concentration) for the chemical in each media (air, water, soil /plants, sediment, suspended solids, aerosols, depositions, food). Recently, mathematical modelling to predict environmental fate of chemical has been developed as an effective tool to assess exposure of biota in media. Modelling is a low-cost method to expand and complete the information from expensive analyses to make practical environmental risk management for large number of possibly hazardous substances feasible. The risk management scheme for an anthropogenic chemical emitted to environment is illustrated in Figure 1.

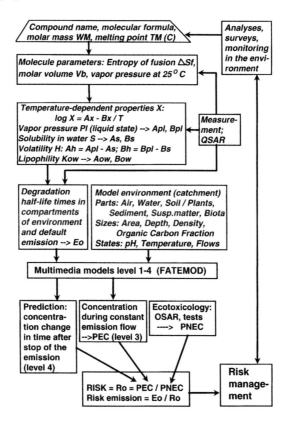

Figure 1:   Scheme of the use of modelling in the environmental risk estimation and management of chemicals

For preliminary estimation of environmental risk, the exposure assessment part (Figure 1) produces PEC in air, water, soil/vegetation and sediment. While obtained by multimedia modelling on the environmental fate of the chemical [1,2], PEC is practical for evaluating instant (acute harmful effect) risk to biota. Risk estimation of long-term (chronic) effects needs additional analyses, modelling and QSAR estimations of bioaccumulation processes [3-5]. The risk assessment result can be computed by fate model using default emission as input parameter Eo. Then risk ratio is Ro = PEC/PNEC and defined risk emission RE = Eo/Ro. The lowest value for significant risk could be RE = 1, although a guideline for tolerable emission can be some lower RE value - for safety.

The modelling of PEC requires input data of the certain properties of the model environment and chemical. They can be written as a database to the model program for their fluent use. Such predictive models for chemical exposure prediction were evaluated and completed in a series of international European projects, our research team as partners, during 1989-2002 [6-8]. We learnt that ambient temperature was an essential factor in all model predictions of the fate of chemicals in the environment, but automatic temperature corrections of physical and chemical properties of the target substances were absent in their program codes. These procedures were, first time, taken in to our FATEMOD where all temperature dependent physical (Pl, S, H and Log Kow) and chemical (degradation life-times) properties of chemical (Figure 1) were instantly corrected to the ambient temperature [9-11]. Development of these temperature adjustments and their applications in PEC modelling is reviewed in this paper.

2   Temperature coefficients of the compound properties

2.1 Development

Our first paper on necessary temperature correction coefficients in fate model, paper on vapour pressures, solubilities and Henry's law coefficients as function of temperature for 72 POPs was published in 1999 [8]. Secondly, we published assessment rules for degradation rates of very persistent POPs [9]. Thirdly, our research group contributed in the evaluation and chromatographic validation of the temperature-dependent properties, including also LogKow, of twelve environmentally important synthetic musks. There, also an example of automatic use of temperature correction coefficients in risk estimation by modelling was presented [10].

2.2 Vapour pressure: VPLEST

Temperature dependence of vapour pressure has been known over 150 years. It is applied in PEC modelling by simple integrals of Clausius-Clapeyron equation (1) and (2), where T is the ambient temperature in degrees Kelvin. In case that Aps and Bps only are known, FATEMOD converts them to Apl and Bpl by eqn. (3) and eqn. (4). ΔS_f = entropy of fusion (J K^{-1} mol^{-1}), TMK = melting point (K), and 19.1444 = Ln10 * R (gas constant in J K^{-1} mol^{-1}).

Solid state vapour pressure: $Log\ Ps = Aps - Bps\ /\ T.$ (1)

(Subcooled) liquid state vapour pressure: $Log\ Pl = Apl - Bpl\ /\ T.$ (2)

Conversion of intercept: $Apl = Ap\ s - \Delta S_f\ /\ 19.1444$ (3)

Conversion of slope: $Bpl = Bps - \Delta S_f * TMK\ /\ 19.1444.$ (4)

For organic molecular substances, only Pl (liquid state vapour pressure) is needed for modelling of PEC. Determination of coefficients Apl and Bpl can be performed by gas chromatography [11]. However, good results can be obtained also by QSAR methods, as well. Method of Crain [12] is capable to produce eqn. (2) for liquid state vapour pressure Pl from one known value (Pl_1 at T_1) with no other approximations needed than structure parameter K_F. It was adopted in our program VPLEST to derive Apl and Bpl coefficients [10] by regression at given temperature (from 0 to +30 $^\circ$C as the usual environmental range). Comparison of Pl assessments by VPLEST and gas chromatography [13] are shown in Figure 2.

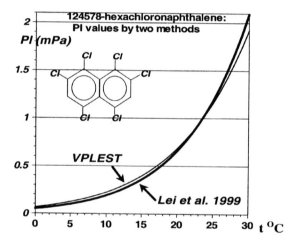

Figure 2:   Comparison of Pl values determined by QSAR program VPLEST by author and gas chromatography by Lei et al. [13].

2.3 Solubility in water: WATSOLU

For organic molecules, there exists only one kind of solubility S which is relevant in environment [6,8,14,15]. Experimental determination of S for hydrophobic compounds are very inaccurate – in these cases water solubility estimated by mobile order thermodynamics presented by Ruelle and his coworkers [15-17] are more reliable. Unfortunately, all their results are only for one temperature (25°C). WATSOLU program is our modification of Ruelle's equations [16,17] to consist of separate temperature non-dependent and dependent parts for calculation of As and Bs coefficients to use in eqn (5) and first applied to the As and Bs evaluation for selected POPs [8] and synthetic musks [10].

WATSOLU calculations:

$$Solubility\ in\ water\ (mol\ m^{-3}):\qquad Log\ S = As - Bs\ /\ T. \qquad (5)$$

$$As=\Delta S_f/(R*Ln10)-0.036*Vb-0.217*LnVb+\Sigma N_{OH}*(2+boh)/Ln10$$
$$+aAcc+aDon+5.154. \qquad (6)$$

$$Bs = \Delta S_f *TMK/(R*Ln10)+(DB-20.5)^2*Vb/(19.1444*(1+MAXW/18.1)). \qquad (7)$$

$$Association\ terms:\qquad aAcc = \Sigma vAcc(i)*Log(1+KAccW(i)/18.1). \qquad (8)$$

$$aDon = \Sigma vDon(i)*Log(1+KDonW(i)/18.1). \qquad (9)$$

Terms vAcc(i) and vDon(i) are numbers of active sites, and KAccW(i) and KDonW(i) are stability constants for proton acceptor and donor groups in water. R (gas constant) = 8.3143. R*Ln10 = 19.1444.Vb = liquid state molar volume $cm^3\ mol^{-1}$ = sum of increments [17]. N_{OH} = sum of the hydroxyl groups. Term boh = 1, 2 or 2.9 or prim., sec. or tert. OH, respectively. 18.1 is molar volume of pure water. TMK = melting point (K). DB = solubility parameter of Ruelle [16]. MAXW = the greatest value of KAccW(i) or KDonW(i) in water.

Together with melting point (TKM in Kelvins), thermodynamic parameter ΔS_f is most important for evaluation temperature coefficients As and Bs. It is ratio of the heat of fusion ΔH_f (J mol^{-1}) and TKM (K). An accurate method to determine ΔH_f is thermoanalysis of the pure crystals by differential scanning calorimetry (DSC) [18]. By measuring some members of a congener group by DSC to obtain accurate heats of fusion values, the results can be expanded using multiple regression with molecular descriptors to other compounds in the congener group [19]. A QSAR method to estimate ΔSf is to calculate it as sum of molecular increments [20]. The results are often equal between isomers, and therefore less accurate than from method above.

HPLC at different temperatures can be used for validation the completely theoretical results from WATSOLU. This has been successfully done by us in the musk study /10/. Another example of comparison of WATSOLU and HPLC-derived values of S for 4,5,6-trichloroguaiacol (Figure 3).

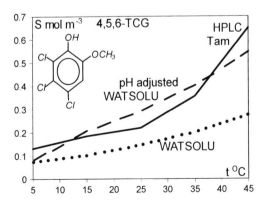

Figure 3:   Solubility of 4,5,6-trichloroguaiacol in water at different temperatures [21] with HPLC compared to WATSOLU results: lower curve without and upper curve with correction to pH of the solvent in HPLC.

Temperature-controlled HPLC study of chloroguaiacols had been done in the University of Toronto by Tam et al. [21]. The HPLC and WATSOLU values without considering pKa of the substrate and pH of the elution solvent were not in agreement. But after making correction the WATSOLU values of the substrate (phenol) to pH of the HPLC solvent, the values coincide.

2.4 Octanol-water partition coefficient: TDKLOW program

Ruelle applied the mobile order thermodynamics for Log Kow at 25 °C [16]. The procedure was to subtract solubility equations for Log S in water-saturated n-octanol and in pure water from each other, which resulted to equation of Log Kow. We isolated temperature independent and dependent parts from these equations to produce coefficients Aow and Bow in eqn (10). Most temperature dependence was eliminated by subtraction, but some remained by polyfunctional molecules. The first application was presented in paper from our musk study [10].

TDLKOW calculations:

$$\textit{Octanol water partition:Log Kow} = Aow - Bow \, / \, T. \tag{10}$$

$$Aow = \Delta B + \Delta F + \Delta Acc + \Delta Don. \tag{11}$$

$$\Delta B = (0.5*Vb*(1/124.2 - 1/18.1) + 0.5*Ln*(18.1/124.2))/Ln10. \tag{12}$$

$$\Delta F = (Vb*(rw/18.1 - ro/124.2) - \Sigma N_{OH}(j)*(boh + rw - ro))/Ln10. \tag{13}$$

$$\Delta Acc = \Sigma vAcc(i)*(Log((1 + KaccO(i)/124.2)/(1 + KaccW(i)/18.1)). \tag{14}$$

$$\Delta Don = \Sigma vDon(i)*log((1 + KdonO(i)/124.2)/(1 + KdonW(i)/18.1)). \tag{15}$$

$$Bow = (Vb/19.1444)*((db - 20.5)^2/(MAXW/18.1)$$
$$-(db - 16.38)^2/(1 + MAXO/124.2)). \tag{16}$$

Most molecular descriptors are defined in previous chapter (**2.3**). Here 124.2 is the reduced molar volume of water-saturated n-octanol, rw is structuration factor of water = 2.0, ro is structuration factor for water-saturated n-octanol (= 1.275). KaccO(i) and KdonO(i) are stability constants of proton acceptor and donor groups in octanol, and MAXO the greatest value of KoO(i) and KohO(i) for solute in n-octanol [16].

2.5 Volatility: Henry's law function

Henry's function is used for volatilisation the estimate in FATEMOD. Its value is temperature-dependent, expressed by eqn (17). At environmental temperature range the values can be approximated as H = Pl / S for cases of low solubility compounds which are not miscible in water [22]. Then, Ah and Bh are obtained by simple subtractions (eqns (18) and (19)). For more hydrophilic compounds other methods to calculate H in the fate modelling are needed [22].

$$\textit{Temperature dependence of Henry's law function: Log H} = Ah - Bh \, / \, T. \tag{17}$$

$$\textit{Approximation of coefficients:} \quad Ah = Apl - As. \tag{18}$$

$$Bh = Bpl - Bs. \tag{19}$$

2.6 Estimation of the degradation lifetimes

Degradation rates $R(i)$ h^{-1} and half-life times $HL(i)$ (hours) of the compound in each of the four major compartments $(i = 1-4)$ are necessary factors for fate prediction [9]. The values of $R(i)$ and $HL(i)$ depend with different weight of significance in different compartments on rate constants of the photodegradation $(k(p)$, hydrolysis $k(h)$ and biodegradation $k(b)$:

$$\textit{In Air:} \qquad R(1) = k(p) \qquad HL(1) = Ln\ 2\ /\ R(1). \qquad (20)$$

$$\textit{In Water:}\ \ R(2) = k(p)/12 + k(h) + k(b)/12. \qquad HL(2) = Ln\ 2\ /\ R(2). \quad (21)$$

$$\textit{In Soil/Plants:}\ \ R(3) = k(h)/12 + k(b). \qquad HL(3) = Ln\ 2\ /\ R(3). \qquad (22)$$

$$\textit{In Sediment:}\ \ R(4) = k(h)/12 + k(b)/12. \quad HL(4) = Ln\ 2\ /\ R(4) \qquad (23)$$

There are "thumb rules" for temperature correction of the $HL(i)$ values: $10^{O}C$ decrease of temperature (T) from reference value HLT increases degradation lifetime in air by a factor of 1.2, and in other major compartments by a factor of 2 [9]. The corresponding correction equations are:

$$HL(1)corr\ =\ HL(1)ref * 1.2^{((HLT-T)/10)}. \qquad (24)$$

$$HL(2)corr\ =\ HL(2)ref * 2^{((HLT-T)/10)}. \qquad (25)$$

$$HL(3)corr\ =\ HL(3)ref * 2^{((HLT-T)/10)}. \qquad (26)$$

$$HL(4)corr = HL(4)ref * 2^{((HLT-T)/10)}. \qquad (27)$$

3 FATEMOD applications: role of temperature

3.1 Model program

FATEMOD is a practical tool for PEC predictions for relatively persistent chemicals or their more persistent metabolites in environment. The results, together with effect potency (PNEC) data could give a reasonably realistic risk estimate (Figure 1) for management of pollution (e.g. source detection and discharge guidelines), need of restrictions or remediation, groundwater quality control, obstacles for use of the chemical in agriculture, and assessment of potential ecosystem damages. In addition, it is a realistic tool for fast prediction of environmental risks of chemicals for the planned EU-wide management.

The model environment in FATEMOD consist of single box (catchment) of six compartments (parts in Figure 1). Air, Water, Soil/plants and Sediments contribute in the mass balance calculation. For exposing concentrations, also compartments Suspended solids and Fish are included. Examples of environments are: SWF = catchment areas of rivers flowing to the Bothnian Sea (South-West Finland) and KemR = Kemijoki River catchment area (North Finland).

Database of compounds in FATEMOD lists following property parameters: CAS Nr, molecular mass WM, melting point TM (in Celsius), entropy of fusion ΔS_f (in $J\ K^{-1}\ mol^{-1}$), pKa value, temperature coefficients (see above) for vapour pressure (Apl, Bpl or Aps, Bps), water solubility (As, Bs), LogKow (Aow, Bow), and degradation half-life times HL at reference temperature HLT in parts Air, Water, Soil/plants and Sediment.

3.2 Risk estimation of dimethoate application in Finland

Dimethoate is the most used insecticide in Finland while it possesses low toxicity to man and mammals. However, its estimated PNEC value to fish is about 0.3 mg L^{-1}. By FATEMOD model run with Eo = 500 mg/ha h^{-1} to soil/plants, PNEC to fish was exceeded, but recovered in five months in South-West Finland but in North Finland not before winter (Figure 4).

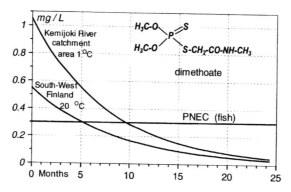

Figure 4:   FATEMOD concentrations (PEC) of the insecticide dimethoate in the water at mean spring-summer temperature after stop of emission of 500 mg/ha h^{-1} to soil/plants.

Figure 5:   Waste chemicals discharged to purification ponds (waste stream WS) and further to the recipient sea area (RSA).

3.3 Determination of guidelines for discharge

The risk based guidelines were needed for the waste chemicals of a factory led through a purification plant (WS) to a coastal recipient sea area (RSA). The wastes were solvent/reagent chlorobenzene (CBz) and herbicide isoxaflutole (IFT). The latter is melabolized in water to a more persistent diketonitrile (DKN; Figure 5). PECs of these compounds were estimated by FATEMOD modelling with default emission to WS Eo = 1 Kg h^{-1} Output from WS was used as emission to RSA (two successive modellings). PNEC values were taken from the acute ecotoxicity test results with *Daphnia*, Fish and Algae. Results as Ro and RE values and discharge guidelines led from them are shown in Table 1.

Observe that the mean water temperatures in winter and summer in these water flows were 5 and 20 °C, respectively. However, the risk emission values did not differ very much. This is due the dominance of water the compartment in the model systems.

Table 1:   Risk levels of the chemical discharges at RSA sea area.

Values of ecotoxicity		Daphnia		Fish		Algae	
Code	Toxic level --> Compound	LC50 mg L^{-1}	PNEC µgL^{-1}	LC50 mg L^{-1}	PNEC µg L^{-1}	LC50 mg L^{-1}	PNEC µg L^{-1}
CBz	Chlorobenzene	5.8	580	22.0	2200	12.5	1250
IFT	Isoxaflutole			1.7	170	0.33	33
DKN	Diketonitrile			1.7*	170	0.33*	33
RE determination		Daphnia		Fish		Algae	
Code	PEC t °C	µg L^{-1}	Ro	RE Kg h^{-1}	Ro	RE Kg h^{-1}	Ro	RE Kg h^{-1}
CBz	5	2.319	0.004	250	0.0011	949	0.0019	539
CBz	20	2.229	0.0038	260	0.0010	987	0.0018	561
IFT	5	2.529			0.0149	67	0.0766	13
IFT	20	1.566			0.0092	109	0.0475	21
DKN	5	3.458			0.0203	49	0.1048	10
DKN	20	3.386			0.0199	50	0.1026	10

* Assumed toxicity for DKN was the same as for its precursor IFT
Risk ratio Ro = PEC / PNEC (for discharge from factory Eo = 1 Kg h^{-1})
Risk emission RE = 1 / R (Kg h^{-1} to the first waste basin)
Guidelines: RE / 10 (safety factor) for CBz = 25 and for IFT 1 Kg h^{-1}

References

[1]   Mackay, D., Multimedia Environmental Models. The Fugacity Approach. Lewis, Chelcea, MI, USA, **L-242,** 1991.

[2]   Paasivirta, J., *Chemical Ecotoxicology.* Lewis, Chelsea, MI, USA, **L366,** 1991.

[3]   Gobas, F.A.P.C., & J.A. McCorquodale (eds.), *Chemical Dynamics in Fresh Water Ecosystems.* Lewis, Chelsea, MI, USA, **L-511,** 1992.

[4]   Paasivirta, J., Long-term effects of bioaccumulation in ecosystems, Chapter 3. In: *The Handbook of Environmental Chemistry, Vol. 2 Part J. bioaccumulation,* ed. B. Beek, Springer, Berlin, pp. 201-233, 2000.

[5]   Calamari, D., (ed.) *Chemical Exposure Predictions.* Lewis, **L-852,** 1993.

[6]   Trapp, S. & Matthies, M. *Chemodynamics and Environmental Modeling, An Introduction.* Annex: Baumgarten, G., Reiter, R., Scheil, S., Schwartz, S. & Wagner, J.O., *CemoS User's Manual and included program diskette.* Springer, Berlin, 1998.

[7]   Pacyna, J.M., Brorström-Lunden, E., Runge, E., Paasivirta, J., Münch, J., Fudala, J. Calamari, D. & Broman, DS., *Environmental Cycling of Selected Persistent Organic Pollutants (POPs) in the Baltic Region.* ENV4-CT96-0214 / EU. Executive Final Summary Report, NILU, Kjeller, Norway, 1999.

[8]   Paasivirta, J., Sinkkonen, S., Mikkelson, P., Rantio, T. & Wania, F., Estimation of vapour pressures, solubilities and Henry's law constants of selected persistent organic pollutants as functions of temperature, *Chemosphere* **39**, pp.811-832, 1999.

[9]   Sinkkonen, S. & Paasivirta, J., Degradation half-life times of PCDDs, PCDFs and PCBs for environmental fate modelling. *Chemosphere* **40**, pp.943-949, 2000.

[10] Paasivirta, J., Sinkkonen, S., Rantalainen, A-L., Broman, D. & Zebühr, Y., Temperature Dependent Properties of Environmentally Important Synthetic Musks. *Environ. Sci. & Pollut. Res.* **9(5)**, pp. 345-355, 2002.

[11] Hinckley, D.A., Bidleman, T.F. & Foreman, W.T., Determination of vapour pressures for nonpolar and semipolar organic compounds from gas chromatographic retention data, *J. Chem. Eng. Data* **35**, pp. 232-237, 1990.

[12] Crain, C.F., Tables 14-4 and 14-5, Eqn 14-25 (Chapter 14): Vapor Pressure In: W. J. Lyman, W. F. Reehl and D. H. Rosenblatt, ed. *Handbook of Chemical Property Estimation Methods*, ACS, Washington, DC, 1990.

[13] Lei, Y.D., Wania, F. & Shiu, W.Y., Vapour pressures of polychlorinated naphthalenes, *J. Chem. Eng. Data* **44**, pp. 577-582, 1999.

[14] Schwarzenbach, R.P., Cschwend, P.M. & Imboden, D.M., *Environmental Organic Chemistry*, Wiley, New York, pp. 80-81, 1993.

[15] Ruelle, P & Kesselring, U.W. Aqueous solubility prediction of environmentally important chemicals from the mobile order thermodynamics, *Chemosphere* **34**, pp.275-298, 1997.

[16] Ruelle, P., The n-octanol and n-hexane/water partition coefficient of environmentally relevant chemicals predicted from the mobile order and disorder (MOD) thermodynamics, *Chemosphere* **40**, pp.457-512, 2000.

[17] Ruelle, P., Farina-Cuendet, A. & Kesselring, U.W., The mobile order solubility equation applied to polyfunctional molecules: The non-hydroxy steroids in aqueous and non aqueous solvents, *Int. J. Pharm.* **157**, pp. 219-232 1997.

[18] Plato, C., Differential scanning calorimetry as a general method for determining purity and heat of fusion of high-purity organic chemicals. Application to 64 compounds. *Anal Chem.* **44**, pp1531-1534, 1972.

[19] Lahtinen, M., Paasivirta, J. & Nikiforov, V.A., Evaluation of entropies of fusion of polychlorinated naphthalenes by model congeners: a DSC study, *Thermochim. Acta*, in print, 2006.

[20] Chickos, J.S., Braton, C.M. & Hesse, D.G., Estimating entropies and enthalpies of fusion of organic compounds J.Org.Chem. **56**, pp. 927-938,1991.

[21] Tam, D.,Varhanickova, D., W-Yu. Shiu, W. Yu. & Mackay, D., Aqueous solubility of chloroguaiacols. J. *Chem. Eng. Data* **39**, pp. 83-86 1994.

[22] Boethling, R.S., Howard, P.H. & Meylan W.M., Finding and estimating chemical property data for environmental assessment. *Environ. Toxicol. Chem.* **23**, pp. 2290-2308, 2004.

Human fluorosis related to volcanic activity: a review

W. D'Alessandro
Istituto Nazionale di Geofisica e Vulcanologia – Sezione di Palermo, Palermo, Italy

Abstract

Fluorosis is a widespread disease related to ingestion of high levels of fluorine through water and food. Although sometimes of anthropogenic origin, high levels of fluorine are generally related to natural sources. One of the main sources is represented by volcanic activity, which releases magmatic fluorine generally as hydrogen fluorine through volcanic degassing. For example, Mt. Etna in Italy is considered the greatest point source at the global scale, releasing on average 70 Gg of HF each year. But the impact of fluorine on human health is highly dependent on its chemical state, which means that high rates of release do not necessarily point to high impacts. The major pathway of magmatic fluorine to humans is in the form of fluoride ion (F^-), through consumption of contaminated vegetables and drinking water. Contamination can happen either through direct uptake of gaseous HF or through rainwaters and volcanic ashes. Furthermore hydrogen fluoride, being one of the most soluble gases in magmas, exsolves only partially ($< 20\%$) during volcanic activity. Volcanic rocks thus contain high levels of fluorine, which are transferred to groundwaters through water-rock interaction processes in the aquifers. Large magmatic provinces, like for example the East African Rift Valley, are therefore endemic for fluorosis. Finally a literature review of volcanic related fluorosis is given.
Keywords: fluorosis, magmatic fluorine, volcanic activity, groundwater.

1 Introduction

The influence of the geological environment on human health has long been known with the first links being recognized probably by Chinese physicians in the 4th century [1]. Such influence is related not only to excessive intake of particular elements but also to their deficiency. For fluorine, like for a few other elements (e.g. I, Se), human health status depends upon a delicate balance between excess and deficiency. Low intake of this element is related to dental caries while exposure to high chronic levels can lead to dental or skeletal fluorosis [2].

Many natural geological processes, sometimes exacerbated from anthropogenic influences (mining activities, fuel combustion, etc.), can be responsible of the impact of harmful compounds. One of the most important is volcanic activity. Apart from the obvious impact of eruptive, especially explosive, activity, in recent times the scientific community has become aware of the effects on the environment and particularly on human health deriving from geochemical processes acting in quiescent periods and even in volcanic systems considered extinct. The importance of this new scientific branch brought specialists of different disciplines to join in the International Volcanic Health Hazard Network (IVHHN) with the common aim of trying to better define the health effects of volcanic emissions [3].

1.1 Magmatic fluorine

Fluorine is the 13th most abundant element in the earth's crust [4]. Several natural and anthropogenic sources contribute to the geochemical cycling of fluorine. The most important industrial processes that release fluorine compounds are: aluminium smelting, coal burning, phosphate fertiliser and cement production, and brick and ceramic firing [5].

Volcanoes represent the main natural persistent source of fluorine [6,7]. Fluorine is emitted by volcanoes mostly in the form of $HF_{(g)}$ [6], but emissions also contain much lower amounts of gaseous NH_4F, SiF_4, $(NH_4)_2SiF_6$, $NaSiF_6$, K_2SiF_6, KBF_4 and organo-fluorides [4, 8, 9]. Volcanic emissions of fluorine take the form of either sluggish permanent release from quiescent volcanoes (passive degassing) or rarer but more impacting discharges during short-lived volcanic eruptions. Estimates of the global volcanogenic fluorine flux range 50 to 8600 Gg/a [6,7,10], with the former figure being probably an underestimate. Total anthropogenic emissions are in the same order of magnitude with the highest emissions are due to chlorofluorocarbon production (300 Gg/a) and coal burning (200 Gg/a) [5].

It has been estimated that passive degassing, like that existing at Mt. Etna (Italy) and Masaya (Nicaragua) volcanoes, accounts for about 90% of the volcanic fluorine release. In particular Mt. Etna is the largest known point source of atmospheric fluorine, contributing for about 70 Gg/a [11], even stronger than today's total estimated anthropogenic release over Western Europe [12]. The influence of these emissions on the surrounding environment and in particular on vegetation has been investigated by several authors [12–14].

Although responsible on average for the emission of lower amounts, the impact of fluorine emitted during explosive volcanic eruptions has been better studied [15–17]. Fluoride was certainly the agent responsible for the death of sheep after the volcanic eruption described in the Icelandic sagas. Ashes emitted during explosive activity scavenge very effectively the erupted volcanic gases including HF thus enhancing their deposition around the erupting volcano up to distances of hundreds of km [15]. Acute and chronic fluorosis on grazing animals has been described for many explosive eruptions all around the world (Mt. Hekla – Iceland [18], Lonquimay – Chile [19], Nyamuragira – Democratic Republic of Congo [20], Mt. Ruapehu – New Zealand [17]). Consequences on livestock are due either to direct ingestion of F-rich ashes deposited on the grass or to grazing of grass or drinking water that are F-contaminated. The problem is widespread in Iceland were the magmas are particularly F-rich. Since it's settling in 9th century many eruptions on Iceland were responsible of F-poisoning of livestock. The first account of this problem was made after the 1693 eruption of Mt. Hekla by the farmer Eiriksson and the clergyman Petursson, which described the deformed teeth in sheep, cattle and horse calling them "ash-teeth" [21]. Death due to F-poisoning of livestock caused serious famines among Icelanders who were totally dependent on them. The worst episode followed the Laki eruption in 1783 causing the death of half of the population of Iceland [21].

Chronic fluorosis on grazing animals was also related to passive degassing (Ambrym, Vanuatu [22]) and to geothermal activity in recent volcanic areas (Yellostone, U.S.A. [23]).

1.2 Human fluorosis

Fluorine is an essential element for human growth being incorporated in the mineral part of bones and in teeth in the form of fluoroapatite. Dental fluorosis is characterised by discoloured, blackened, mottled or chalky-white teeth. These symptoms are connected with an overexposure to fluoride during childhood when teeth were developing [5]. Fluorine intake above safe limit for very long time or in very high amounts can lead to skeletal fluorosis, with severe and permanent bone and joint deformations [5].

The main route of intake for humans is through drinking water. Only rarely high fluorine intake may derive from F-rich vegetables due either to natural accumulation (tea) or to anthropogenic contamination [24]. Absorption through the skin is very low but HF causes severe burns. Also the contribution of inhalation to the body burden of fluorine is very low. Workers exposed to high HF concentrations have higher probabilities to develop asthma or chronic lung diseases rather than fluorosis [5].

Fluorine, which is accumulated by vegetation as fluoride ion, is readily adsorbed by herbivorous animals. Animals accumulate this element as fluoroapatite that is relatively inert so that fluorine will not be biomagnificated in flesh-eating animals.

Whether dental or skeletal fluorosis are irreversible and no treatment exists, the only remedy being prevention by keeping fluoride intake within safe limits. The World Health Organization fixed a maximum admissible limit of 1.5 mg/l for drinking water [5], although for hotter climates this limit should be lowered to 1.0 mg/l due to higher water consumption.

Third world populations are more prone to develop fluorosis because of bad nutritional status. Clinical data indicate that to low protein, calcium and vitamin C and D intake raises the dental fluorosis risk [25, 26].

A conservative estimate indicates that fluorosis affects tens of millions of people and is endemic in at least 25 countries across the globe [25]. Among these, Japan, New Zealand, Mexico and the countries along the east African Rift Valley (Eritrea, Ethiopia, Uganda, Kenia and Tanzania) display widespread volcanic activity.

2 Review of human fluorosis related to volcanic activity

The problem of fluorosis related to volcanic activity was first recognised in Japan were this pathology was called "Aso volcano disease" [27] due to the fact that fluorosis was widespread in the population living at the foot of this volcano.

Water intake being the main route of fluorine into the human body, fluorosis in volcanic areas is generally associated to elevated fluoride content in surface- and ground-waters. Contamination of vegetation, which is the main agent of volcanic-related fluorosis in herbivorous animal, is only of secondary importance for human health.

High fluorine content in waters derive either from water-rock interaction (WRI) processes in volcanic aquifers (groundwaters) or to contamination due to wet or dry deposition of magmatic fluorine (surface waters - reservoirs).

Furthermore, paleopathologic studies on human skeletons found in Herculaneum, referable to victims of 79 AD eruption of Mt. Vesuvius, evidenced that fluorosis in this area had the same incidence as in modern times, pointing to the constancy of the geochemical processes responsible for fluorine enrichment of the drinking water in the area over at least the last 2000 years [28].

2.1 Water-rock interaction

Volcanic rocks are often enriched in fluorine. Hydrogen fluorine is, in fact, one of the most soluble gases in magmas and exsolves only partially during eruptive activity. Burton et al. [29] estimated for example that Etnean magmas exsolve only about 20% of their initial HF content during effusive activity. Furthermore fluorine behaves as incompatible element being highly enriched in differentiated volcanic products [30].

In volcanic aquifers elevated temperatures and/or strong acidic conditions enhance WRI processes. Such conditions often lead to high concentrations of harmful elements [31]. Therefore fluorine concentrations in volcanic aquifers above safe drinking limits are rather the rule than the exception. Values as high as tens of mg/l of fluorine are often achieved in groundwaters, which if used for human consumption can easily lead to skeletal fluorosis.

Dental fluorosis due to groundwaters enriched by WRI in recent or active volcanic areas has been assessed in many parts of the world. Many articles illustrate such cases. Some of them refer to limited areas like Gölcük – SW Turkey [32], Mt. Aso volcano, Japan [27], Island of Tenerife – Spain [33],

Furnas volcano, São Miguel – Azores, Portugal [34], Albano Lake – Italy [35], while other evidence a widespread problem throughout entire countries like Mexico [36], Ethiopia [37], Kenya [38], Tanzania [26]. In these areas, populations as high as 200,000 people could be at risk to develop fluorosis like for example the inhabitants of the Los Altos the Jalisco region in Mexico [39].

Particularly high fluorine concentrations (thousands of mg/l) can only be achieved under extreme conditions, partially or totally ascribable to volcanic activity. Lakes along the East Africa Rift Valley display values as high as 1980 mg/l (Lake Magadi – Kenia [40]). Such high contents originate from geothermal weathering of the F-rich volcanic rocks further concentrated by evaporative processes in arid climate. Calcium concentrations in these lakes, which could limit fluorine concentrations through fluorite precipitation, are very low due to precipitation of carbonate phases in a highly alkaline environment.

Very acidic lakes in active volcanic systems (pH values ≈ 0) can also achieve extreme fluorine concentrations not only due to intense WRI processes but also to direct input of F-rich volcanic gases. Lakes like Poas – Costa Rica [41] and Ijen Crater Lake – Indonesia [42] reach concentrations far above 1000 mg/l.

Seepage or effluent rivers from these extremely F-rich lakes can easily contaminate ground- or surface waters. It has been estimated that the Ijen Crater Lake discharges daily in the surface and groundwaters of the highly populated area of Asembagus about 2800 kg of fluorine [42], which is responsible of the widespread occurrence of fluorosis in the area. Furthermore the fluorine contained in the salts extracted from the shores of the East African Rift Valley lakes and used for cooking purposes represent an additional fluorine source for the local population [26].

2.2 Degassing activity

Persistent open conduit degassing being the main source of fluorine to the atmosphere can be also the source of contamination of water resources close to volcanoes. In two volcanic systems, high magmatic fluorine deposition has been correlated to human fluorosis.

Stromboli Island – Italy releases 2 Gg/a of fluorine of which about 1% is deposited on its flanks [43]. Deposition decreases exponentially from the summit craters (70 mg/m^2/d) to the periphery of the island (0.5-1 mg/m^2/d). Up to few decades ago the inhabitants of Stromboli derived their water resources mainly from rainwater collection. A recent study highlighted volume weighted mean values of fluorine in rainwater ranging from 0.4 to 1.9 mg/l in the inhabited area of the island were rainwater was collected [43, 44]. Such concentrations are high enough to cause dental fluorosis and actually elderly people suffer from this pathology. Nowadays people all drink bottled water and most of the water for domestic use is brought by ship from the Sicilian mainland so that young people do not show any sign of dental fluorosis.

Recent studies assessed that Ambrym volcano - Vanuatu releases enormous amount of HF (up to 400 Gg/a) with severe environmental consequences [22]. Fluorosis is widespread on the island and is due to contamination of drinking water resources (mainly rainwater) and vegetables through abundant fluorine deposition [22].

2.3 Explosive activity

During explosive activity huge quantities of fluorine are deposited with ashes around the volcano up to distances of hundreds of km. Contamination is in this case generally of short duration because of permanent adsorption of fluorine in soils and dilution through rainwaters, and generally does not bring to chronic expositions of humans. The most exposed to the transient high fluorine levels are herbivorous animals.

The only known eruption that had also consequences on humans was that of Laki – Iceland in 1783. The Icelandic clergyman and historian Jon Steingrimsson, who described in detail all consequences of fluorine intoxication on livestock, tells us also that some of the people living in the area most affected by ashfall developed the same bone and teeth deformations like the animals [21]. In this case fluorosis affected people who had to live on food and to drink water that were contaminated by fluorine. The consequences were probably worsened by the bad nutritional status deriving from the food shortage that followed the eruption. Furthermore this was a long-lasting eruption, exposing people to very high fluorine levels for many months.

3   Remediation

There are basically two approaches for treating water supplies to remove fluoride: flocculation and adsorption [45]. In the former method fluoride is removed through reaction with chemicals (generally hydrate aluminium salts) that coagulating into flocs settle at the bottom of the container. The other method is to filter water down through a column packed with a strong adsorbent, such as activated alumina, activated charcoal or ion exchange resins. Both methods can be used for community or household treatment plants but often their exercise costs are to high for third world countries.

Resent research highlighted the strong fluoride sorption properties of volcanic soils, which are readily available in volcanic areas [45, 46]. The high content of amorphous phases of aluminium (allophane, imogolite), clay minerals and organic bound aluminium enhance fluoride adsorption properties.

Volcanic soils exert their defluoridation properties also naturally. Bellomo et al. [43] evidenced that the soils of Mt. Etna adsorb about 70% of the magmatic fluorine deposited on its flanks protecting the huge groundwater resources and maintaining the fluoride concentration always below the safe drinking water level.

4   Concluding remarks

Fluorosis is generally not considered in the volcanic hazard evaluation of volcanic systems. Nevertheless fluorosis related to volcanic activity affects probably, at the global level, nearly some million of people. In the majority of cases this brings only little sufferance or even only aesthetic problems (dental fluorosis) but in the worst cases it can bring to complete inability (skeletal fluorosis).

Problems of magmatic fluorine contamination have to be managed in different ways. Contamination due to WRI processes has to be resolved with methodologies attaining correct water resource management, i.e. searching alternative (low-F) water sources or applying water treatment methods. It has to be highlighted that while in the more evolved nations fluorosis is generally declining because there are enough economic resources to find alternative water sources (i.e. Stromboli and Furñas) or to apply correct water treatments, in the less evolved nations the problem is still increasing. This depends on many reasons. One is the constant population growth in third world nations, particularly in the highly fertile volcanic areas [47], exposing an increasing number of people to the risk of fluorosis (and other volcanic risks). Furthermore in some of these countries, which often suffer for arid climate and consequent water shortage problems, emphasis is usually on water availability rather than quality. But while in some cases, discoloured teeth may be an acceptable side effect of the overriding need to provide microbially clean, easily accessible and cheap drinking water, in some country of the East African Rift Valley maximum admissible concentrations in drinking waters (3 mg/l in Ethiopia and 8 mg/l in Tanzania) expose people to the more serious consequences of skeletal fluorosis.

Fluorine contamination due to eruptive activity often requires emergency management. One of the major tools is quantitative modelling that gives important information for the identification of exposure pathways and for the fluorosis risk management. The problem was discussed in two recent papers. One deals with the contamination of rainwater tanks and open drinking water reservoirs through volcanic ashes [48]. The risk connected to this type of contamination depends on the quantity of leachable fluorine adsorbed by ashes and the ash/water mass ratio in the contaminated reservoir. The other paper applies mathematical models, developed by the U.S. Environmental Agency, to different volcanic activity scenarios [49]. The risk assessment was applied to quantitatively estimate exposure pathways and daily average intake of fluoride in hypothetical fluoride-contaminated volcanic areas.

References

[1]   Appleton, J.D., Fuge, R., McCall, G.J.H. (eds.) *Environmental Geochemistry and Health*, Geological Society Special Publ. 113, 1996.
[2]   Edmunds, W.M., Smedley, P.L., Groundwater geochemistry and health: an overview. Appleton, J.D., Fuge, R., McCall, G.J.H. (eds.) *Environmental Geochemistry and Health*, Geological Society Special Publ. 113, 1996.
[3]   http://www.ivhhn.org/.
[4]   Weinstein, L.H., Davison, A., *Fluoride in the Environment*, CABI Publishing, 2003.

[5]   WHO, *Fluorides*, Geneva, World Health Organization. Environmental Health Criteria 227, pp. 268, 2002.

[6]   Symonds, R.B., Rose, W.I., Reed, M.H., Contribution of Cl- and F-bearing gases to the atmosphere by volcanoes, *Nature*, **334**, 415-418, 1988.

[7]   Halmer, M.M., Schmincke, H.U., Graf, H.F., The annual volcanic gas input into the atmosphere, in particular into the stratosphere: a global data set for the past 100 years, *J. Volcanol. Geoth. Res.*, **115**, 511-528, 2002.

[8]   Francis, P., Chaffin, C., Maciejewski, A., Oppenheimer, C., Remote determination of SiF_4 in volcanic plumes: A new tool for volcano monitoring, *Geophys. Res. Lett.*, **23(3)**, 249-252, 1996.

[9]   Schwandner, F.M., Seward, T.M., Gize, A.P., Hall, P.A., Dietrich, V.J., Diffuse emission of organic trace gases from the flank and the crater of a quiescent active volcano (Volcano, Aeolian Islands, Italy), *J. Geophys. Res.*, **109**, D04301, 2004.

[10]   Cadle, R.D., A comparison of volcanic with other fluxes of atmospheric trace gas constituents, *Rev. Geophys.*, **18**, 746-752, 1980.

[11]   Francis, P., Burton, M.R., Oppenheimer, C., Remote measurements of volcanic gas compositions by solar occultation spectroscopy, *Nature*, **396**, 567-570, 1998.

[12]   Aiuppa, A., Bellomo, S., Brusca, L., D'Alessandro, W., Di Paola, R., Longo, M., Major ion bulk deposition around an active volcano (Mt. Etna, Italy), *Bull. Volcanol.*, **68**, 255-265, 2006.

[13]   Garrec, J.P., Plebin, R., Faivre-Pierret, R.X., The Influence of volcanic fluoride emissions on surrounding vegetation, *Fluoride*, **10**, 152-156,1984.

[14]   Notcutt, G., Davies, F., Accumulation of volcanogenic fluoride by vegetation: Mt. Etna, Sicily, *J. Volcanol. Geotherm. Res.*, **39**, 329-333, 1989.

[15]   Oskarsson, N., The interaction between volcanic gases and tephra: fluorine adhering to tephra of the 1970 Hekla eruption, *J. Volcanol. Geotherm. Res.*, **8**, 251-266, 1980.

[16]   Thorarinsson, S., On the damage caused by volcanic eruptions with special reference to tephra and gases, In: Sheets, P.D., Grayson, D.K. (eds.), *Volcanic activity and human ecology*, Academic Press, New York, p. 125-159, 1979.

[17]   Cronin, S.J., Neall, V.E., Lecointre, J.A., Hedley, M.J., Loganathan, P., Environmental hazards of fluoride in volcanic ash: a case from Ruapehu volcano, New Zealand, *J. Volcanol. Geotherm. Res.*, **121**, 271-291, 2002.

[18]   Georgsson, G., Petursson, G., Fluorosis of sheep caused by the Hekla eruption in 1970, *Fluoride*, **5(2)**, 58-66, 1972.

[19]   Araya, O., Wittwer, F., Villa, A., Ducom, C., Bovine fluorosis following volcanic activity in the southern Andes, *Vet. Rec.*, **126**, 641-642, 1990.

[20]   Casadevall, T.J., Lockwood, J.P., Active volcanoes near Goma, Zaire: hazard to residents and refugees, *Bull. Volcanol.*, **57**, 275-277, 1995.

[21]   Fridriksson, S., Fluoride problems following volcanic eruption, In: Shupe, J.L., Peterson, H.B., Leone, N.C., (Eds.) *Fluorides, - Effect on vegetation, animals and humans*, Pearagon Press, UT, 339-344, 1983.

[22]   Crimp, R., Cronin, S., Charley, D., Oppenheimer, C., Bani, P., Dental fluorosis attributed to volcanic degassing on Ambrym, Vanuatu, *Cities on Volcanoes 4, Quito, Ecuador, 23-27 January 2006, Abstract book*, 2006.

[23]   Garrot, R.A., Eberhardt, L.L., Otton, J.K., White, P.J., Chaffee, M.A., A geochemical trophic cascade in Yellowstone's geothermal environments, *Ecosystems*, **5**, 659-666, 2002.

[24]   Ando, M., Todano, M., Yamamoto, S., Tamura, K., Asanuma, S., Watanabe, T., Kondo, T., Sakurai, S.J.R., Liang, C., Chen, X., Hong, Z., Cao, S., Health effects of fluoride caused by coal burning, *Sci. Total Environ.*, **271**, 107-116, 2001.

[25]   Qian, J., Susheela A.K., Mudgal, A., Keast, G., Fluoride in water: an overview, *Waterfront*, **13**, 11-13, 1999.

[26]   Nanyaro, J.T., Aswathanarayana, U., Mungure, J.S., Lahermo, P.W., A geochemical model for the abnormal fluoride concentration in waters in parts of northern Tanzania, *J. African Earth Sci.*, **2**, 129–140, 1984.

[27]   Kawahara, S., Odontological observations of Mt. Aso-volcano disease, *Fluoride*, **4**, 172-175, 1971.

[28]   Morettini, L., Ciranni, R., *Herculaneum – other mysteries unearthed*, C.N.R., Bologna, 2000.

[29]   Burton, M., Allard, P., Murè, F., Oppenheimer, C., FTIR remote sensing of fractional magma degassing at Mt. Etna, Sicily, In: *Volcanic degassing*, Oppenheimer, C., Pyle, D., Barclay, J. (eds) Geol. Soc. London Spec. Publ. 213, 281-293, 2003.

[30]   Sigvaldason, G.E., Oskarsson, N., Fluorine in basalts from Iceland, *Contr. Mineral. Petrol.*, **94**, 263-271, 1986.

[31]   Aiuppa, A., D'Alessandro, W., Federico, C., Palumbo, B., Valenza, M., The aquatic geochemistry of arsenic in volcanic groundwaters from southern Italy, *Appl. Geochem.*, **18**, 1283-1296, 2003.

[32]   Pekdeger, A., Özgür, N., Schneider, H.J., Hydrogeochemistry of fluoride in shallow aqueous systems of the Gölcük area, SW Turkey, In: Kharaka, Y.K., Maest, A.S. (eds) *Proc. 7th Intern. Symp. on Water Rock Interaction*, Utah, 821-824, 1992.

[33]   Hardisson, A., Rodriguez, M.I., Burgos, A., Diaz Flores, L., Gutierrez, R., Varela, H., Fluoride levels in publicly supplied and bottled drinking water in the island of Tenerife, Spain, *Bull. Environ. Contam. Toxicol.*, **67**, 163-170, 2001.

[34]   Baxter, P.J., Buabron, J.C., Coutinho, R., Health hazards and disaster potential of ground gas emissions at Furnas volcano, São Miguel, Azores, *J. Volcanol. Geotherm. Res.*, **92**, 95-106, 1999.

[35]   Beccari, M., Dall'Aglio, M., Nuove frontiere nell'approvvigionamento di acque per usi civili ed agricoli alla luce della loro accettabilità, Atti del Convegno dell'Accademia dei LINCEI pp. 99-127, "Accettabilità delle acque per usi civili ed agricoli". Roma, 5 Giugno 2002.

[36]   Soto-Rojas, A.E., Ureña-Cirett, J.L., Martínez-Mier E.A., A review of the prevalence of dental fluorosis in Mexico, *Pan Am. J. Public Health*, **15**, 9-17, 2004.

[37]   Kloos, H., Tekle Haimanot, R., Distribution of fluoride and fluorosis in Ethiopia and prospects for control, *Tropical Medicine and International Health*, **4**, 355-364, 1999.

[38]   Nyaora Moturi, W.K., Tole, M.P., Davies, T.C., The contribution of drinking water towards dental fluorosis: a case study of Njoro Division, Nakuru District, Kenya, *Environ. Geochem. Health*, **24**, 123-130, 2002.

[39]   Hurtado, R., Gardea-Torresdey, J., Tiemann, K.J., Fluoride occurrence in tap water at "Los Altos de Jalisco" in the central Mexico region, *Proc. of the 2000 Conf. on Hazardous Waste Research*, 211-219, 2000.

[40]   Jones, B.F., Eugster, H.P., Reitig, S.L., Hydrogeochemistry of the Lake Magadi Basin, Kenya, *Geochim. Cosmochim. Acta*, **41**, 53-72, 1977.

[41]   Rowe, G.L., Brantley, S.L., Fernandez, J.F., Borgia, A., The chemical and hydrologic structure of Poas Volcano, Costa Rica, *J. Volcanol. Geotherm. Res.*, **64**, 233-267, 1995.

[42]   Heikens, A., Sumarti, S., van Bergen, M., Widianarko, B., Fokkert, L., van Leeuwen, K., Seinen, W., The impact of the hyperacid Ijen Crater Lake: risks of excess fluoride to human health, *Sci. Total Environ.*, **346**, 56-69, 2005.

[43]   Bellomo, S., D'Alessandro, W., Longo, M., Volcanogenic fluorine in rainwater around active degassing volcanoes: Mt. Etna and Stromboli island, Italy, *Sci. Total Environ*, **301**, 175–185, 2003

[44]   Liotta, M., Brusca, L., Grassa, F., Inguaggiato, S., Longo, M., Madonia, P., Geochemistry of rainfall at Stromboli volcano (Aeolian Islands): isotopic composition and plume-rain interaction, *Geochem. Geophys. Geosyst.*, in press, 2006.

[45]   Bjorvatn, K., Reimann, C., Østvold, S.H., Tekle-Haimanot, R., Melaku, Z., Siewers, U., High-fluoride drinking water. A health problem in the Ethiopian Rift Valley. 1. Assessment of lateritic soils as defluoridating agents, *Oral Health & Preventive Dentistry* **1**, 141-148, 2003.

[46]   Zevenberger, C., van Reeuwijk, L.P., Frapporti, G., Louws, R.J., Schuiling, R.D., A simple method for defluoridation of drinking water at village level by adsorption on Ando soil in Kenya, *Sci. Total Environ.*, **188**, 225-232, 1996.

[47]   Small, C., Naumann, T., The global distribution of human population and recent volcanism, *Environ. Hazards*, **3**, 93-109, 2001.

[48]   Stewart, C., Johnston, D.M., Leonard, G.S., Horwell, C.J., Thordarson, T., Cronin, S.J., Modelling contamination of water supplies by volcanic ashfall, *J. Volcanol. Geotherm. Res.*, in press, 2006.

[49]   Delmelle, P., A quantitative look at fluoride exposure and intake of residents in volcanic areas using a health risk assessment approach, *Cities on Volcanoes 4, Quito, Ecuador, 23-27 January 2006, Abstract book*, 2006.

Toxicity of atmospheric particulate matter using aquatic bioassays

C. Papadimitriou[1], V. Evagelopoulos[2], P. Samaras[1],
A. G. Triantafyllou[2], S. Zoras[2] & T. A. Albanis[3]
[1]*Department of Pollution Control Technologies,*
Technological Educational Institute of W. Macedonia, Kozani, Greece
[2]*Department of Geotechnology and Environmental Engineering,*
Laboratory of Atmospheric Pollution and Environmental Physics,
Technological Educational Institute of W. Macedonia, Kozani, Greece
[3]*Department of Chemistry, University of Ioannina, Panepistimioupolis,*
Ioannina, Greece

Abstract

The investigation of ecotoxicological effects of particulate matter and the increasing interest in correlating chemical composition to biological impacts is of paramount importance. A complete assessment of biological impacts will require a broad look at the effects of human health and the terrestrial and aquatic environments. The objective of this work was to assess the effects of various sizes of particulates on aquatic organisms. Samples were collected from filters and the filter contents were extracted with DCM using a Soxhlet extractor. Then concentration DCM extracts were condensated to 2 ml by rotary evaporation and each millilitre of extract was exchanged with 1 ml of DMSO and then diluted to freshwater. The samples were then analyzed for their ecotoxicological properties with *Vibrio fischeri* and bioassay. Five consecutive dilutions of the sample were tested in order to determine the EC50 values as a result of different size of particulates.
Keywords: PM10, PM2.5, particulates, toxicity, bioassays.

1 Introduction

The knowledge of the distribution of airborne particulate matter (PM) into size fractions has become an increasing area of focus during the examination of the

WIT Transactions on Biomedicine and Health, Vol 10, © 2006 WIT Press
www.witpress.com, ISSN 1743-3525 (on-line)
doi:10.2495/ETOX060041

effects of particulate pollution. Particle size distribution is important for human exposure and risk assessment, as well as for understanding the mechanisms of atmospheric processes [1]. Several studies have shown correlation between respiratory symptoms and heart diseases with the inhalable particles [2, 3]. Particles with a size less than 10 mm (PM10) have long been implicated in causing adverse health effects and increased mortality [4] whereas fine (PM2.5) and ultrafine particles impose even higher risk [5–7]. Thus, it is of paramount importance the ability to correlate the sources and chemical composition of aerosols with their biological impact. Complete assessment of the biological impact of aerosols would require a broad look at the effects on human health and the terrestrial and aquatic environments. Several analytical methods have been developed for the determination of the detailed chemical composition of atmospheric aerosol (gas chromatography coupled with mass spectroscopy GC-MS, inductively coupled plasma ICP mass spectroscopy, X-ray fluorescence, etc) [13]. These techniques have been used to demonstrate the extreme diversity of the composition of atmospheric aerosols as a function of temporal and spatial factors. Similarly, biological characterization of atmospheric and emission source particulates has been conducted on a variety of species, with rodent inhalation studies [14, 15] and bacterial assays [16–19] representing the bulk of the *in vivo* research. In addition, *in vitro* work with human and rat cell lines has further expanded the collective knowledge of the biological response mechanisms to particulates [20–22]. However, due to the complex nature of analytical monitoring and especially of biological indicators, a limited number of studies have presented potential correlations of chemical parameters and biological characteristics, focusing mainly on bulk or physical characterization (elemental and organic carbon, inorganic elements, particle size, etc.).

The area of Kozani – Ptolemais Basin (KPB) is a heavy industrialized area in the north-western part of Greece, which is characterized by a complex topography. Within the basin, four lignite thermal power stations (PS) are operated by the Greek Public Power Corporation (GPPC) with a total installed power capacity of more than 4 GW. These power stations contribute to about 70% of the total electrical energy produced in Greece. The lignite that is combusted in the power stations is mined in the nearby open-pit mines. Sulphur content in feedstock is low (<1%); in addition the high content of calcium oxides in the mineral matter of the raw lignite contributes to negligible releases of sulphur oxides in the flue gases. Thus, dust emissions seem to be the most significant air pollution problem in the area; the measured ambient concentrations of suspended particles are often at high levels exceeding local and international standards [4, 5]. The high particulate concentration in this area has been attributed to two parameters: (a) particulates released in the flue gases of the thermal power stations. The required amount of lignite for the operation of one 1200 MW power station is estimated to about 54000 tons per day. About 15% of raw fuel is transformed to ash, corresponding to about 8100 tons of ash produced per day. Electrostatic precipitators have been installed in the power stations, aiming to the removal of fly ash from flue gases; an optimum removal capacity of about 99.9% results to about 8.5 tones of fly ash emitted per day in the atmosphere as primary suspended particulate pollutants; (b) another significant source of inhalable particulates is the dust generated from mining operations, including excavation, transportation by uncovered trucks, deposition of lignite and ash and potential subsequent re-suspension due to air currents. Monitoring of air quality in the area has started in 1983 by the installation and operation of a network of sampling stations that has been upgraded in 1997. However the ecotoxicological effects of the particulate matter have not been investigated, in order to assess the environmental impact associated to the particulate matter.

The aim of this work was the evaluation of the ecotoxic properties of particulates emitted in a highly industrialized area, the assessment of potential parameters affecting the toxicity and the correlation to sampling conditions.

2   Materials and methods

2.1 Sample collection

Size-segregated particulate samples (PM10 and PM2.5) were collected in the industrial area of the opencast mines at the village Klitos and the urban area of Kozani in north-western Greece (Figure 1), for a period of three months (January to March). The samples were collected by Andersen Reference Ambient Air Sampler (RAAS), and dichotomous sampler with a PM10 inlet probe for the collection of particles in the PM2.5 and PM10 size range for gravimetric analysis. The samples were collected by passing air through a 37 mm PTFE filter. The technique for monitoring the size distribution of airborne aerosols with a dichotomous sampler was based on inertial impaction of particles into a void (virtual surface). The total flow and the coarse particle flow were adjusted by the use of certain rotameters on the sampler.

RAAS PM2.5/10 Sequential Filter Sampler System specifically designed to comply with requirements of the National Ambient Air Quality Standards (NAAQS) for Particulate Matter (40 CFR Part 50). According to the new standard a fixed flow rate sampler is required, operating at 16.67 liters/minute (one cubic meter per hour), using a specified inlet, tubing, secondary size selective impactor, and filter holder. The inlet should be a 10-micrometer (nominal) cut-point providing wind speed and direction independent sampling to remove particles larger than 10μm. The connecting tubing was a 12-inch length anodized aluminum tubing with a precisely specified inner diameter and surface finish. The secondary size selective element was the WINS impactor that further reduced the passed particulate to exclude particles larger than 2.5μm. The particulates were then collected on a 47mm PTFE filter.

Before sampling, the filter was equilibrated to constant temperature and relative humidity conditions and weighed. After sampling, the filter was equilibrated to the constant temperature and humidity conditions and weighed. The concentration was calculated by dividing the weight of the particulates captured on the filter by the volume of air (at ambient conditions) that passed through the sampler. The flow rate was required to be maintained within 5% of 16.67 liters per minute with a coefficient of variation less than 2%.

Figure 1:   Map of the sampling area [20].

2.2 Treatment of samples

Each filter was Soxhlet extracted in 300 ml dichloromethane (DCM, provided by Merck) followed by rotary evaporation to 1 ml. The concentrated extract was then exchanged with 1 ml dimethylsulfoxide (DMSO) and diluted to 300 ml with deionised water. Toxicity testing was performed using various dilutions of the final sample volume of 300 ml, that were prepared by the appropriate amounts of deionised water.

2.3 Toxicity Testing with *Vibrio fischeri*

Toxicity testing included the measurement of bioluminescence inhibition of the marine bacteria *Vibrio fischeri* within a short exposure time. The bacteria were in freeze-dried form and were activated prior to the use. The salinity of the samples was adjusted with the addition of a 2% NaCl solution. The light emitted from a control sample and the various samples were measured after exposure of the bacteria to the sample at certain times, 5-15 and 30 min, by the Microtox model 500 analyzer (Azur Environmental) (Microtox Manual, 1998). The percentage bioluminescence inhibition and the EC_{50} value (the percentage of sample concentration that causes 50 % effect on the test organism) of each sample were calculated as an end point.

3 Results and discussion

During the sampling period, 12 samples were collected from the sampling point of Kozani, including six filters of fine particulates and 6 filters of coarse particulates, while 6 samples were collected from Klitos sampling point corresponding to 3 filters of PM 10 and 3 of PM2.5. Sample symbols and the corresponding collection dates are given in Table 1.

Table 1:   Sample symbols and the corresponding sampling dates for the toxicity assessment of particulates.

Kozani		Klitos	
Symbol	Date	Symbol	Date
A_1	6/1/2006	B_1	6/1/2006
A_2	14/1/2006	B_2	2/2/2006
A_3	2/2/2006	B_3	18/3/2006
A_4	27/2/2006		
A_5	15/3/2006		
A_6	18/3/2006		

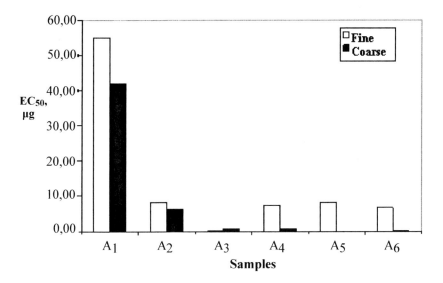

Figure 2: EC_{50} values for the fine and coarse particulate matter collected from Kozani sampling site.

The EC_{50} values obtained from the toxicity testing of the particulate matter collected from Kozani and Klitos sampling points are shown in Figures 2 and 3, respectively.

EC_{50} values of samples collected from the Kozani sampling point were ranging from 0 to 55 µg. Coarse particulates exhibited EC_{50} concentrations ranging from 0 to 42 µg, while the respective concentrations for fine particulates were ranging from 8 to 55 µg. In general, coarse particulates exhibited increased toxic effects to *Vibrio fischeri*, and the EC_{50} values were lower that those of the fine particulate matter.

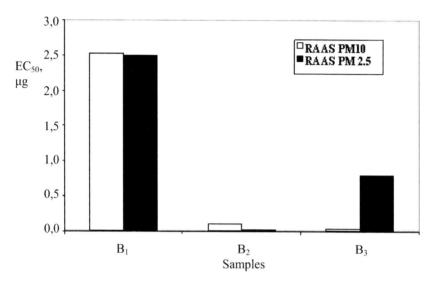

Figure 3: EC_{50} values for PM 10 and PM 2.5 collected from Klitos sampling site.

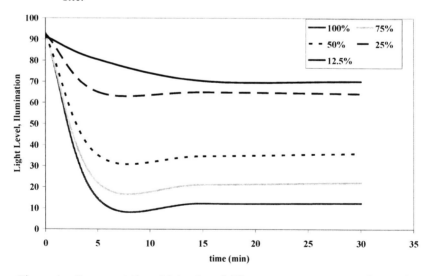

Figure 4: Representative Light Level–Time response curves for various dilutions of sample A2.

EC_{50} values of samples collected from Klitos sampling site were ranging from 0 to 2.5 µg for both PM10 and PM 2.5 samples. PM2.5 samples exhibited higher toxicity to *Vibrio fischeri* than PM10 samples. The EC_{50} values found for Klitos were lower that those of Kozani, suggesting differences in the physical and chemical characteristics of particulate matter.

For all samples, the light - level response curves presenting the relation of the exposure time of the organisms to the various dilutions of samples were investigated, in order to evaluate potential factors responsible for the toxicity of the samples. Typical light response curves from various dilutions of a representative sample are shown in Figure 4; all samples presented almost similar response curves.

As shown in this figure, a sharp decrease of emitted bioluminescence was observed within 5 min of exposure, and remained almost constant at higher contact times. In addition, the higher the concentration of the sample extract, the higher was the initial decrease of bioluminescence. This pattern of light – level response curve is typical of samples containing mainly organic compounds.

Ambient samples may be used for the investigation of the effect of pollutants on specific organisms, by using bioassay techniques. However, the estimation of potential factors responsible for the observed toxicity is a complicated process, as several parameters may affect the response of test species. In the current study, increased toxic properties were observed for samples collected from a rural area (Klitos), than samples collected from an urban one (Kozani). However, as it was mentioned, the work area is an industrial location with significant environmental problems, attributed both to the excavation and transportation of coal resources and to the flue gas releases from the thermal power plants. For the urban area of Kozani, low toxicity effects were observed that could be attributed to sources such as exhaust particulates, gasoline powered vehicles and even biogenic particulates.

Acknowledgement

This work was funded by the Greek Ministry of Education and Religion Affairs, within the framework of "Archimedes II", enhancement of research activities in Technological Educational Institutes, Project 2.6.1IA, "Study of the adsorption and transport processes from particulate matter to urban and rural areas situated close to coal mines and electricity production power plants".

References

[1]   Fishbein, L. Sources, nature and levels of air pollutants, in: L. Tomatis Ed., Air Pollution and Human Cancer, Monographs, European School of Oncology, Springer-Verlag, New York, 1993, pp. 9–34.

[2]   Lewtas, J. Experimental evidence for the carcinogenicity of air pollutants, in: L. Tomatis Ed., Air Pollution and Human Cancer, Monographs, European School of Oncology, Springer-Verlag, New York, 1993, pp. 49-61.

[3]   Triantafyllou, A.G.: 2003, 'Levels and trend of suspended particles around large lignite power stations', Environmental Monitoring and Assessment 89, 15-34.

[4]   Triantafyllou, A.G., Zoras, S., Evagelopoulos, V.: 2005, 'Particulate matter over a seven year period in urban and rural areas within, proximal and far from mining and power station operations in Greece, *Environmental Monitoring and Assessment*, 159-165.

[5]   Sasaki, Y. Kawai, T. Ohyama, K.-I. Nakama, A. Endo, R. Carcinogenicity of extract of airborne particles using newborn mice and comparative study of carcinogenic and mutagenic effect of the extract, Arch. Environ. Health 42 1987 14–18.

[6]   Crebelli, R. Fuselli, S. Meneguz, A. Aquilina, G. Conti, L. Leopardi, P. Zijno, A. Baris, F. Carere, A. In vitro and in vivo mutagenicity studies with airborne particulate extracts, Mutat. Res. 204 1988 565–575.

[7]   Crebelli, R. Fuselli, S. Baldassarri, L. Turrio Ziemacki, G. Carere, A. Benigni, R. Genotoxicity of urban air particulate matter: correlations between mutagenicity data, airborne micropollutants, and meteorological parameters, Int. J. Environ. Health Res. 5 1995 19–34.

[8]   De Raat, W.K. Polycyclic aromatic hydrocarbons and mutagens in ambient air particles, Toxicol. Environ. Chem. 16 1988 259–279.

[9]   Hadnagy, W. Seemayer, N.H. Tommingas, R. Ivanfy, K. Comparative study of sister-chromatid exchanges and chromosomal aberrations induced by airborne particulates from an urban and a highly industrialized location in human lymphocyte cultures, Mutat. Res. 225 1989 27–32.

[10]   Viras, L.G. Athanasiou, K. Siskos, P.A. Determination of mutagenic activity of airborne particulates and of the benzo *a* pyrene concentrations in Athens atmosphere, At mos. Environ. B 24 1990 267–274.

[11]   Viras, L.G. Siskos, P.A. Samara, C. Kouimtzis, T. Athanasiou, K. Vavatzandis, A. Polycyclic aromatic hydrocarbons and mutagens in ambient air particles sampled in Thessaloniki, Greece, Atmos. Environ. 10 1991 999–1007.

[12]   Fuselli, S. Benigni, R. Conti, L. Carere, A. Crebelli, R.. Volatile organic compounds VOCs and air mutagenicity: results of one year monitoring at an urban site, Int. J. Environ. Health Res. 5 1995 123–132.

[13]   Pagano, P. De Zaiacomo, T. Scarcella, E. Bruni, S. Calamosca, M. Mutagenic activity of total and particle-sized fractions of urban particulate matter, Environ. Sci. Technol. 30 1996 3512–3516

[14]   Kim Oanh, N.T. Upadhyay, N. Zhuang, Y.-H. Hao, Z.-P. Murthy, D.V.S. Lestari, P. Villarin, J.T. Chengchua, K. Co, H.X. Dung, N.T. Lindgren, E.S. 2006 Particulate air pollution in six Asian cities: Spatial and temporal distributions, and associated sources, Atmospheric Environment 40 3367-3380

[15]   Pershagen, G. Simonato, L. Epidemiological evidence on air pollution and cancer, in: L. Tomatis Ed. , Air Pollution and Human Cancer, Monographs, European School of Oncology, Springer-Verlag, New York, 1993, pp. 64–74

[16] Obot, C. J.; et al. A comparison of murine and human alveolar macrophage responses to urban particulate matter. *Inhalation Toxicol.* 2004, *16* (2), 69-76.

[17] Hannigan, M. P.; et al. Bacterial Mutagenicity of Urban Organic Aerosol Sources in Comparison to Atmospheric Samples. *Environ. Sci. Technol.* 1994, *28* (12), 2014-2024.

[18] Oanh, N. T. K.; Nghiem, L.; Phyu, Y. L. Emission of polycyclic aromatic hydrocarbons, toxicity, and mutagenicity from domestic cooking using sawdust briquettes, wood, and kerosene. *Environ. Sci. Technol.* 2002, *36* (5), 833-839.

[19] Pedersen, D. U.; et al. Human-cell mutagens in respirable airborne particles in the Northeastern United States. 1. Mutagenicity of Fractionated Samples. *Environ. Sci. Technol.* 2004, *38* (3), 682-689.

[20] Google Earth, http://earth.google.com.

[21] B. B. Hicks, R. A. Valigura, F. B Courtright, The role of the atmosphere in coastal ecosystem declines Future research directions. *Estuaries*, **2000** *23* (6), 854-863.

[22] Sheesley, R. J.; et al. , Toxicity of ambient atmospheric particulate matter from the Lake Michigan (USA) airshed to aquatic organisms. *Environ. Toxicol. Chem.*, in press

Influence of ozone air contamination on the number of deaths ascribed to respiratory causes in Madrid (Spain) during 1990–1998

P. Fernández [1], R. Herruzo[2], A. Justel[3] & F. Jaque[1]
[1]Facultad de Ciencias, Universidad Autónoma de Madrid, Spain
[2]Departamento de Medicina Preventiva y Salud Pública,
Facultad de Medicina, Madrid, Spain
[3]Departamento de Matemáticas, Universidad Autónoma de Madrid,
Madrid, Spain

Abstract

In this paper the influence of the air quality on the mortality associated with respiratory causes in the Madrid (Spain) area during 1990-1998 is analysed. Air pollutants and mortality show that during this period there were two different effects; periodic annual fluctuations plus a monotonic evolution. It has been found that the slope values and signs of the monotonic components are different for each pollutant and mortality ascribed to respiratory causes; negative for particles, SO_2, and NO_x and positive for the O_3 concentration and mortality values. A casual relationship between the mortality ascribed to respiratory causes and ozone concentration during the 1900-1998 period has been found. The variation detected in the Ultraviolet Index and traffic in Madrid City during the 1990-1998 period is also considered
Keywords: air pollution, ozone and nitrogen oxides trends, ultraviolet radiation, health effects.

1 Introduction

In the last years, much effort has been given to the health effects of air contamination, detecting a correlation between air pollution and daily mortality realised in the U.S and Europe (Brimblecombe [1] and Brunekreef and Holgate [2]). A large study concerning the correlation between air pollution and mortality was carried out in 20 U.S cities during the 1987-1994 period (Samet *et al* [3]). In Europe, work in this topic has been developed in the frame of the APHEA project (Air Pollution and Health: an European Approach) (Katsouyami *et al* [4] and Le Trete *et al* [5]). It has been recently suggested that small increases in air pollution levels could affect the mortality rate. In particular, an increase in deaths from cardiovascular and respiratory causes has been correlated with enhancement in air contamination (Robert *et al* [6]). This conjecture is generally accepted, even considering the very low contamination levels of exposure, which are far from being levels that are normally considered unsafe. Among all air pollutants, particles (PM_{10}) and sulphur dioxide (SO_2) have been established as having a strong association with health effects. However, as the concentration of both pollutants in air has decreased strikingly, attention has shifted to ozone (O_3), nitrous oxides (NO_x) and small particles $PM_{2.5.}$ In fact, weak evidence that increases in ozone levels increased the relative rates of deaths during the summer period has been found in some U.S. cities [3].

In this work, a study of air pollution and the mortality rate due to respiratory causes in the area of Madrid (Spain), made over the 1990-1998 period, is presented.

2 Experimental procedure

The study was carried out in the area of Madrid (Spain). Its population is 5.423.384, according to the 2001 population census.

The information related to the deaths associated with respiratory causes was supplied by the National Institute of Statistics. Table I shows the codes of the respiratory diseases. Information regarding the daily values of surface air pollutant concentrations was supplied by the Community of Madrid Air Pollution Monitoring Network (MAPMN). The basic parameters, which this network deals, are the following: particles in suspension (PM_{10}), SO_2, NO_x, CO, and O_3. The analytical methods are that established within the context of the European Union.

Table 1:   Codes of the respiratory diseases.

Code	
J00-J99	All respiratory causes
J10,J11	Influenza
J12-J18	Pneumonia
J40-J44,J45-J47	Chronic illness and asthma
Rest of J00-J99	Other respiratory causes

The concentration values were obtained according to the European norms 85/203/CEE and 92/72/CEE respectively.

Unfortunately, only the monthly mortality data were available in the 1990-1998 period. Therefore, from the daily data the monthly concentration values for all pollutants were calculated for their posterior comparison with the mortality data. It is evident that working with monthly means, the effects associated with short term events are minimised and some information could be masked.

The current network has 27 fixed monitoring sets for measuring the levels of air pollutants in the Madrid area.

Transport is considered to be one of the principal sources of pollutant emission, taking into account that there is not a great deal of heavy industry located in the area of Madrid. Madrid parking was supplied by the Traffic Office of Madrid.

The Ultraviolet Index (UVI) data were supplied by the National Institute of Meteorology.

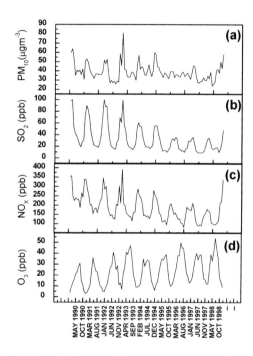

Figure 1:   Evolution of the monthly mean concentration values of PM$_{10}$ (a), SO$_2$ (b), NO$_x$ (c) and O$_3$ (d).

3   Results and discussion

Fig. 1 shows the monthly mean concentration values of PM$_{10}$ (a), SO$_2$ (b), NO$_x$ (c) and O$_3$ (d) during the 1990-1998 period. The experimental data indicate that the tropospheric concentration of PM$_{10}$, SO$_2$, CO$_2$ and NO$_x$ show, in addition to a periodic annual fluctuations (PAF) with maximum values around the winter session, a monotonic descending component (tend component (TC)). The evolution of the O_3 content in the air shows in comparison with the former pollutants two different facts; the maximum values are centred on the summertime and the trend component (TC) shows a positive slope.

Fig. 2(a) shows the evolution in the same period of 1990-1998 the number of monthly deaths associated with all respiratory causes (a) and other respiratory causes (b) (see Table 1). As can be seen, together with the PAF effect, a monotonous increase (TC) in the number of deaths is observed. It is important to point out that in the average of the PAF due to respiratory causes, calculated by using the Statistical Analysis Software Program (SPSS), two peaks are perceived; an intense peak in winter and another less intense centred in summertime.

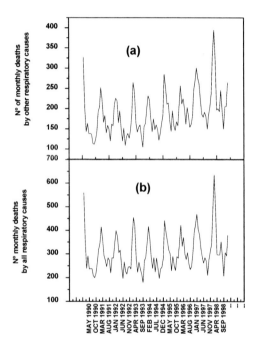

Figure 2:   Monthly deaths associated with other respiratory causes (a) and all respiratory causes (b).

Considering these experimental results it is reasonable to split the experimental set data in the PAF and TC components. In addition, as the O_3 is the unique pollutant that presents a positive TC slope, the study has been restrict between the mortality associated with respiratory causes and the ozone concentration, taking into account only the trend components.

Fig. 3 shows the TC components obtained using the Statistical Analysis Software Program (SPSS) for the Ozone concentration (a), monthly deaths associated with other respiratory causes (b) and the monthly deaths ascribed to all respiratory causes (c) during the 1990-1998 period. As can be observed the monthly mean ozone values fluctuate between $8 - 35$ μgm^{-3} and the number of deaths associated with respiratory causes between 200-400 (all respiratory causes) and 120-250 (others causes). It is important to remark that these ozone values are very far below to those considered as potentially hazardous to health. In fact, the "safe" ozone level protection for the population is considered to be 120 μgm^{-3}.

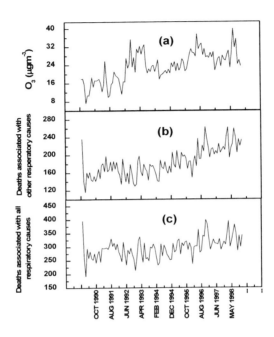

Figure 3:   Tends components (TC): (a) ozone concentration (b) and (c) monthly deaths associated with other and all respiratory respectively.

In a first approach, the scatter-plot diagram of ozone concentration versus mortality by all respiratory and other respiratory causes indicates that despite the considerable scatter of the data, there is still a significant correlation with $R = 0.465$ and $p < 0.0001$ between the number of deaths associate with other respiratory causes and the ozone concentration. When all respiratory causes are considered, the correlation is less significant ($R < 0.163$, $p < 0.19$). Never the less, it has been found that the monthly rate from other respiratory causes increases by ~11% respectively for each 10 $\mu gr/m^3$ augmentation of the O_3 concentration level. These values are considerably higher than those reported for the U.S cities [3] with the consideration that this value was obtained considering all death causes and not only respiratory causes as in the present work. In an early work (Diaz *et al* [7]) completed in Madrid City a value of 12% for an augmentation of 25 $\mu gr/m^3$ has been reported which is in acceptable accordance with the datum calculated in this work.

WIT Transactions on Biomedicine and Health, Vol 10, © 2006 WIT Press
www.witpress.com, ISSN 1743-3525 (on-line)

The Box-Jenkins metrology has been used for obtaining multivariate, autoregressive integrated moving average (ARIMA) model of the time–based series taken into consideration. Multivariate models show a 8% (95% CI: 1.4 to 14.8) increase in mortality for each $10\mu g/m^3$ increase in the ozone concentration.

It is generally accepted that the formation of ozone in the troposphere includes a complex series of ultraviolet photolysis reaction involving nitrogen dioxide and hydrocarbons. The major source of anthropogenic emission of NO_x is the combustion of fossil fuels: heating, power generation and motor vehicles. In particular, in the Madrid area, transport is the basic source of the NO_x contamination. In this point it is important to remark that in the period 1995-2001 the Madrid parking increased ~33%.

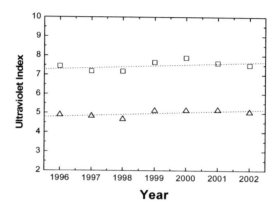

Figure 4: Variation of the Ultraviolet Index during the 1996-2002 period in Madrid.

Fig. 4 shows the variation of the Ultraviolet Index (UVI) (annual and summertime) during the 1996-2002 period, which were the only available data from the National Institute of Meteorology of Madrid. The experimental points correspond to the average of all daily Ultraviolet Index (UVI) maximum for a specific year (□) or summer season (Δ). As can be observed, a positive and linear trend with a slope ~ 7% is found for both sets of data. This value is in good concordance with the European Environmental Agency [8] that reports an increase of the UV radiation intensity between 7-9% in the Madrid area during the 1980-1997 period. More explicitly, it has been published an enhancement of the UV radiation intensity in Salonika City (Greece) of ~5% in the 1990-1997 period (Zeferos et al [9]). Therefore, the coincidence of these two phenomena in Madrid City, the increase in the UV radiation level and traffic can explain the rise in ozone concentration.

As has been commented before, a week peak in the number of deaths due to respiratory causes during the summer period was observed, which appears much closer to the maximum value of the O_3 concentration. It is clear that the application of time series model covering only the summer months during the

1990-1998 period is not fully correct. However, a Pearson regression could add light over the Ozone effect in the mortality associated to respiratory causes. Following this argument, fig. 5 shows the Pearson regression between the monthly deaths associated with other respiratory causes and the monthly mean values of the O_3 concentration for the summer period. Considering this set of data, a value of R=0.729 is reached, which is significantly higher to that reported in this work when all months are considered. An inspection of fig. 5 reveals that the monthly rate for other respiratory causes increases by a ~17% for each 10 $\mu gr/m^3$ augmentation of O_3 concentration level. This value is higher than that obtained for the TC in this work when all months are used and considerably higher than that reported for the U.S cities [3] with the consideration that this value was obtained using all death causes and not only respiratory causes, as in the present work.

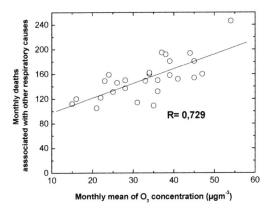

Figure 5:   Regression between the monthly deaths associated with other respiratory causes and the monthly mean values of O_3 concentration for the summertime.

In summary, this work supports that there is a casual relationship between air pollution associated with the Ozone content and mortality ascribed to respiratory causes associated with the monotonic component (TC) during the 1990-1998 period.

Acknowledgements

The authors would like to express their gratitude to Professors G. Lifante and P.D. Townsend for helpful discussions and continuous encouragement. F.J. would like to acknowledge the financial support of MAT2005- 05950 project.

References

[1] Brimblecombe P (eds). *Air pollution and health history.* 5-18 Eds Academic Press, San Diego 1990.

[2] Brunekreef. B., Holgate S. T. *Air pollution and health (review).* Lancet 360, pp. 1233-1242, 2002.

[3] Samet J.M. et al. *Fine particulate air pollution and mortality in 20 US cities 1987,1994.* N. Engl. J. Med, pp.1742-1749, 2000.

[4] Katsouyami K., Zmirou D., Spix C. *Short-term effects of air pollution on health: a European approach using epidemiological time series.* Eur. Respir. J. 8, pp. 1030-1038, 1995.

[5] Le Tertre A. et al. *Short-term effects of particulate air pollution on cardiovascular diseases in eight European cities.* J. Epidemol. Community Health 56, pp.773-779, 2002.

[6] Robert D. et al. *Air pollution and cardiovascular disease: a statement for healthcare professional from the Expert Panel On Population and Prevention Science of the Amerial Heart Association.* AHA Scientific Statement Circulation. 109, pp. 2655-2671, 2004.

[7] Diaz J. et al. *Modelling of air pollution and its relationship with mortality and morbidity in Madrid Spain.* Int. Arch. Occup. Environ. Health. 72, pp. 366-376, 1999.

[8] Europe's Environment: The Second Assessment 1998. Elsevier Science Ltd Luxemburg ,1998.

[9] Zeferos C., Meleti C., Balis D., Tourpali K., Bais A. F. Quasil-biennial and longer-term changes in clear sky UV-B solar irradiance. Geophys. Res. Letters. 25, 4345-4348, (1008).

Risk assessment of toxic *Cyanobacteria* in Polish water bodies

J. Mankiewicz-Boczek[1], K. Izyorczyk[1] & T. Jurczak[2]
[1]*International Centre for Ecology, Polish Academy of Sciences, Łódź, Poland*
[2]*Department of Applied Ecology, University of Łódź, Poland*

Abstract

Cyanobacteria and cyanotoxins constitute a serious problem for water supply systems and recreation. The most common in Polish water bodies are microcystin-producing *Cyanobacteria*. Polish regulations regarding the quality of drinking water indicate the limit of 1 µg/l for microcystin-LR. Regulations for recreational water are addressed through a guideline that the water should only be inspected visually. Poland's regulations, unfortunately, do not recommend any specific standardized methods, equipment or programmes for the risk assessment resulting from *Cyanobacteria* exposure. Therefore, this study is aimed at developing and evaluating a toxicity monitoring programme that would be based on the early identification of microcystin-producing genera, followed by examination of their toxic potential and toxicity. The risk of cyanobacterial blooms was assessed in six eutrophic water bodies located in different regions of Poland in the summers of 2004 and 2005. The first step of the proposed monitoring programme included a microscopic analysis of seasonal variation in phytoplankton composition and measurement of the cyanobacterial biomass. Then the toxigenic strains of cyanobacteria were determined by amplification of selected *mcy* genes in the microcystin biosynthesis pathway using PCR. The presence of the *mcy* genes in sampling water were detected at the beginning of the July, which coincided with a low cyanobacterial biomass (0.1 mg/l). In the last step, the water samples toxicity and microcystins concentration were examined by a screening method, called protein phosphatase inhibition assay, and an analytical method, called high-performance liquid chromatography, respectively. The greatest concentration of microcystins were found to be 31.22 µg/l (October) and 11.25 µg/l (September) in samples dominated by *Planktothrix agardhii* and *Microcystis aeruginosa,* respectively. Since a relationship between amplification of *mcy* genes and production of microcystins was not always found, it seems to be necessary to use a microscopic analysis and molecular method in parallel with toxicity analysis for providing more complete information on the cyanobacterial risk in water.
Keywords: cyanobacteria, toxigenic strains, mcy genes, microcystin, toxicity.

WIT Transactions on Biomedicine and Health, Vol 10, © 2006 WIT Press
www.witpress.com, ISSN 1743-3525 (on-line)
doi:10.2495/ETOX060061

1 Introduction

In many countries, including Poland, the problem of toxicity caused by cyanobacterial blooms has important consequences [1–5]. Cyanobacteria and cyanotoxins are dangerous for water supply systems and recreation [5, 6]. The most common in Polish water bodies are microcystin-producing cyanobacteria. At present, the toxins (cyanotoxins) that are produced by cyanobacteria can be classified into four main groups: hepatotoxins, neurotoxins, cytotoxins and dermatotoxins [4, 7]. Microcystins, including hepatotoxins, are the best-known group. They are synthesized nonribosomally by a peptide synthetase polyketide synthase enzyme complex encoded by the microcystin synthetase gene cluster (*mcyA-J*), which contains 55 kb of DNA [8, 9]. The microcystin biosynthesis *mcy* genes have been used to establish molecular techniques for the detection of toxigenic cyanobacteria. Microcystins are one of the most toxic and dangerous cyanotoxins. Contact with microcystin-producing cyanobacteria in bathing water can cause skin irritations, allergic reactions and gastrointestinal symptoms [4, 10]. Moreover, chronic exposure to low microcystins concentration in drinking water can lead to cancer promotion [11, 12].

According to WHO [13, 14] and Pilotto *et al* [15] the concentrations of microcystins above 2–4 µg/l in the samples indicated a first alert level for health risks. Irritative or allergenic effects were observed on 30% of exposed people on microcystin-producing cyanobacteria [15]. The first level of risk means a relatively low probability of adverse health effects. The second level of alert for recreational water was proposed if the concentration of microcystins is in the region of 20–50 µg/l for the bloom consisted of *Microcystis* occured. However, toxic levels may approximately double if *Planktothrix agardhii* dominates. If the first or second level of alert appears, information to the relevant health authorities is required. Additionally, the second level of alert requites an introduction of bathing restriction and further investigation of the toxicity risk. The third alert level includes an intensive scum forming by cyanobacterial water bloom. In this case, swimming and other water-contact activities must be strictly forbidden.

The Polish regulation regarding the quality of drinking water indicates the limit of 1 µg/l for microcystin-LR [16]. Regulation for recreational water is addressed through a guideline that the appearance of cyanobacterial blooms should be controlled by observation of water colour, turbidity and/or odour [17]. Poland's regulations, unfortunately, do not recommend any specific standardized methods, equipments or programmes for the risk assessment resulting from *Cyanobacteria* exposure.

Therefore, this study is aimed at developing and evaluating a toxicity monitoring programme that would be based on the early identification of microcystin-producing genera, followed by an examination of their toxic potential and toxicity. The risk of cyanobacterial blooms was assessed in six eutrophic water bodies (105 water samples) located in different regions of Poland in the summers of 2004 and 2005. The presence and identification of toxigenic strains was studied by PCR amplification of a *mcy* genes. Sample toxicity and microcystins production were established by protein phosphatase inhibition assay (PPIA) and high-performance liquid chromatography (HPLC).

2 Material and methods

2.1 Study sites

Six water bodies from three different regions of Poland were studied: Lake Jeziorak (Northern Poland); lakes Bnińskie and Bytyńskie (Western Poland) and reservoirs Sulejów, Jeziorsko and Zegrzyn (Central Poland). The water bodies were selected because of the high intensity of recreational activities. Additionally, Zegrzyn Reservoir is a source of drinking water for Warsaw, the capital city of Poland and Sulejów Reservoirs is an alternative source of drinking water for the city of Łódź (1 million inhabitants).

2.2 Sampling and phytoplankton analysis

Water samples of the surface layer (0–0.5 m) were collected every two weeks from studied water bodies, from July to October 2004 and 2005. The chemical parameters of the monitored lakes and reservoirs were determined according to standard methods [18]. For DNA analysis, 100–200 ml of water were filtered and filters were stored at -20°C until DNA extraction. For the PPIA and HPLC analysis, water samples containing cyanobacterial material were filtered immediately on a GF/C (Whatman, England) filter and stored at 4°C for a maximum of 24 h. After that, toxicity and microcystins concentration were determined.

Water samples for phytoplankton estimation were preserved in Lugol's solution and sedimented in the laboratory. Phytoplankton was counted using a Fusch-Rosenthal counting cell. The phytoplankton biomass (fresh weight) was determined based on a volumetric analysis of cells using geometric approximation. Biomass computed in volume units was transposed to freshmass (FM) assuming the specific mass of phytoplankton as a unit (=1) [19].

Concentrations of selected nutrients were assessed by spectrophotometer DR 2010 HACH.

2.3 PCR – sample preparation and amplification

Nucleic acid extraction from the filters was performed as described in Giovannoni et al [20]. The PCR reactions for the amplification of selected *mcy* genes were carried out as described previously [21–23]. The part of *mcyA* (291–297 bp) was amplified with Cd1F (5'-aaaattaaaagccgtatcaaa-3') and Cd1R (5'-aaaagtgttttattagcggctcat-3') primers [21]. An 758 bp region of the *mcyB* was amplified with FAA (5'-ctatgttatttatacatcagg-3') and RAA (5'-ctcagcttaacttgattatc-3') primers [22]. For amplification of the *mcyD* fragment (818 bp), the primers of mcyDF (5'-gatccgattgaattagaaag-3') and mcyDR (5'-gtattccccaagattgcc-3') were used [23]. An 809-812 bp region of the *mcyE* was amplified with the mcyE-F2 (5'-gaaatttgtgtagaaggtgc-3') and mcyE-R4

(5'-aattctaaagcccaaagacg-3') [23]. PCR products were separated by 1.5% agarose gel electrophoresis, using a constant voltage (85 V) and visualized using ethidium bromide.

2.4 Estimation of microcystins toxicity

2.4.1 Protein phosphatase inhibition assay (PPIA)
The inhibition of protein phosphatase type 1 (PP1) from rabbit skeletal muscle (BioLabs, New England) by microcystins was estimated according to An and Carmichael [24] with minor modification. The samples were incubated with 2.5 U of PP1 and 15 mM p-nitrophenol phosphate (Fluka, UK) as a substrate was used. PPIA enables detection of microcystins at the 0.125 µg/l level, without concentrating the sample.

2.4.2 High performance liquid chromatography (HPLC)
The preparation of cyanobacterial material and determination of microcystins by HPLC-DAD was performed according to previous description [6]. Total microcystins concentration (extra and intracellular) was measured. The HPLC method enables the detection of microcystin at the 0.01 µg/l level after concentrating the samples.

3 Results and discussion

The risk of toxic cyanobacterial blooms was measured in six water bodies (lake Jeziorak and lake Bnińskie as well as the Bytyńskie, Sulejów, Jeziorsko and Zegrzyn reservoirs), which represent three different regions of Poland. These water bodies are the most intensively used for recreation in each region. Additionally, Sulejów and Zegrzyn reservoirs are a source of drinking water for the city of Łódź and Warsaw, respectively. We studied 105 of the water samples collected in summer season from July to October in 2004 and 2005.

The chemical conditions in water bodies supported the occurrence of cyanobacterial blooms. The average concentration of total phosphorus (TP) and total nitrogen (TN) almost exceeded the limits for eutrophication of reservoirs (0.1 mg P/l and 1.5 mg N/l) [25] (Tab.1). The low TN/TP ratio is a factor responsible for the dominance of bloom-forming cyanobacteria [26, 27].

Microscopic analysis as a first step of the proposed monitoring programme enabled description of phytoplankton composition and measurement of the cyanobacterial biomass (Tab. 1 and 2). Hepatotoxic cyanobacteria with domination by *Microcystis* were detected in Sulejów Reservoir, Zegrzyn Reservoir and Jeziorsko Reservoir in 2004 and 2005, also these genera were detected in Lake Jeziorak. Lakes Bnińskie and Bytyńskie were dominated by *Planktothrix agardhii* with the highest biomass over 50 mg/l. Additionally, *Cylindrospermopsis raciborskii* occurred in Western lakes, in 2005.

The inability to differentiate between toxic and nontoxic strains of cyanobacteria by morphological analysis lead to applying sensitive molecular methods, which enable identification of microcystin-producing strains [21, 28, 29]. The toxigenic strains of cyanobacteria were determined by amplification of selected *mcy* genes (*mcyA*, *mcyB*, *mcyD* and *mcyE*) in the microcystin biosynthesis pathway. Genetic analysis of water samples containing cyanobacteria collected from monitored water bodies indicated that the potential to produce toxins was determined early in July and persisted throughout summer monitoring, until October 2004 and 2005 (Fig. 1). The presence of the *mcyA, B, D* and *E* genes in sampling water at the beginning of the July coincided with a low cyanobacterial biomass (0.1 mg/l) (Fig. 1 and Tab. 2). The *mcyE* gene fragment was amplified in all water samples collected, irrespective of source or date, with the exception of the Jeziorsko reservoir. In this case, the *mcyE* gene was amplified less often in 75% in summer season 2005. The *mcyD* was also amplified in all water samples but only in summer 2005. The other *mcy* genes were amplified less often, depending on the water body (Fig. 1). These results suggested that the *mcyA, mcyB, mcyD* and *mcyE* genes were different in being able to predict microcystin toxigenicity. The molecular analysis indicated that the *mcyE* gene that encode the glutamate-activating adenylation domain is the best molecular marker for the determination of toxigenic strains of cyanobacteria in different water samples even if their biomass is very low (0.1 mg/l).

Table 1:   The average concentration of nutrients and microcystin- producing cyanobacteria in studied water bodies in the summers of 2004 and 2005.

Water body	Year	TN	TP	*TN/TP	Cyanobacteria
Sulejów	2004	1.00	0.14	9.70	*Microcystis aeruginosa*
	2005	1.18	0.10	10.85	*Microcystis aeruginosa*
Jeziorsko	2004	1.20	0.10	11.60	*Microcystis aeruginosa*
	2005	2.84	0.59	9.48	*Microcystis aeruginosa*
Zegrzyn	2004	1.40	0.20	8.50	*Microcystis aeruginosa, Microcystis viridis*
	2005	1.02	0.21	6.18	*Microcystis aeruginosa*
Bnińskie	2004	2.30	0.21	12.40	*Planktothrix agardhii,*
	2005	2.21	0.10	16.06	*Planktothrix agardhii, Cylindrospermopsis raciborskii*
Bytyńskie	2004	0.80	0.11	10.10	*Planktothrix agardhii*
	2005	1.84	0.12	18.5	*Planktothrix agardhii*
Jeziorak	2004	1.50	0.08	20.90	*Microcystis aeruginosa*
	2005	1.46	0.11	12.63	*Microcystis aeruginosa*

* - TN/TP – the ratio between total nitrogen (TN) and total phosphorus (TP).

WIT Transactions on Biomedicine and Health, Vol 10, © 2006 WIT Press
www.witpress.com, ISSN 1743-3525 (on-line)

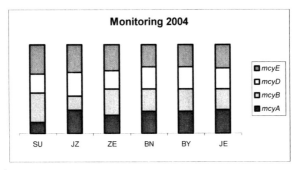

A.

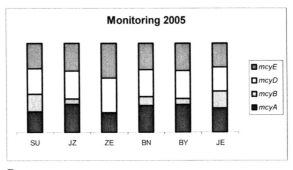

B.

Figure 1:   PCR results of *mcy* genes amplification in summer A) 2004 and B) 2005.

However, although genetic markers provided information about potential toxicity they did not indicate what the real risk to health was. Therefore, toxicological analysis of water as the last step of the proposed monitoring programme was needed. Microcystins production and their toxicity was measured by biochemical (PPIA) and analytical (HPLC) methods. The presented data indicated variations in microcystin concentrations in water samples collected from different water bodies (Tab. 1). Additionally, seasonal variations were observed also. Samples collected in 2004 were more toxic than samples from 2005. The greatest concentration of microcystins were found to be 31.22 µg/l (Lake Bnińskie, October 2004) and 11.25 µg/l (Jeziorsko reservoir, September 2004) in samples dominated by *Planktothrix agardhii* and *Microcystis aeruginosa,* respectively (Tab. 1 and 2). In Sulejów reservoir the highest microcystins concentration (4.67 µg/l) was found in September 2004. The low microcystins concentration with a maximum of 0.63 µg/l was detected in Lake Jeziorak. These variations detected in the water body might be a result of population dynamics altering the proportion of toxic genotypes within the population of cyanobacteria [30]. Additionally, physicochemical conditions that were not studied could be an influence on the dynamics of cyanobacteria growth and their toxigenicity (microcystins production). Microcystins were almost not detected in the Zegrzyn reservoir in spite of *mcy* genes amplification. Non-toxic strains do not generally contain the *mcy* genes. However, some strains may have fragments of microcystin synthetase genes or mutations within these genes. Therefore, these strains can be amplified with *mcy* primers, although they are not able to produce toxins [29]. Nishizawa *et al* [31] suggested that non-toxic *Microcystis* strains comprise two groups: those with and those without *mcyABC*.

Table 2: The maximum of the microcystins concentration and biomass of hepatotoxic cyanobacteria in the summers of 2004 and 2005.

Water body		July 2004	July 2005	August 2004	August 2005	September 2004	September 2005	October 2004	October 2005
Sulejów	*Biomass	0.12	43.64	0.87	29.58	9.70	8.12	1.30	2.31
	PPIA	3.96	2.83	3.48	2.27	>4.00	0.76	0.94	1.09
	**HPLC					4.67			
Jeziorsko	*Biomass	0.90	0.97	52.55	0.76	27.3	34.22	0.61	0.93
	PPIA	0.36	0.72	3.67	0.85	>4.00	1.55	0.41	0.56
	**HPLC					11.25			
Zegrzyn	*Biomass	0.11	0.57	3.62	0.41	1.33	0.62	2.23	0.12
	PPIA	<0.13	0.32	<0.13	<0.13	<0.13	<0.13	0.42	<0.13
	**HPLC	n.d.		n.d.	n.d.	n.d.	0.05		n.d.
Bnińskie	*Biomass	50.50	0.03	32.75	0.06	51.13	1.99	33.71	3.66
	PPIA	>4.00	0.66	>4.00	0.56	>4.00	0.82	>4.00	2.19
	**HPLC	10.98		10.67		11.13		6.98	
Bytyńskie	*Biomass	31.5	0.18	40.18	0.453	44.55	1.08	53.54	1.813
	PPIA	>4.00	0.60	>4.00	0.48	>4.00	0.63	>4.00	0.59
	**HPLC	12,61		16.06		31.22		18.62	
Jeziorak	*Biomass	1.20	5.9	2.30	3.39	0.10	1.04	4.10	3.78
	PPIA	0.34	0.50	0.31	0.52	0.47	0.17	0.23	0.63
	**HPLC								

* - biomass of hepatotoxic cyanobacteria [mg/l] in samples with maximum microcystins concentration.
** - if microcystins concentration was below 0.13 µg/l or exceeded 4 µg/l, the high performance liquid chromatography (HPLC) technique was used.

The highest microcystins concentration (>20 µg/l) corresponds to the second alert level for recreational water as determined for the lakes of Western Poland. In Central Poland the concentration of microcystins reached the first alert level (>2 µg/l). In the lake in Northern Poland the occurrence of microcystin did not exceed 1 µg/l in the whole sampling period.

In conclusion, the microscopic analysis of water samples from six Polish water bodies indicated an appearance of microcystin-producing cyanobacteria. The most common was *Microcystis* and *Planktothrix* genera. The microscopic detection of cyanobacteria and determination of toxigenic strains by molecular markers (*mcy* genes) was an early signal of potential toxicity of cyanobacterial blooms, and acts as an effective alert to possible health risk. However, only biochemical and analytical methods enabled determination of microcystins concentration in studied water bodies. Therefore, it seems to be necessary to use a microscopic analysis and molecular method in parallel with toxicity analysis for providing complete information on the cyanobacterial risk in water.

Acknowledgments

These studies were supported by State Committee for Scientific Research 2PO4F 044 27.

References

[1]   Fleming, L.E., Rivero, C., Burns, J., Williams, Ch., Bean, J.A., Shea, K.A. & Stinn, J., Blue green algal (cyanobacterial) toxins, surface drinking water, and liver cancer in Florida. *Harmful Algae*, **1**, pp. 157-168, 2002.

[2]   De Figueiredo, D.R., Azeiteiro, U.M., Esteves, S.M., Goncalves, F.J.M. & Pereira, M.J., Microcystin-producing blooms – a serious global public health issue. *Ecotoxicology and Environmental Safety,* **59**, pp. 151-163, 2004.

[3]   Pawlik-Skowrońska, B., Skowroński, T., Pirszel, J. & Adamczyk A., Relationship between cyanobacterial bloom composition and anatoxin-a and microcystin occurrence in the eutrophic dam reservoir (SE Poland). *Polish Journal of Ecology*, **52**, pp. 479-490, 2004.

[4]   Mankiewicz, J., Tarczyńska, M., Walter, Z. & Zalewski, M., Natural Toxins from Cyanobacteria (Blue-Green Algae). *Acta Biologica Cracoviensia,* **45(2)**, pp. 9-20, 2003.

[5]   Mankiewicz, J., Komarkova, J., Izydorczyk, K., Jurczak, T., † Tarczyńska, M. & Zalewski M., Hepatotoxic cyanobacterial blooms in lakes of North Poland. *Environmental Toxicology*, **20**, pp. 499-506, 2005.

[6]   Jurczak, T., Tarczyńska, M., Izydorczyk, K., Mankiewicz, J., Zalewski, M. & Meriluoto J., Elimination of microcystins by water treatment process – examples from Sulejow Reservoir, Poland. *Water Research,* **39**, pp. 2394-2406, 2005.

[7]   Codd, G.A., Morrison, L.F. & Metcalf, J.S., Cyanobacterial toxins: risk management for health protection. *Toxicology and Applied Pharmacology*, **203**, pp. 264-272, 2005.

[8]   Tillett, D., Dittmann, E., Erhard, M., von Döhren, Börner, T. & Neilan, B.A., Structural organization of microcystin biosynthesis in Microcystis aeruginosa PCC7806 : an integrated peptide-polyketide synthetase system. *Chemistry & Biology,* **7**, pp. 753-764, 2000.

[9]   Christiansen, G., Fastner, J., Erhard, M., Borner, T. & Dittman E., Microcystin Biosynthesis in *Planktothrix*: Genes, Evolution and Manipulation. *Journal of Bacteriology*, **185**, pp. 564-572, 2003.

[10]   Chorus, I. & Bartram J., *Toxic Cyanobacteria in Water. A Guide to their Public Health Consequences, Monitoring and management*, E.&F. N. Spon: London, 1999.

[11]   Carmichael, W.W., Heath Effects of Toxin-Producing cyanobacteria: The CyanoHABs. *HERA*, **7**, pp. 1393-1407, 2001.

[12]   Briand, J.F., Jacquet, S., Bernard, C. & Humbert, J.F., Health hazards for terrestrial vertebrates from toxic cyanobacteria in surface water ecosystems. *Veterinary Research*, **34**, pp. 361-377, 2003.

[13]   Falconer, I.R., An overview of problems caused by toxic Blue-Green Algae (Cyanobacteria) in drinking and recreational water. *Environmental Toxicology and Water Quality*, **14**, pp. 5-12, 1999.

[14]   WHO, Guidelines for safe recreational water environments. Coastal and fresh waters, **1**, pp.150, 2003.

[15]   Pilotto, L.S., Burch, M.D., Douglas, R.M., Cameron, S., Roach, G.J., Cowie, C.T., Beers. M., Robinson, P., Kirk, M., Hardman, S., Moore, C. & Attewell, R.G., Health effect of exposure to cyanobacteria (blue-green algae) during recreational water activities. *Australian and New Zealand Journal of Public Health*, **21**, pp. 562-566, 1997.

[16]   Ministry of Health regulation, Regulation of the Minister of Health of 19/11/02 on the requirements concerning the quality of drinking water. *Journal of Laws* № 203/2002, item 1718, 2002a.

[17]   Ministry of Health regulation, Regulation of the Minister of Health of 16/10/02 on the requirements concerning the quality of bathing water. *Journal of Laws* № 183/2002, item 1530, 2002b.

[18]   Golterman, H.L., Clynio, R.S. & Ohstand M.A.M., *Methods for physical and chemical analysis of freshwater*, IBP Hand Book, 1988.

[19]   Komárková, J., Vyhnalek, V. & Kubecka J., Impact of fishstock manipulation on the composition of net phytoplankton in the Rimov Reservoir (Czech Republic). *Water Science and Technology*, **32**, pp. 207-216, 1995.

[20]   Giovannoni, S.J., DeLong, E.F., Schmidt, T.M. & Pace N.R., Tangential flow filtration and preliminary phylogenetic analysis of marine picoplankton. *Applied and Environmental Microbiology*, **568**, pp. 2572-2575, 1990.

[21]   Hisbergues, M., Christiansen, G., Rouhiainen, L., Sivonen, K. & Borner T., PCR-based identification of microcysin-producing genotypes of different cyanobacterial genera. *Archives of Microbiology*, **180**, pp. 402-410, 2003.

[22]   Bittencourt-Oliveira, M., Detection of potential microcystin-producing cyanobacteria in Brazilian reservoirs with a *mcyB* molecular marker. *Harmful Algae*, **2**, pp. 51-60, 2003.

[23]   Rantala, A., Fewer, D.P., Hisbergues, M., Rouhiainen, L., Vaitomaa, J., Börner, T. & Sivonen, K., Phylogenetic evidence for the early evolution of microcystin synthesis. *PNAS,* **101(2)**, pp. 568-573, 2004.

[24]   An, J.S. & Carmichael, W.W., Use a colorymetric protein phosphatase inhibition assay and enzyme linked immunosorbent assay for study of microcystins and nodularins. *Toxicon,* **32**, pp. 1495-1507, 1994.

[25]   OECD, Eutrophication of water. Monitoring assessment and control. Technical Report. *Environment Directorate*, OECD, Paris, pp. 154, 1983.

[26]   Rapala, J., Sivonen, K., Lyra, C. & Niemela, S. I., Variation of Microcystin, Cyanobacterial hepatotoxins, in Anabaena spp. as a function of growth stimulate. *Applied and Environmental Microbiology,* **63**, pp. 2206-2212, 1997.

[27]   Kaebernick, M. & Neilan, B.A., Ecological and molecular investigations of cyanotoxin production. *FEMS Microbiology Letters,* **35**, pp. 1-9, 2001.

[28]   Kurmayer, R., Dittmann, E., Fastner, J. & Chorus, I., Diversity of microcystin genes within a population of the toxic cyanobacterium Microcystis spp. in Lake Wannsee (Berlin, Germany). *Microbiol Ecology,* **43**, pp.107-118, 2002.

[29]   Vaitomaa, J., Rantala, A., Halinen, K., Rouhiainen, L., Tallberg, P., Mokelke, L. & Sivonen K., Quantitative Real-Time PCR for Determination of Microcystin Synthetse E Copy Numbers for Microcystis and Anabaena in Lakes. *Applied and Environmental Microbiology*, **69**, pp. 7289-7297, 2003.

[30]   Dittmann, E. & Börner, T., Genetic contributions to the risk assessment of microcystin in the environment. *Toxicology and Applied Pharmacology,* **203**, pp. 192-200, 2005.

[31]   Nishizawa, T., Ueda, A., Asayama, M., Fujii, K., Harada, K., Ochi, K. & Shirai, M., Polyketide Synthase Gene Coupled to the Peptide Synthetase Module Involved in the Biosynthesis of the Cyclic Heptapeptide Microcystin. *Journal of Biochemistry,* **127**, pp. 779-789, 2000.

Bacteriological monitoring of ships' ballast water in Singapore and its potential importance for the management of coastal ecosystems

V. Ivanov
Maritime Research Centre, School of Civil and Environmental Engineering, Nanyang Technological University, Singapore

Abstract

The bacteriological quality of ships' ballast water was studied to evaluate the risk of invasion by alien bacterial species in Singapore. Guidelines of the International Maritime Organization (IMO) are currently in place concerning the threat posed by invasive pathogens due to uncontrolled discharge of ballast water and sediment from ships. The bacteriological quality of ship's ballast water was studied to evaluate the risk of invasion by alien bacterial species in Singapore. The concentrations of total bacteria, enterobacteria, *Vibrio* spp., and *Escherichia coli* have been compared for ballast water samples taken from ships in Singapore Harbour. The share of facultative-anaerobic bacteria, which are often the agents of water-borne diseases, in a ship's ballast water was usually higher than in seawater, especially for the cases when the ballast water tank was filled to the top. Samples of ballast water contained from 0.7 to 39.5% of eubacteria, 0 to 2.5% of enterobacteria, 0.2 to 35.8% of *Vibrio* spp., and from 0 to 2.5% of *E. coli*. The significant percentage of *Vibrio* spp. in some samples of ballast water shows possibly increased risk of invasion of microbial pathogens as ballast water will be discharged in coastal area. Simple self-manufactured chromogenic test kit for coliforms in water showed semi-quantitatively the level of faecal pollution in water. The regular monitoring of ballast water for selected microbial pathogens must be performed continuously for reliable conclusions. Disinfection of ballast water, discharged in the coastal area of Singapore, can be recommended as mandatory treatment if some microbial pathogens are frequently detected in the ship's ballast water.
Keywords: ballast water, bacteriological monitoring, coliforms, Vibrio sp.

WIT Transactions on Biomedicine and Health, Vol 10, © 2006 WIT Press
www.witpress.com, ISSN 1743-3525 (on-line)
doi:10.2495/ETOX060071

1 Introduction

Microbiological monitoring of discharged in the port or coastal area ships'
ballast water could diminish the risk of invasive species of microorganisms and
microbial water-borne infectious diseases, It is also can be used to verify quality
of discharged ballast water treatment/disinfection. Risk of invasion by microbial
pathogens due to discharge of ballast water in coastal area is evaluated using the
regulation D-2 of International Convention for the Control and Management of
Ships Ballast Water & Sediments which was adopted by consensus at a
Diplomatic Conference at IMO in London on Friday 13 February 2004. The
indicator microbes but are not be limited to: (a) toxicogenic *Vibrio cholerae* (O1
and O139) with less than 1 colony forming unit (cfu) per 100 milliliters or less
than 1 cfu per 1 gram (wet weight) zooplankton samples; (b) *Escherichia coli*
less than 250 cfu per 100 milliliters; (c) intestinal enterococci less than 100 cfu
per 100 milliliters. The conventional methods are often not appropriate for the
analysis of ballast water. For practical ballast water analysis, new methods have
been developed for simple monitoring of bacteriological quality of water. These
methods are able to provide rapid results, be easily performed by an analyst with
little or no microbiological training, and give an accurate enumeration of harmful
microorganisms. Some commercial probe assay kits for the rapid detection and
enumeration of the indicator bacteria *E. coli* are known but all existing standards
for microbiological water quality are still based on cultivation methods.

Flow cytometry is a useful tool for assessing many parameters of
microorganisms including enumeration, viability, and identification.
Fluorescence *in situ* hybridization (FISH) methods with oligonucleotide probes
(matched to the genetic sequences of particular organisms) are useful for the
identification of individual microbial cells by flow cytometry. Specific
oligonucleotide probes for the detection of *Escherichia coli*, *Vibrio*,
dinoflagellate algae producing paralytic shellfish toxins, marine nanoplanktonic
protists, photosynthetic eukaryotic microorganisms, marine nanoflagellates,
marine cyanobacteria, and other indicator or pathogenic microorganisms may be
used for monitoring the microbiological quality of a ship's ballast water.

The aim of this research was to develop simple and rapid methods of
detecting microbes in discharged ballast water that are applicable to the
monitoring of ballast water treatment quality in ships. The applicability of a flow
cytometry-based method has been currently under examination for ballast water
quality monitoring involving the enumeration and identification of
microorganisms.

2 Methods

Methods of the sampling and analysis of ballast water by conventional
microbiological approaches and by flow cytometry combined with molecular
probes have been described earlier (Tay et al. [5]; Joachimsthal et al. [2–4]).

3   Results

Total concentration of bacteria in ballast water varied significantly, from 30 to 2100 thousand cells/mL. It shows variability of bacterial pollution of ballast water (Table 1).

Table 1:   Variability of bacterial numbers in the samples of ballast water.

Sample	Target bacteria	Concentration of specifically stained cells (cells/mL)
Ballast 1	Eubacteria	2100×10^3
Ballast 2	Eubacteria	50×10^3
Ballast 3	Eubacteria	2300×10^3
Ballast 4	Eubacteria	1400×10^3
Ballast 5	Eubacteria	69×10^3
Ballast 6	Eubacteria	30×10^3

It was a reliable correlation between the number of particles with size from 1 to 4 μm (determined by portable particle counter) and the number of bacterial cells (determined by flow cytometer). Therefore, concentration of bacterial cells in ballast water can be determined in the port laboratory or even on board by portable particle counter. However, it cannot differentiate dead and viable bacterial cells in the sample.

The majority of microorganisms in the seawater samples were either obligate aerobes or microaerophilic. For the ballast water, the proportion of aerobic microorganisms was 11%. However, the proportion of facultative anaerobic microorganisms increased, from 7% in the seawater samples, up to 50% in the ballast water samples. Domination of facultative anaerobes increases the risk of transfer of water-borne diseases.

The number of enterobacteria in ballast water samples varied from 0 to 15 thousand cells/mL. It shows high risk of water borne-diseases in some samples of ballast water (Table 2). Because of high variability of enterobacteria, *E. coli* and *Vibrio spp.* cell numbers, which are indicators of faecal pollution of ballast water and risk of water-borne disease due to the discharge of ballast water, cells with β-galactosidase activity can be detected as an indicator of ballast water pollution.

Table 2:   Variability of enterobacterial numbers in the samples of ballast water.

Sample	Target bacteria	Concentration of specifically stained cells (cells/mL)
Ballast 1	Enterobacteria	15×10^3
Ballast 2	Enterobacteria	11×10^3
Ballast 3	Enterobacteria	14×10^3
Ballast 4	Enterobacteria	0
Ballast 5	Enterobacteria	0
Ballast 6	Enterobacteria	14×10^3

4   Conclusions

The number of coliforms in almost all samples of ballast water taken in Singapore exceeded the values of regulation D-2 of International Convention for the Control and Management of Ships Ballast Water & Sediments. Therefore, the disinfection of ballast water discharged in the port or coastal area of Singapore can be recommended for the cases when the regulation D-1 (ballast water exchange shall do so with an efficiency of 95% volumetric exchange of ballast water or by the pumping through three times the volume of each ballast water tank) was not performed. In this case, instead of measurement does ships' ballast water exceed IMO–recommended levels, the detection of alive bacterial cells in ballast water must be performed using fast and simple specially designed immunochemical test kits to evaluate efficiency of ballast water disinfection on the ships discharging ballast water in the port or coastal area.

References

[1] Ivanov V., Joachimsthal E.L., S.T.L. Tay and J.H. Tay. (2004) Bacteriological quality of ship's ballast water. Singapore Maritime and Port Journal: 101 -108.
[2] Joachimsthal E. L., V. Ivanov, J.-H. Tay, S.T.-L. Tay (2002). Flow cytometry and conventional enumeration of microorganisms in ships ballast water and marine samples. Marine Pollution Bulletin, 46 (3): 308 – 313.

[3] Joachimsthal E. L., V. Ivanov, J.-H. Tay, S.T.-L. Tay (2003). Quantification of whole-cell in situ hybridization with oligonucleotide probes by flow cytometry of Escherichia coli cells. World Journal of Microbiology and Biotechnology, 19 (5): 527-533.
[4] Joachimsthal E.L., V. Ivanov, S.T.L. Tay and J.H. Tay. (2004) Bacteriological examination of ballast water in Singapore harbour by flow cytometry with FISH. Marine Pollution Bulletin, 49(4): 334- 343.
[5] Tay Joo Hwa, Volodymyr Ivanov, Xiaoge Chen, Eva L. Joachimsthal, Maszenan Bin Abdul Majid, and Stephen Tiong Lee Tay. (2002) Monitoring of bacteriological quality of ballast water in Singapore by flow cytometry. Singapore Maritime & Port Journal: 87-99.

Section 2
Integrated approach
in risk assessment
(Special session
organised by T.-S. Chon)

Modeling of denitrification rates in eutrophic wetlands by artificial neural networks

K. Song[1], M.-Y. Song[2], T.-S. Chon[2] & H. Kang[1]
[1]Department of Environmental Science and Engineering,
Ewha Womans University, Seoul, Korea
[2]Division of Biological Sciences, Pusan National University,
Busan, Korea

Abstract

Eutrophication can be controlled by denitrification which is complex microbial processes converting nitrate to nitrogen gas in water body. Various environmental factors such as oxygen, available carbon and pH are known to regulate denitrification rate. However, those controlling variables affect denitrification rates nonlinearly. Further, interactions between those factors hinder a good prediction on the rate. As such, conventional mechanistic modeling of denitrification often fails to fit with data collected from fields.

In this study, we applied artificial neural networks to elucidate complex relationships between denitrification rate and environmental factors. A Multi-Layer Perceptron network based on the back-propagation algorithm was utilized for prediction of denitrification rate. High predictability of denitrification was achieved with $R=0.910$ with the trained network. Subsequently, sensitivity analysis was carried out to evaluate factors predominantly controlling denitrification rates. A sensitivity analysis exhibited that DO exert a dominant controlling effect on denitrification rate over other environmental factors.

Keywords: denitrification, eutrophic wetland, artificial neural networks.

1 Introduction

Aquatic ecosystems near agricultural or urban systems can be polluted by a high loading of nitrogen, which often results in eutrophication in summer season. Since eutrophication causes oxygen depletion, fish kill and odor, many studies have focused on economical methods to control eutrophication from nitrogen loading [1, 2].

Wetlands, placed between upland and aquatic systems, are known to control eutrophication by removal of N through various mechanisms including denitrification [3, 4]. Denitrification has been considered as a major mechanism for N removal, and wetlands are known to be a hotspot of denitrification [5, 6]. In a wetland where anaerobic conditions are easily introduced, many facultative microorganisms, called denitrifying bacteria, can use nitrate as an electron acceptor and remove the nitrogen from water [5, 7]. High organic matter content, nitrate input and development of rhizosphere with anaerobic condition also induce high denitrification rate [4, 8–10]. Denitrifying Enzyme Activity (DEA) reflects the amount of denitrifying enzymes in soil, and it is often correlated with denitrification rate positively [6, 11].

In wetlands, denitrification rates often exhibit extremely high variations both temporally and spatially, due to heterogeneity of environmental conditions. Although many controlling variables of denitrification have already been reported, models for denitrification are still empirically based. Several studies have proposed mechanistic models but they require a large amount of data sets as well as many assumptions [12–15]. As such, conventional modeling often fails to provide a powerful explanation for denitrification in wetlands.

An Artificial Neural Networks' (ANNs) approach based on human brain performance has been applied to various types of ecological data successfully (e.g., [16–19]). In most of the previous studies, ANNs have been applied to biological indicator with environmental factors such as the occurrence of river bird distributions [20], benthic insect species [18], trout density [16], and phytoplankton production [19]. Recently, the applications of ANNs have been accelerated to environmental science. ANNs have been successfully applied to climate change simulation [21], nutrient cycling [15, 22] and water quality [23–25]. According to the previous studies, ANNs have often shown better predictions for ecological data than classical linear or logistic regression models [20, 26, 27]. A Multi-Layer Perceptron (MLP), which is one of the most popular neural networks, operates under a supervised learning procedure to minimize the errors between the actual and target values based on the back-propagation algorithm [15, 26]. Sensitivity analysis in neural networks such as the 'Weight' and 'Profile' methods can additionally explain the inter connection weights of each input-hidden-output layer and the responses of the output according to the changes of input variables, respectively [16, 17, 27].

In this study, we aimed 1) to apply ANNs to simulate denitrification rates in the eutrophic constructed-wetland, and 2) to determine key controlling variables for denitrification.

2 Materials and methods

2.1 Mesocosm operation

To investigate denitrification in eutrophic states in the constructed wetlands, mesocosm-scale wetlands were built at Ewha Womans University in South

Korea. Wetlands were composed of two types, namely a marsh characterized with shallow water and vegetation, and a pond with deep water without vegetation [1]. We planted 30 *Phragmites australis* in marsh mesocosm which contains 608 L of water, while pond was filled with 1301 L of water. The retention times were ca. 3 days in marsh and *ca.* 5 days in pond due to differences in water volume. Contaminated inflow water was prepared by adding ammonia-nitrate (10 mg-N L^{-1}).

2.2 Sampling and analysis

Water and soil samplings and subsequent analyses were conducted once a week for 2 years except winter to early spring seasons. Water temperature, pH and dissolved oxygen (DO) were measured *in situ*. Soil samples were collected to 5 cm depth from the surface in each mesocosm. After removing the large fragments of detritus and roots, all soils were maintained at 4°C until analysis of denitrification. Denitrification rate and DEA were determined by an acetylene blocking method. Approximately 10 *g* of soil for denitrification and 5 g for DEA with glucose and nitrate were placed in a 100 ml sterilized glass vial and incubated for 20 min with oxygen-free N$_2$ gas purge at 20°C. After acetylene (10%, v/v) addition in the head space, cumulative nitrous oxide concentration in head space collected within 2 hours was measured by gas chromatography equipped with ECD (HP-6890) [28].

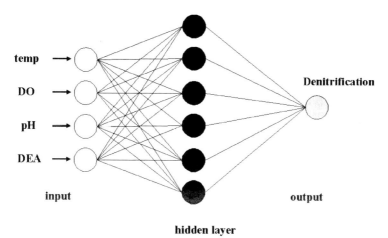

Figure 1:   Conceptual structure of MLP used for predicting denitrification (modified from [17, 29]).

2.3 Application of Artificial Neural Networks

Basic structure of MLP in this study was present in Fig. 1. Water temperature (°C), DO (mg L^{-1}), pH and DEA were chosen as input data (4 nodes). Denitrification rate was used as an output (1 node). In this study, 6 hidden neurons were chosen for the learning procedure, and iteration was performed 3000 times. For training the momentum (0.95 in this study) was used for the learning procedure, while the learning rate was initially set to 0.75 and was gradually decreased as convergence was achieved. The trained MLP was tested by the 'leave one out' method using MATLAB ver.6.1. For sensitivity analysis, the 'Weight' and 'Profile' methods were used [16, 27]. The 'Weight' method involves the weight calculations in hidden-output neuron connections during the whole iteration procedure. The 'Profile' method considers the influence of the certain variations of each input variables between minimum and maximum values while the remaining input variables are fixed [16].

3 Results

3.1 Mesocosm operation

Table 1 presents the environmental characteristics and denitrification in wetlands from May, 2004 to November, 2005. Water temperature varied between 2.5 and 28.1°C while the median value was 22.7°C during the observation period. pH in wetlands was neutral and slightly alkaline ranging 6.3–9.8 with the median value of 7.0. The pH peaked in June, 2004 and 2005 in concomitant with algal blooming. DO ranged between 0.4 to 9.8 mg L^{-1}. DO was dropped to 0.4 on August, 2005, while the highest value was recorded in May, 2005 (Fig. 2). Denitrification rates showed high variations from 0.3 to 1559.6 ng N_2O hr^{-1} g^{-1}. Denitrification rates exhibited the highest in July while it was the lowest in the spring and early winter. The median value of DEA was 4072 ng N_2O hr^{-1} g^{-1}.

Table 1: Environmental characteristics and denitrification rate in the eutrophic constructed-wetlands during the whole operation period in 2004 and 2005.

	Median	Min.	Max.
Temp (°C)	22.7	2.5	28.1
DO (mg L^{-1})	4.3	0.4	12.3
pH	7	6.3	9.8
DEA (ng N_2O h^{-1} g^{-1})	4072	7.3	16261
Denitrification rate (ng N_2O h^{-1} g^{-1})	18.2	0.3	1559.6

3.2 Application of ANNs

The MLP showed high predictability after training. The estimated and observed values were correlated with $R = 0.910$ (P<0.001) after 3000 iterations based on the test of the 'leave one out' method (Fig. 3).

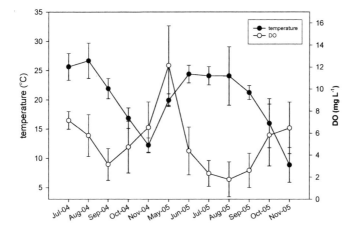

Figure 2:   Seasonal variations of averaged temperature and DO during the wetland operation period. The data chosen as input for the MLP training are exhibited.

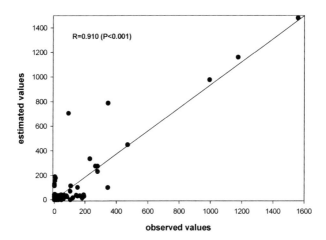

Figure 3:   Estimated and observed values tested by MLP using the 'leave one out' method (n=95). The solid line represents the 1:1 between estimated and observed values.

In order to elucidate the importance of input variables, sensitivity analyses based on 'Weight' and 'Profile' were carried out. According to the 'Weight' analysis, DO was the predominant factor being followed by pH, while temperature and DEA weakly contributed to determination of denitrification. Percents of contribution of DO and pH were 45% and 30%, respectively, whereas percents of contribution of temperature and DEA were *ca.* 10% for both factors (Fig. 4). Fig. 5 shows the contribution profile of the input variables in determining changes in denitrification according to the 'Profile' analysis. The degree of contribution of Oxygen to denitrification decreased monotonically along with the increase in the levels of Oxygen. In contrast, the degree of contribution of pH increased rapidly until pH reached the middle range. From the middle point on, contribution of pH remained in the same range.

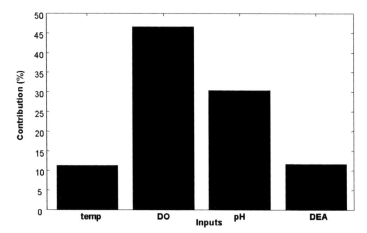

Figure 4: Relative contribution of input variables according to the 'Weight' sensitivity analysis.

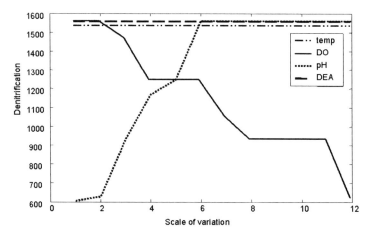

Figure 5: Contribution profile over the scale of the different input variables in determining changes in denitrification.

4 Discussion and conclusion

High predictability of denitrification in eutropic wetlands was achieved with the trained artificial neural networks. This study demonstrated that MLP could be efficiently utilized for analyzing complexity residing in the data related to eutropic states in wetlands.

The sensitivity analyses could provide a comprehensive view on the overall scope of input-output relationships (i.e., profile) representing the whole network procedure and specific information on the degree of contribution (i.e., weight) according to local changes in the input variables as well [16]. According to the 'Weight' analysis, DO and pH predominantly contributed to changes in denitrification (Fig. 4), whereas contribution of the other input variables was minimal. The results were coincident with the results from the 'Profile' method. Denitrification exhibited steep changes according to the changes in DO and pH (Fig. 5). The high contribution of DO could be explained from the fact that denitrification may be discernable when DO is low, regarding that denitrification would be strongly inhibited under high DO levels [30, 31].

The result of pH on the 'Profile' analysis demonstrated that denitrification might be inhibited under low pH but will be stimulated along with the increase in pH. However, this positive relation between pH and denitrification became constant from middle range of pH since neutral or slightly alkaline conditions were in the optimum range for denitrification [4, 32]. Some researchers reported that denitrification decreased in acidic condition [33] while denitrification is facilitated in alkaline condition [34, 35]. Considering these reports, denitrification would not be further enhanced over the optimum range of pH, while denitrification would greatly increase as pH increased from low to medium values. The results from our study confirmed the reports in previous studies on the relationships between pH and denitrification.

The low contribution of temperature shown in the sensitivity analyses (Figs. 4 and 5), on the other hands, implies that temperature might not be a key variable in controlling the states of denitrification in the eutrophic wetlands tested in this study. This low contribution of temperature can be explained as follows. Firstly, we did not consider the condition of highly low temperatures in winter for this study. Generally, denitrification ceased below 5°C, while it increased with higher levels of temperature [30, 33, 36]. However, temperatures measured in this study were in the narrow range because we only operated the wetlands from late spring to autumn (for 2 years) (Fig. 2). Since the input data only covered the given range, the 'Weight' and 'Profile' sensitivity methods could not reveal the effect of temperature beyond the range of temperatures used for the test in this study. Secondly, the effect of temperature in previous studies might be over-estimated. Some researchers reported the highest denitrification rate in winter [37, 38]. The reviews of Q_{10} of the treatment wetlands near 1.0 [3, 39] exhibited that temperature effect can be estimated to be too high. The effect of temperature was also often confounded with other seasonal effects [6, 39] involving light intensity, vegetation growth, etc. Therefore, low contribution of temperature in the limited range indicates that denitrification is not highly sensitive to seasonal variables such as DOC supply, microbial activity or plant growth in this study. It indicates that substrates [8, 40] for denitrification could not be limited in our wetlands due to eutrophication. Eutrophication was also influential in keeping DEA from contributing to denitrification. DEA is highly dependent on nitrate and carbon availability [6, 10]. However, the nutrient sources such as nitrate or carbon could not be a limiting factor in eutrophication. Consequently, denitrification could not be influenced by DEA in the eutrophic wetlands tested in this study.

In summary, denitrification in mesocosm-scale eutrophic wetlands was well predicted using the trained MLP. The results of sensitivity analysis exhibited that, under the condition of the limited range in temperature, DO and pH exert a dominant controlling effect on denitrification rate over other environmental factors in this study. In addition, the sensitivity analysis such as the 'Profile' and 'Weight' methods appeared to be useful for understanding both the overall scope of the environment-denitrification relationships and specific information on local contribution of input variables in determining denitrification. The results of this study suggest ANNs could be a powerful modeling tool to explain complexity residing in ecological data.

Acknowledgements

This subject is supported by Ministry of Environment as The Eco-Technopia 21 project and Advanced Environmental Biotechnology Research Center.

References

[1]   Mitsch W.J. Mitsch R.H., & Turner R.E., Wetlands of the old and new worlds: ecology and management. In Mitsch, W.J. (ed), Global Wetlands: Old World and New. Elsevier Science, pp. 3-56. 1994.
[2]   Rezania T., Cicek N. & Oleszkiewicz, J.A., Kinetics of hydrogen-dependent denitrification under varying pH and temperature conditions. Biotechnology and Bioengineering 92, pp. 900-906. 2005.
[3]   Kadlec R.H & Knight R.L., Treatment wetlands, CRC/Lewis Publishers, Boca Raton, FL. 1996.
[4]   Mitsch W.J & Gosselink J.G., Wetlands. John Wiley and Sons Inc; New York, pp. 107-118. 2000.
[5]   Reddy K.R & D'Angelo E.M., Soil Processes regulating water quality in wetlands. In Mitsch, W.J. (ed), Global Wetlands: Old World and New. Elsevier Science, pp. 309-324. 1994.
[6]   White J.R & Reddy K.R., Influence of nitrate and phosphorus loading on denitrifying enzyme activity in Everglades wetland soils. Soil Science Society of America Journal 63, pp. 1945-1954. 1999.
[7]   Zumft, W.G., Cell biology and molecular basis of denitrification. Microbiology and Molecular Biology Reviews 61, pp. 533-616. 1997.
[8]   Brix H., 1997. Do macrophytes play a role in constructed treatment wetlands? Water Science and Technology 35, pp. 11-17. 1993.
[9]   Groffman P.M. & Hanson G.C., Wetland denitrification: Influence of site quality and relationships with wetland delineation protocols. Soil Science Society of America Journal 61, pp. 323-329. 1997.
[10]   Hume N.P. & Fleming M.S., Horne A.J., Denitrification potential and carbon quality of four aquatic plants in wetland microcosms. Soil Science Society of America Journal 66, pp. 1706-1712. 2002.

[11] Smith M.S & Tiedje J.M., Phases of denitrification following oxygen depletion in soil. Soil Biology and Biochemistry 11, pp. 261-267. 1979.

[12] Halling-Sørensen B. & Nielsen S.N., A model of nitrogen removal from waste water in a fixed bed reactor using simultaneous nitrification and denitrification (SND). Ecological Modelling 87, pp. 131-141. 1996.

[13] Heinen M. Simplified denitrification models: Overview and properties. Geoderma, in press. 2005a.

[14] Heinen M. Application of a widely used denitrification model to Dutch data sets. Geoderma, in press. 2005b.

[15] Holmberg M., Forsius M., Starr M. & Huttunen M. An application of artificial neural networks to carbon, nitrogen and phosphorus concentrations in three boreal streams and impacts of climate change. Ecological Modelling 195, pp. 51-60. 2006.

[16] Gevrey M., Dimopoulos I. & Lek S., Review and comparison of methods to study the contribution of variables in artificial neural network models. Ecological Modelling 160, pp. 249-264. 2003.

[17] Lek S., Delacoste M., Caran P., Dimopoulos I., Lauga J. & Aulagnier S., Application of neural networks to modelling nonlinear relationships in ecology. Ecological Modelling 90, pp. 39-52. 1996.

[18] Park Y.S, Ce´re´ghino R., Compin A. & Lek S., Applications of artificial neural networks for patterning and predicting aquatic insect species richness in running waters. Ecological Modelling 160, pp. 265-280. 2003.

[19] Scardi M., Artificial neural networks as empirical models for estimating phytoplankton production. Marine Ecology Progress Series 139, pp. 289-299. 1996.

[20] Manel S., Dias J-M & Ormerod S.J. Comparing discriminant analysis, neural networks and logistic regression for predicting species distributions: a case study with a Himalayan river bird. Ecological Modelling 120, pp. 337-347. 1999.

[21] Seginer, I., Boulard T. & Bailey B.J., Neural network models of the greenhouse climate. Journal of Agricultural Engineering Research 59, pp. 203-216. 1994.

[22] Melesse AM. & Hanley RS., Artificial neural network application for multi-ecosystem carbon flux simulation. Ecological Modelling, 189, pp. 305-314. 2005.

[23] Nour M.H., Smith D.W., El-Din M.G. & Prepas E.E., The application of artificial neural networks to flow and phosphorus dynamics in small streams on the Boreal Plain, with emphasis on the role of wetlands. Ecological Modelling 191, pp. 19-32. 2006.

[24] Acharya C., Mohanty S., Sukla L.B. & Misra V.N., Prediction of sulphur removal with *Acidithiobacillus sp.* using artificial neural networks. Ecological Modelling 190, pp. 223-230. 2006.

[25] Karakitsios S.P., Papaloukas C.L., Kassomenos A. & Polodis G.A., Assessment and prediction of benzene concentrations in a street canyon using artificial neural networks and deterministic models: Their response to "what if" scenarios . Ecological Modelling 193, pp. 253-270. 2006.

[26]   Lek S. & Guégan J.F., Artificial neural networks as a tool in ecological modeling, an introduction. Ecological Modelling 120, pp. 65-73. 1999.
[27]   Özemi S.L, Tan C.O. & Özemi U., Methodological issues in building, training, and testing artificial neural networks in ecological applications. Ecological Modelling 195, pp. 83-93. 2006.
[28]   Beauchamp E.G. & Bergstrom D.W., Denitrification. In Carter M.R. (ed), Soil Sampling and Methods of Analysis. Lewis Publishers, pp. 351-355.
[29]   Goh A.T.C., Back-propagation neural networks for modeling complex systems. Artificial Intelligence in Engineering 9, pp. 143-151. 1995.
[30]   Richardson W., Strauss, E., Bartsch L., Monroe E., Cavanaugh J., Vingum L. & Sovalle D., Denitrification in the upper Mississippi River: rates, controls, and contribution to nitrate flux. Canadian journal of fisheries and aquatic sciences. 61, pp. 1102-1112. 2004.
[31]   Sirivedhin T. & Gray K.A., Factors affecting denitrification rates in experimental wetlands: Field and laboratory studies. Ecological Engineering 26, pp. 167-181. 2006.
[32]   Etherington J.R., Wetland Ecology. Edward Arnold, London. p.67. 1983.
[33]   Bremner A.M., Shaw R.L. & Macalady D.L., Denitrification in soil. Factors affecting denitrification. Journal of Agricultural Science 51, pp. 40-52. 1954.
[34]   Ellis, S., Howe, M.T., Goulding K.W.T., Mugglestone, M.A. & Dendooven, L., Carbon and nitrogen dynamics in a grassland soil with varying pH: Effect of pH on the denitrification potential and dynamics of the reduction enzymes. Soil Biology and Biochemistry 30, pp. 359-367. 1998.
[35]   Simek M. & Hopkins D.W., Regulation of potential denitrification by soil pH in long-term fertilized arable soils. Biology and Fertility of Soils 30, pp. 41-47. 1999.
[36]   Stanford G., Dzienia S., & Van der Pol R., Effect of temperature on denitrification rate in soils. Journal of Soil Science Society of America 39, pp. 867-870. 1975.
[37]   Hasegawa T. & Okino T., Seasonal variation of denitrification rate in Lake Suwa sediment. Limnology 5, pp. 33-39. 2004.
[38]   Pinay, G., Roques, L. & Fabre, A., Spatial and temporal patterns of denitrification in a riparian forest. Journal of Applied Ecology 30, pp. 581-591. 1993.
[39]   Bachand P.A.M. & Horne, A.J., Denitrification in constructed free-water surface wetlands: Effects of vegetation and temperature. Ecological Engineering 14, pp. 17-32. 2000.
[40]   Platzer C., Enhanced nitrogen elimination in subsurface flow artificial wetlands- a multi stage concept. Preprints of the 5th International Conference on Wetland Systems for Water Pollution Control. University für Bodenkultur Wien. Vienna, Austria. pp. 1-8. 1996.

Recurrent Self-Organizing Map implemented to detection of temporal line-movement patterns of *Lumbriculus variegatus* (Oligochaeta: Lumbriculidae) in response to the treatments of heavy metal

K.-H. Son[1], C. W. Ji[2], Y.-M. Park[1], Y. Cui[3], H. Z. Wang[3], T.-S. Chon[2] & E. Y. Cha[1]

[1]*Department of Computer Science and Engineering,*
Pusan National University, Republic of Korea
[2]*Division of Biological Sciences,*
Pusan National University, Republic of Korea
[3]*Institute of Hydrobiology,*
Chinese Academy of Sciences, People's Republic of China

Abstract

Measurement of behavioral responses have been recently considered as an important method for monitoring risk assessment. Computational processing could be applied to continuous data for automatic determination of changes in behavioural states of indicator specimens. Behavioral monitoring could be used as an alternative tool to fill the gaps between large (e.g., ecological survey) and small (e.g., molecular analysis) scale methods for risk assessment. While the points were conventionally used for indicating movement of test specimens, the line shapes of blackworms, *Lumbriculus variegatus,* were trained by Artificial Neural Networks in this study. We proposed an unsupervised temporal model, Recurrent Self-Organizing Map (RSOM), to detect sequential changes in the line-movement of blackworms after the treatments of a toxic substance, copper, in this study. RSOM was feasible in addressing the stressful behaviors of indicator specimens such as body contraction, high degree of folding, etc. We demonstrated that the unsupervised temporal model is efficient in classifying temporal behavior patterns and could be used as an alternative tool for the real-time monitoring of toxic substances in aquatic ecosystems in the future.
Keywords: Recurrent Self-Organizing Map, response behavior, temporal sequence processing, Lumbriculus variegates.

 WIT Transactions on Biomedicine and Health, Vol 10, © 2006 WIT Press
www.witpress.com, ISSN 1743-3525 (on-line)
doi:10.2495/ETOX060091

1 Introduction

Considering the urgency of water contamination, development of methods for assessing toxicity has been regarded as an important issue in maintaining sustainable ecosystem health. Toxicity exposed to ecosystems can be assessed in various scales. Previous practices in assessment, however, were skewed to either extremely large (e.g., biodiversity evaluation in communities) or small (e.g. molecular or chemical analyses) scales.

Recently, continuous behavioral monitoring of indicator organisms [1–4] has been considered as one of the efficient tools filling the gaps between the large and small-scale assessment methods. Monitoring of the locomotory behavior has been introduced as an efficient means of evaluation of contaminated ecosystems with toxic chemicals [5]. A numerous accounts of research on effects of toxic chemicals on behaviors of organisms have been reported in various taxa, including crustaceans [6, 7], snails [8], fish [9, 10] and insects [11, 12]. However, these studies have been mostly limited to observation of single or combinations of single behaviors mainly with qualitative descriptions. Quantitative characterization of behaviors, however, is difficult for analysis due to complexity residing in the behavioral data.

Theoretical study on behavior has been initiated with research on biological motion regarding random walk [13], correlation function [14, 15], movelength analysis [16], fractal dimension [17–19], etc. However, these parameters are highly condensed and tend to emphasize the totality of the movement states of indicator organisms. Addressing behavioral states in a compressed form (i.e., as a parameter), however, may not be suitable for uniquely characterizing various behavioral patterns. Local information on movement patterns may also be critical in determining various states of animal behaviors [1, 2].

In this study we used Artificial Neural Networks (ANNs) to extract information from complex behavioral data. ANNs has been regarded as an efficient non-linear filter, and have been widely used for forecasting and data organization in ecological sciences [20–22]. Recently ANNs have been applied to behavioral monitoring. Kwak et al [1] implemented Multi-Layer Percepton (MLP) for detecting response behaviors of medakas treated with Diazinon. Wavelets and ANNs have been also used in combination to detect changes in response behaviors of chironomids for water quality monitoring [23]. In addition to supervised learning by MLP, Self-Organizing Map (SOM) was implemented to patterning the movement tracks of indicator organisms in an unsupervised manner in response to the treatments of toxic substances. SOM was efficient in classifying different states of response behaviors of indicator organisms such as cockroach [3] and medaka [24].

In the previous studies with ANNs, however, the static patterns were mainly considered as input, and location of the specimens (i.e., points) was the main source for training the movement tracks in the networks. In this study we demonstrated the feasibility of SOM in revealing the patterns of sequential movement of indicator specimens in a recurrent manner. In addition, a species in oligochaetes was selected as indicator organisms, and the line shapes of the specimens were used as input data for training.

Recent models of neural networks have been used for temporal sequence processing (TSP). The temporal network was reported to be more feasible in learning time series data than conventional methods based on linear (e.g., AR and ARMA) and non-linear (e.g., NARMAX and MARS) statistical analyses [25]. In ecology, the Elman [26] and recurrent [27] networks have been used for predicting the time-series data for community dynamics. These networks, however, were mainly used for training with the templates (i.e., supervised learning). In real situations, however, there are numerous patterns in behavioural data. Consequently it is difficult to have all the pre-determined patterns be ready for training, considering that a huge amount of behavioural data could be accumulated through the real-time, continuous recording. Data mining would be desired to provide the overall scope of the behavioral data. In this regard, we incorporated the unsupervised network to accommodate the sequential line-movement data of specimens collected from continuous recording.

Temporal Kohonen Map (TKM) [28], being derived from the Kohonen Self-Organizing Map [29, 30], has been regarded as an efficient learning tool for TSP. In the TKM the involvement of the earlier input vectors in each unit is represented by using recursive difference. An unsupervised temporal model, Recurrent Self-Organizing Map (RSOM), was further proposed to provide more flexibly in dealing with the sequential data. RSOM was originally designed by Varsta et al [31] and can be presented as an enhancement of the TKM algorithm. While TKM does not directly use the temporal contextual information of input sequences in weight updating [31], direct learning of the temporal context is possible with RSOM. It allows model building using a large amount of data with only a little a *priori* knowledge. RSOM provided promising results in dealing with classification of temporal data with simple property [31–33].

In this study, we proposed RSOM for detecting temporal response behaviors of *Lumbriculus variegatus* treated with copper. We demonstrated that RSOM was feasible in patterning the sequential line-movement of oligochaetes after the treatments of toxic substances and efficiently characterize the stressful behaviors of the specimens.

2 Materials and methods

2.1 Observation system

The body shape of the test specimens of *Lumbriculus variegatus* was recorded by using an observation system consisting of an observation aquarium, a camera and software for an image recognition system. During the observation period, groups and individuals of *Lumbriculus* were placed in a glass aquarium (diameter: 9 cm), and their position was scanned from top view in 0.25 sec intervals using a CCTV camera (Kukjae Electronics Co. Ltd.; IVC-841®) for two days (one day before the treatments and one day after the treatments). The analog data captured by the camera were digitized by using a video overlay board

(Sigmacom Co., LTD.; Sigma TVII®), and were sent to the image recognition system to digitize the line-movement of tested specimens. The software for detecting the specimens and other supporting mathematical programs were developed by the Neural Network and Real World Application Laboratory, Department of Computer Science and Engineering, Pusan National University. The stock populations were maintained in a glass tank, and were reared with artificial dry diet (Tetramin®) under the light condition (back light; twenty five 0.2W green diodes vertically located underneath the observation cage 10 cm apart) of 24 hours.

2.2 Computational method

Self-Organizing Map (SOM) is a vector quantization method to map patterns from an input space V_i onto lower dimensional space V_M of the map such that the topological relationships between the inputs are preserved [29, 30] to find the best matching unit b in time step t in the following equation:

$$\|x(t) - w_b(t)\| = \min_i \{\|x(t) - w_i(t)\|\} \qquad (1)$$

where $i \in V_M$, $x(t)$ is an input vector, and $w_i(t)$ is a weight vector of the unit i in the map. Subsequently the weight vector of the best matching unit b is updated towards the given input vector $x(t)$ according to

$$w_b(t+1) = w_b(t) + \gamma(t)h_b(t)(x(t) - w_b(t)) \qquad (2)$$

where $\gamma(t)$, $0 < \gamma(t) \le 1$, is a learning rate, and $h_b(t)$ is the neighbourhood function.

RSOM [31] is similar to the SOM except for the following difference equation, fig. 1:

$$y_i(t) = (1 - \alpha)y_i(t-1) + \alpha(x(t) - w_i(t)) \qquad (3)$$

where $0 < \alpha \le 1$ is a leaking coefficient, $y_i(t)$ is a leaked difference vector, $w_i(t)$ is the reference or weight vector in the unit i, and $x(t)$ is the input pattern in time step t. The best matching unit b at time step t is searched by

$$y_b = \min_i \{\|y_i(t)\|\} \qquad (4)$$

where $i \in V_M$. The process of updating weight is the same to SOM. However, the input sequence should be noticed before learning in a recurrent manner, fig. 1. The property of RSOM is described in reference articles in detail [31–33].

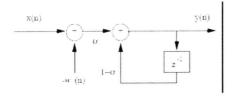

Figure 1:   Schematic picture of an RSOM unit acting as a recurrent filter [32].

For input data, we obtained 13 *x-y* coordinates of the line shape of blackworm specimens through computer recognition system measured in every 0.25 s interval. The coordinates were converted to a line consisting of 12 sub-segments with 12 lengths and 11 angles, fig. 2. These features carried information on the line shape of *Lumbriculus*. The general difference of the body shape was observable before and after the treatments, fig. 3. The lengths of the body segments were similar, being consistently shortened after the treatments, fig. 3(a). In contrast, the angles were smaller close to the center of the body, fig. 3(b). The angles in all sub-segments also decreased consistently after the treatments. This indicated that the body of the treated blackworms tended to contract and fold strongly after the treatments. For training RSOM, the data for 12 lengths and 11 angles were provided as input. The whole sequence of the line movement in two days (one day before the treatments and one day after the treatments) was divided into 100 sections with equal intervals (duration of ca 28.8 minutes). The line-data (12 lengths and 11 angles for the body segments, fig. 2) of the specimens at the beginning point was selected for the initial data for each section. Subsequently eleven more line data were selected in every 25 s interval in each section merged to the initial line-data. In total the 12 sequential line-data for 5 minutes (25 s X 12 = 300 s) were regarded as a sample unit provided to RSOM as input. Twenty specimens were observed for recording. Due to difficulty of detection of the line-shapes, however, some portion of data segments were not properly recorded due to noise. In this study we selected 6 specimens with the full records of the sequence movement during the whole observation period. The overall movement patterns in the specimens not selected for training were in general similar to the patterns of the specimens used for training according to the preliminary studies.

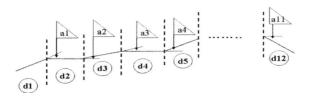

Figure 2:   The body (bold line) of *Lumbriculus variegatus* consisting of 12 sub-segments with the lengths (e.g., d1, d2, etc) and the angles (e.g., a1, a2, etc)

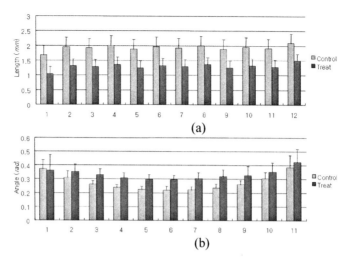

Figure 3:   Comparison of lengths and angles of the sub-segments of *Lumbriculus* before and after the treatments (a) Length, (b) Angle.

3   Results

For the purpose of patterning the sequential line-movements of *Lumbriculus*, we initially checked angle and length separately for training RSOM. Subsequently length and angle were combined, and the whole data were trained with the network.

3.1   Movement patterns based on angles

The angle changes in the line-movements of the test specimens were accordingly grouped on RSOM, fig. 4(a), (b). According to the Ward's linkage method calculated under MATLAB environment [34], the patterns were largely grouped into 4 patterns, fig. 4(a), (c). The horizontal gradient of the angle changes was observed from right to left area of the map. Angle changes in the body segments were minimal in the right area, while the specimens' body shape was more folded in the left area as shown in, fig. 4(b).

The samples were divided to two main clusters on the map. In the right area, cluster B occupied a large area. In this cluster, the stretched line shape was mainly observed in the movement of the specimens, fig. 4(b). The segments before the treatments (white circle) were dominantly grouped in cluster B. The other cluster A occupied a broad area of the map at the left hand side. The folded body shapes were abundantly observed in this cluster. The cluster A was sub-clustered to smaller groups, AI and AII. The gradients were also observed between the sub-clusters. The highly folded ones appeared in the sub-cluster AI at the bottom left area of the map, while the data segments with less folded shapes were grouped in subcluster AII, fig. 4(b). The cluster AII was further divided to sub-sub-clusters, AIIa and AIIb. The segments in cluster AIIa in the top area were presented by partial folding in the body segments, while the body segments in AIIb were characterized by the "U" shape, fig. 4(b). In cluster AI, the segments after the treatments (dark circle) were dominantly grouped. The segments in cluster AII were mixed between 'before' and 'after' the treatments.

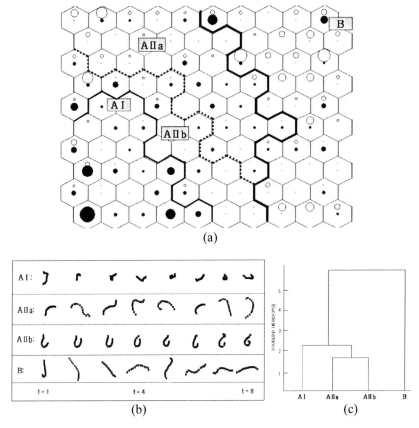

(a)

(b)

(c)

Figure 4:   (a) Grouping of the sequential line-movements of *Lumbriculus* specimens after training with RSOM based on angles of body segments (Clustering carried out on the patterned nodes by the Ward linkage method. White and black circles indicate the segments obtained from 'before' and 'after' the treatments respectively. Size of the circles indicates the number of line segments grouped in the RSOM units relatively (Max. number of the samples grouped in one unit; 120)). (b) Time sequence of the line-movements of *Lumbriculus* in different clusters listed on fig. 4(a) (c) Dendrogram of the RSOM units on fig. 4(a), by the Ward's linkage method.

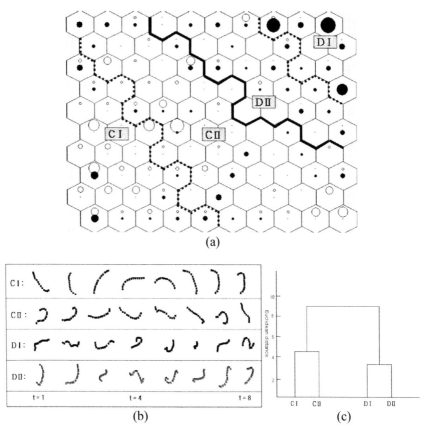

(a)

(b)   (c)

Figure 5:   (a) Grouping of the sequential line-movements of *Lumbriculus*
             specimens after training with RSOM based on lengths of body
             segments (Clustering carried out on the patterned nodes by the
             Ward linkage method. White and black circles are explained in
             fig. 4 (Max. number of the samples grouped in one unit; 120)).
             (b) Time sequence of the line-movements of *Lumbriculus* in
             different clusters listed on fig. 5(a) (c) Dendrogram of the RSOM
             units in fig. 5(a) by the Ward's linkage method.

3.2 Movement patterns based on lengths

The changes in length of the body segments of the blackworm specimens were
also accordingly grouped on RSOM, fig. 5(a), (b). According to Ward's linkage
method, the patterns were grouped to 4 patterns, fig. 5(a), (c). In this case the
gradient of the length was observed diagonally. From bottom left to top right on
the map, the length of body segments accordingly decreased, fig. 5(a), (b). The
patterned samples were divided to two large clusters, C and D. In cluster C
occupying a large area in the bottom left area, the samples with long body
segments were characteristically observed. Cluster C was further divided to two

sub-clusters, CI and CII, depending upon the degree of length. In cluster CI at the bottom left corner, the segment length was in the maximum range while the specimen's body was close to the line-shape without much folding, fig. 5(a), (b). The body length in cluster CII was relatively more contracted than in cluster CI, fig. 5(b). In cluster CI, the body segments before the treatments were more abundantly observed. However, some body segments after the treatments also occurred in this cluster. In cluster CII, the body segments before the treatments were abundant but were mixed with the body segments after the treatment in some degree.

In cluster D close to the upper right area of the map, the contracted forms of body segments were strongly grouped. Especially in cluster DI, the specimens highly contracted and folded. In cluster DII, the body segments were less contracted. In cluster DI, the segments after the treatments were strongly grouped. The segments after the treatments were still dominant in cluster DII. It appeared that the degree of grouping was stronger for the body segments after the treatments (i.e., DI and DII) than before the treatments (i.e., CI and CII).

3.3 Movement patterns based on combination of angles and lengths

The data for angles and lengths of the body segments of the blackworm specimens were combined and were subsequently used for training RSOM, fig. 6(a), (b), (c). Clustering appeared in a characteristic manner more diversely compared with separate patterning by angles, fig. 4, and lengths, fig. 5. According to the Ward's linkage method, the samples were divided to 6 groups, with inclusion of sub- and sub-sub-clusters, fig. 6(a), (c).

Although clustering was diverse, the gradient was still observed diagonally from bottom left to top right on the map. While the body segments with larger and less folded bodies were observed in cluster F in the area of bottom right, the shorter and strongly folded bodies were placed in cluster E in the area of top left, fig. 6(a), (b).

Sub-clusters were similarly divided according to degree of contraction and folding. In cluster FI at the bottom right corner, the longer and less curved specimens were observed. The sub-cluster FI was further divided according to the line shape of the specimens. While more straight forms were allowed in the body segments in sub-sub-cluster FIa, folding (e.g., the fifth snap shot in FIb, fig. 6b) was produced in the sequence of the line-movement in the sub-sub-cluster FIb.

Sub-cluster, FII, was differentiated from FI regarding the stronger degree of contraction and folding. The gradient was further observed in division of sub-clusters. In sub-sub-cluster FIIa, the body segments were strongly contracted and folded. In sub-sub-cluster FIIb, however, the shape of the body segments was relatively longer and somewhat similar to the body segments shown in sub-cluster FI. Clear difference of the body segments between these two clusters needs more verification, but the body segments in cluster FIIb appeared to include the more contracted segments (e.g., the sixth snap shot in FIIb, fig. 6(b) during the course of sequential movement.

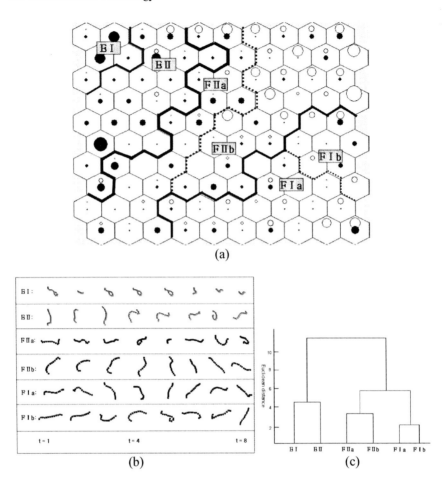

(a)

(b)

(c)

Figure 6:   (a) Grouping of the sequential line-movements of *Lumbriculus* specimens after training with RSOM based on angles and lengths of body segments (Clustering carried out on the patterned nodes by the Ward linkage method. White and black circles are explained in fig. 4 (Max. number of the samples grouped in one unit; 100)). (b) Time sequence of the line-movements of *Lumbriculus* in different clusters listed on fig. 6(a) (c) Dendrogram of the RSOM units in fig. 6(a) by the Ward's linkage method.

In cluster E, the samples showed the strongly contracted and folded bodies, fig. 6(b). In the body segments belonging to sub-cluster EI, the length was in the least range while the degree of folding was in the highest range. Almost all the grouped samples in this cluster were the line-movements after the treatments. In sub-cluster EII, the body was less contracted and the degree of folding decreased, but the samples were still dominated by the line-movement after the treatments. The sequential movement in the contracted and folded forms of the treated specimens was accordingly revealed in clusters EI and EII. This demonstrated that RSOM would efficiently visualize the symptomatic movements of blackworms after exposure to copper. In contrast to cluster E, the samples before the treatments were more abundantly observed in the areas of cluster F (e.g., upper area in FIIb), however the segments after the treatments were also included in some degree. Overall, classification was more discernible with the body segments after the treatments for the combined data of angles and lengths.

4 Discussion and conclusions

In this study we used a species in oligochates, *Limnodurlus variegatus,* for continuous monitoring of temporal line-movement in response behaviors to the treatments of toxic substances. RSOM was feasible in classifying the sequential line-movements of the blackworms. To the best of our knowledge, no computational method has been carried out on patterning behaviors of blackworms with temporal movement, fig. 4, 5 and 6. This type of the line patterning would be useful for monitoring behavioral changes in animals with long body shapes (e.g., other annelids, snakes, eels, etc) in the future. If the lines are detected instead of the points, the scope in movement changes would be broader in providing diverse information on movement.

We analysed the response behaviors based on separate criteria, angle and length. Firstly, patterns of body folding were accordingly revealed by training with the angle data, fig. 4(b), by the trained RSOM. With the data for the lengths subsequently, fig. 5(b), the 'line' part was more clearly identified: the straight lines (e.g., CI in fig. 5(b)) were efficiently grouped. In the data for angles, fig 4(b), however, the straight segments shown in CI were not observed. Conversely, the highly folded ones (e.g., A1 in fig. 4(b)) were not observed in fig. 5(b). RSOM was efficient in extracting information from complex data accordingly to characteristics of input variables.

We used blackworms for indicator species for behaviour monitoring. Aquatic oligochaetes, such as *Lumbriculus*, are important taxa in freshwater aquatic communities. The oligochaetes serve diverse roles as aiding in decomposition of organic materials in the sediment. *Lumbriculus variegatus* has been proposed as a standard organism for sediment bioaccumulation tests [35]. Researchers reported that *Lumbriculus* have several locomotor behaviors such as rapid withdrawal, crawling, body reversal, and helical swimming. There are stereotyped behaviors that can be used for the sub-lethal toxicology [36, 37]. In this study RSOM was feasible in identifying some symptomatic sequence of line-movements of blackworms, fig. 4, 5 and 6, and the results from our study confirmed the toxicological effect of black worms on behaviours.

In this study we used copper as a toxic agent. Copper has commonly been used for fertilizers, and especially pesticides due to its antifungal properties [38]. Excessive levels can contaminate aquatic ecosystems [38, 39] even though it is an essential trace element. Sub-lethal copper exposure alters a number of behaviors in invertebrates. Locomotor behavior can also be adversely affected by copper exposure [40–43]. Bruce et al [44] reported that copper exposure can significantly affect the ability of aquatic oligochaetes to avoid predators. We were able to demonstrate that copper would cause typical symptomatic behaviours in temporal sequence of line-movement of blackworms by using RSOM.

In conclusion, RSOM, was feasible in addressing the sequential line-movements of stressful behaviors of blackworms such as body contraction, high degree of folding, etc. The temporal patterns of body shapes after the treatments were more strongly grouped than the temporal patterns before the treatments. The recurrent unsupervised model in ANNs used in this study would be an efficient tool for identifying temporal behavior patterns of indicator specimens and could be used as an *in-situ* real-time monitoring device in aquatic ecosystems in the future.

Acknowledgement

This subject is supported by Korea Ministry of Environment as "The Eco-technopia 21 project".

Reference

[1] Kwak, I.-S., Chon, T.-S., Kang, H.-M., Chung, N.-I., Kim, J.-S., Koh, S.C., Lee, S.-K. & Kim, Y.-S., Pattern recognition of the movement tracks of medaka (Oryzias latipes) in response to sub-lethal treatments of an insecticide by using artificial neural networks. Environ. Pollut. 120, 671-681, 2002.

[2] Untersteiner, H., Kahapka, J. & Kaiser, H., Behavioral response of the cladoceran Daphnia magna Straus to sublethal copper stress – validation by image analysis. Aquat. Toxicol. 65, 435-442, 2003.

[3] Chon, T.-S., Park, Y.-S., Park, K. Y., Choi, S.-Y., Kim, K. T. & Cho, E. C., Implementation of computational methods to pattern recognition of movement behavior of Blattella germanica(Blattaria: Blattellidae) treated with Ca^{2+} signal inducing chemicals. Appl. Entomol. Zool. 39, 79-96, 2004.

[4] Park, Y.-S., Kwak, I.-S., Chon, T.-S., Kim, J.-K. & S. E. Jorgensen, Implementation of artificial neural networks in patterning and prediction of exergy in response to temporal dynamics of benthic macroinvertebrate communities in streams. Ecolog. Model. 146, 143-157, 2001.

[5] Teather, K., Harris, M., Boszell, J. & Gray, M., Effects of Acrobat MZ and Tatoo C on Japanese medaka (Oryzias latips) development and adult male behavior. Aquat. Toxicol. 51, 419-430, 2001.

[6] Abgrall, P., Rangeley, R.W., Burridge, L.E. & Lawton, P., Sublethal effects of azamethiphos on shelter use by juvenile lobsters (Homarus americanus). Aquaculture 181, 1-10, 2000.

[7] Roast, S.D., Widdows, J., Jones & M.B., Disruption of swimming in the hyperbenthic mysid Neomysis integer (Peracarida: Mysidacea) by the

organophosphate pesticide chlorpyrifos. Aquat. Toxicol. 47, 227-241, 2000.

[8]   Ibrahim, W.L.F., Furu, P., Ibrahim, A.M. & Christensen, N.Ø., Effect of the organophosphorous insecticide, chlorpyrifos (Dursban), on growth, fecundity and mortality of Biomphalaria alexandrina and on the production of Schistosoma mansoni cercariae in the snail. Journal of Helminthology 66, 79-88, 1992.

[9]   Gray, M.A., Teather, K.L. & Metcalfe, C.D., Reproductive success and behavior of Japanese Medaka (Oryzias latipes) exposed to 4-tert-octylphenol. Enviro. Toxicol. and Chem. 18(11), 2587-2594, 1999.

[10] Oshima Y., Kang, I. J., Kobayashi M., Nakayama K., Imada N. & Honjo T, Suppression of sexual behavior in male Japanese medaka (Oryzias latipes) exposed to 17β-estradiol Chemosphere, 50,429-436, 2003.

[11] Chon, T.-S., Park, Y.S. & Ross, M.H., Activity of German cockroach, Blattella germanica (L.) (Orthoptera : Blattellidae), at different microhabitats in semi-natural conditions when treated with sublethal doses of pesticides. Journal of Asia-Pacific Entomology 1, 99-107, 1998.

[12] Chon, T.-S., Park, Y.S. & Ross, M.H., Temporal pattern of local activity at harborage in two strains of the German cockroach, Blattella germanica (L.), in semi-natural conditions after treated with sublethal doses of insecticides. Korean Journal of Entomology 28, 77-83, 1998.

[13] Berg, H. C., Random Walks in Biology. Trends in Neurosciences. 8, pp. 83-84, 1983.

[14] Alt, W., Correlation analysis of two-dimensional locomotion paths. In: Alt, W., Hoffmann, G., (Eds.), Biological Motion. Springer, pp. 254-268, 1990.

[15] Scharstein, H., Paths of carabid beetles walking in the absence of orienting stimuli and the time structure of their motor output. In: Alt, W., Hoffmann, G., (Eds.), Biological Motion. Springer, Berlin, pp. 269-277, 1990.

[16] Tourtellot, M.K., Collins, R.D. & Bell, W.J., The problem of

[17] Johnson, A.R., Milne, B.T. & Weins, J.A., Diffusion in fractal landscapes: simulations and experimental studies of tenebrionid beetle movements. Ecology 73, 1968-1983, 1992.

[18] Weins, J.A., Crist, T.O., With, K.A. & Milne, B.T., Fractal patterns of insect movement in microlandscape mosaics. Ecology 76, 663-666, 1995.

[19] Alados, C.L., Escos, J.M. & Emlen, J.M., Fractal structure of sequential behavior patterns: an indicator of stress. Anim. Behav. 51, 437-443, 1996. movelength and turn definition in analysis of orientation data. J. Theo. Bio. 150, 287–297, 1991.

[20] Sovan L. & J.F. Guegan, Application to Ecology and Evolution. Artificial Neuronal Networks, 2002.

[21] Friedrich R., Ecological Informatics, Springer, 2003.

[22] Sovan L., Michele S., Piet F.M.V., Jean-Pierre D. and Park Y.-S., Modelling Community Structure in Freshwater Ecosystems, Springer, 2005.

[23] Kim, C.-K., Kwak, I.-S., Cha, E.-Y. & Chon, T.-S., Implementation of wavelets and artificial neural networks to detection of toxic response behavior of chironomids(Chironomidae: Diptera) for water quality monitoring. Ecol. Model. 195, 61-71, 2006.

[24] Park, Y.-S., Chung, N.-I., Choi, K.-H., Cha, E. Y., Lee, S.-K. & Chon, T.-S., Computational characterization of behavioral response of medaka (Oryzias latipes) treated with diazinon, Aquat. Toxicol. 71, 215-228, 2005.

[25] N. Gershenfeld & A. Weigend. The future of time series: Learning and understanding. In A. Weigend and N. Gershnfeld, editors, Time Series Prediction: Forecasting the Future and Understanding the Past, page 1- 70. Addison-Wesley, 1993.

[26] Chon, T.-S., Park, Y.-S. & Cha, E.Y., Patterning of Community Changes in Benthic Macroinvertebrates Collected from Urbanized Streams for the Short Time Prediction by Temporal Artificial Neuronal Networks. Artificial Neural Networks, Springer-Verlag , pp. 99-114, 2000.

[27] Chon, T.-S., Kwak, I.-S., Park, Y.-S., Kim, T.-H & Kim Y.S., Patterning and short-term predictions of benthic macroinvertebrate community dynamics by using a recurrent artificial neural network. Ecol. Model. 146, 181-193, 2001.

[28] G.J. Chappell & J.G. Taylor., The temporal Kohonen map. Neural Networks, 6:441~445, 1993.

[29] T. Kohonen., Self-Organization and Associative Memory. Springer-Verlag, Berlin, Heidelberg, 1989.

[30] T. Kohonen, Self-Organization Maps. Springer-Verlag, 1995.

[31] M. Varsta, J. Heikkonen. & J.d.R. Millan, Context learning with the self-organizing map. In Proc. Of Workshop on Self-Organizing Maps, 197-202. Helsinki University of Technology, 1997.

[32] K. Timo, M. Varsta, J. Heikkonen & K. Kaski, Temporal Sequence Processing using Recurrent SOM. In Proc of 2nd International Conference on Knowledge-Based Intelligent Engineering Systems, 1, 290-297, 1998.

[33] Peter A., Time Series Prediction Using RSOM and Local Models. IIT.SRC, 2005.

[34] The Mathworks, Inc. MATLAB Version 5.3, Massachusetts. 1999.

[35] ASTM: American Society for Testing and Materials, Standard guide for conducting sediment toxicity tests with freshwater invertebrates. In: Annual Book of ASTM Standards. Vo. 11.05, E 1393-94a, Philadelphia, PA, USA, pp. 802-834, 1995.

[36] Rogge, R.W. & Drewes, C.D., Assessing sublethal neurotoxicity effects in the freshwater oligochaete, Lumbriculus variegatus. Aquat. Toxicol. 26, 73-90, 1993.

[37] Ding, J., Drewes, C.D. & Hsu, W.H., Behavioral effects of ivermectin in a freshwater oligochaete, Lumbriculus variegatus. Environ. Toxicol. Chem. 20, 1584-1590, 2001.

[38] G.E.R.B.E, Programme de recherché: toxicologie – Ecotoxicologie des pesticides et des metaux lourds. In: Deuxieme rapport d'activites.

Laboratory of Eco-Toxicology, Faculty of Sciences, University of Reims Champagne, Ardenne, France, pp. 39, 1998.

[39] Jurado, R., Introduction a la Toxicologia Veterinaria. Tebar-flores, Madrid, pp. 112, 1983.

[40] Sullivan, B.K., Buskey, E., Miller, D.C. & Ritacco, P.J., Effects of copper and cadmium on growth, swimming and predator avoidance in Eurytemora affinis (Copepoda). Mar. Biol. 77, 299-306, 1983.

[41] Rondelaud, D., Les effects. d'une concentration subletale de molluscicide (CuCl2) sur l'activite reproductice et les deplacements du mollusque hote Lymnaea truncatula Muller. Ann. Rech Vet. 19, 273-278, 1988.

[42] Charoy, C. & Janssen, C.R., The swimming behavior of Brachionus calyflorus (Rotifer) under toxic stress. Chemosphere 38, 3247-3260, 1999.

[43] Dhawan, R., Dusenbery, D.B. & Williams, P.L., A comparison of metal-induced lethality and behavioral responses in the nematode Caenorhabdities eldgans. Environ. Toxicol. Chem. 19, 3061-3067, 2000.

[44] Bruce A.O., V. Kim B., Matthew W.T. & Michael B.S., Copper-induced changes in locomotor behaviors and neuronal physiology of the freshwater oligochaete, Lumbriculus variegatus. Aquat. Toxicol. 69, 51-66, 2004.

Computational analysis of movement behaviors of medaka (*Oryzias latipes*) after the treatments of copper by using fractal dimension and artificial neural networks

C. W. Ji[1], S. H. Lee[2], I.-S. Kwak[3], E. Y. Cha[4], S.-K. Lee[5] & T.-S. Chon[1]
[1]*Division of Biological Sciences,*
Pusan National University, Republic of Korea
[2]*Department of Physics, Pusan National University, Republic of Korea*
[3]*Faculty of Marine Technology,*
Chonnam National University, Republic of Korea
[4]*Department of Electronics Engineering,*
Pusan National University, Republic of Korea
[5]*Toxicology Research Center,*
Korea Research Institute of Chemical Technology, Republic of Korea

Abstract

Response behaviors of medaka were computationally analyzed before and after the treatments of copper at low concentration (1.0 mg/L). Parameters (e.g., speed, stop time, turning rate, etc) of the movement patterns were used as input for training the Multi-Layer Perceptron. Detection rates of the movement patterns such as 'Slow movement' and 'No movement' increased after the treatments. However, a higher degree of variation was observed in detection rates. Fractal dimension calculated from the movement data of individual specimens decreased consistently after the treatments. Higher consistency in fractal dimension was further achieved by using the data for collective rearing. Feasibility of behavioral monitoring was discussed in assessing toxic chemicals in environment.
Keywords: response behavior, medaka, copper, fractal dimension, Artificial Neural Network, behavioral monitoring.

WIT Transactions on Biomedicine and Health, Vol 10, © 2006 WIT Press
www.witpress.com, ISSN 1743-3525 (on-line)
doi:10.2495/ETOX060101

1 Introduction

Recently automatic detection of response behaviors of animals has been considered as an efficient tool for bio-monitoring of aquatic ecosystems [1]. Dutta et al [2] suggested that a behavioral bioassay would be more sensitive than other types of testing methods. A numerous accounts of behavioral research on effects of toxic chemicals at low concentrations have been reported in various taxa, including crustaceans [3, 4], snails [5], fish [6] and insects [7, 8]. Recently Oshima et al [9] observed suppression of sexual behavior in male medaka exposed to estradiol. However, these studies are mostly based on observation of single or combinations of single behaviors mainly with qualitative descriptions. Not much computational research has been carried out for automatically detecting behavioral changes from continuous recording.

Behaviors, however, have been regarded as difficult for analysis due to complexity residing in the data. Theoretical studies have been carried out on analyzing movement data regarding correlation function [10], random walk [11], etc. Recently fractal dimension has been considered as an efficient parameter to quantitatively express behavioral states. Fractal dimension has been widely used for analyzing non-linear phenomena in biological and ecological sciences such as geographical features, morphology, etc, [see 12]. Johnson et al [13] and Weins et al [14, 15] used fractal dimension for analysis of insect movement to quantitatively characterize behavioral states that might not be available through absolute measures of pathway configurations. Alados et al [16] used fractal dimension for detecting response behaviors of parasitic infection in Spanish ibex. In this study we used fractal dimension to reveal behavioral states of indicator specimens in response to toxic substances.

Along with fractal dimension, we also implemented Artificial Neural Networks (ANNs) to address pattern changes in response behaviors. While fractal dimension quantitatively compresses behavioral changes as one parameter, ANNs are useful for dealing with local information and for revealing specific behavioral patterns explicitly. ANNs have been widely used for analyzing complex data in computer and electronics engineering [see 17, 18] and have been recently implemented to ecological sciences in various aspects such as forecasting, input-output relationships, data organization, classification, etc [see 19, 20]. Recently ANNs have been applied to behavioral monitoring. Self-Organizing Map was applied to classification of response behaviors of indicator organisms treated with diazinon [21, 22]. Multi-Layer Perceptron (MLP) was used to automatically detect behavioral changes in organisms such as medakas and chionomids in response to toxic chemicals [23, 24].

In this study we intend to extract local and global information residing in behavioral data and to propose a system to quantitatively characterize response behaviors in both explicit (i.e., MLP) and compressed (i.e., fractal dimension) forms. Initially MLP was applied to detection of changes in specific movement patterns after the treatments of toxic substances. Subsequently we elucidated fractal dimension as a means of minimizing the variability of behavioral data to be a reliable parameter to detect changes in behavioral states.

WIT Transactions on Biomedicine and Health, Vol 10, © 2006 WIT Press
www.witpress.com, ISSN 1743-3525 (on-line)

2 Materials and methods

2.1 Test specimens and observation system

Medakas (*Oryzias latipes*), the "*or*" strain originally developed by Bioscience Center, Nagoya University, Japan, were obtained from Toxicology Research Center, Korea Research Institute of Chemical Technology (KRICT; Taejeon, Korea) for testing. The stock populations were maintained in a glass tank, and were reared with artificial dry diet (Tetramin®) under the light regime of L10:D14 in temperature ranging 25±1°C. In photo-phase, a fluorescent lamp (20 W) was used as the light source and was located above the observation aquarium with 30 cm apart. In scoto-phase, a red light (20 W) was provided at the same position.

The position of the test specimens of medaka (age: 6–12 months) was recorded by using an observation system consisting of an observation aquarium, a camera and software for image recognition. Individuals or groups of medakas were placed in a glass aquarium (volume of water: 40 cm ×20 cm ×10 cm), and their position was scanned from the side view at 0.25 s intervals using a CCTV camera (Kukjae Electronics Co. Ltd.; IVC-841®) for four days (two days before the treatments and two days after the treatments). The analog data captured by the camera were digitized by using a video overlay board (Sigmacom Co., LTD.; Sigma TV II®), and were sent to the image processing system to locate the target organisms in two dimension. The software for recognition of the movement tracks and other supporting mathematical programs were provided according to [23].

During the period of observation, disturbances in experimental conditions were minimized: oxygen, fresh water and food were not supplied to test specimens to simplify observation conditions [23]. Before monitoring, the specimens were acclimated to the observation aquarium for 1–2 days. Environmental factors such as light and temperature were maintained to the same condition for rearing stock populations.

2.2 Experimental procedure

Copper was treated to medaka fishes in this study. The level of LC_{50} for copper to medaka population was reported as 5 mg/L [25]. After two days of observation without treatment, reaction behaviors were also recorded for two days after the treatments of copper (1 mg/L).

Initially, we observed 10 medaka fishes individually before and after the treatments. The parameters were extracted from the segmented data in every 30 s. Based on previous research on the movement tracks [23], the following parameters were selected to characterize the movement patterns and were subsequently used as input data for training MLP (see section 2.3):

1) Speed (mm/s): average in movement distance of the fish during the observation time.

2) *Y*-position (mm/sec^2): the average distance in *Y*-axis measured from the surface during the observation time; as the specimens was located close to the surface, the *Y*-position was decreased

3) Stop number: the total frequency in which specimen did not move.

4) Stop duration (*total time of stops:* s): the total duration in which the specimen did not move.

5) Turning rate (rad/s): the sum of angle changes in radian in absolute values divided by the cumulated time duration of movement.

6) Meander (rad/mm): the sum of angle changes in radian in absolute values divided by the path length.

For calculating statistics of the parameters of the movement tracks, we selected 10 sample segments (30 s) by visual observation for each movement pattern. This process was repeated for 10 specimens. In total 100 segments were obtained for each pattern. Among the selected samples 30 segments were randomly chosen for statistical analysis. Subsequently 10 samples were independently selected for the MLP training by random sampling. For testing the trained MLP, the whole sequence of the movement data for four days (2 days separately for each 'before' and 'after' the treatments) were provided to the trained MLP. The data segments in every 30 s interval were continuously provided to be recognized by the trained network.

Medaka fishes were also reared in groups with four specimens in the same conditions applied to individual rearing in 10 replications. Fractal dimension (see section 2.4) was calculated for the movement data for specimens in individual and group rearing.

2.3 Multi-Layer Perceptron (MLP)

The MLP [26] was trained with the data for the movement tracks. Training proceeds to minimize the mean square error between the actual input and desired output (or target value) according to the back-propagation algorithm (Fig. 1) [18, 26]. In this study the parameters characterizing the movement tracks were used as input data (6 nodes), while the decision of the movement patterns were given in the binary form as matching output (6 nodes).

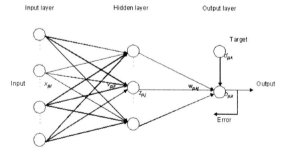

Figure 1:   The schematic diagram of MLP.

The net input ($NET_{p,j}$) to neuron j of the hidden layer for pattern p is calculated as the summation of each input layer output ($X_{p,i}$; input value of parameter) multiplied by weight (v_{ji}). The similar calculation is provided for the neuron k of the output layer being linked by summation of each hidden layer output ($Z_{p,j}$). An activation function (logistic function in this case) is applied to calculate the output of neuron j of the hidden layer ($Z_{p,j}$) and the output of neuron k of the output layer ($O_{p,k}$), according to the following eqn. (1):

$$f(NET) = \frac{1}{1 + \exp(-\lambda NET)}$$

(1)

where λ is the activation function coefficient. NET is expressed either in $Z_{p,j}$ or $O_{p,k}$ as follows, eqn. (2), (3):

$$z_{p,j} = f\left(\sum_i x_{p,i} v_{ji}\right)$$

(2)

$$o_{p,k} = f\left(\sum_j z_{p,j} w_{kj}\right)$$

(3)

where v_{ji} and w_{kj} are the connection weight between neuron i of the input layer and neuron j of the hidden layer, and the connection weight between neuron j of the hidden layer and neuron k of the output layer, respectively.

The back-propagation algorithm adjusts the connection intensities (weights (v_{ji}) and (w_{kj})) of the network in a way that minimizes error. The sum of the errors in each neuron for pattern p, Err_p, is calculated as follows, eqn. (4):

$$Err_p = \frac{1}{2} \sum_k (d_{p,k} - o_{p,k})^2$$

(4)

where $d_{p,k}$ is the target value corresponding to pattern p at neuron k. The value of the activation function coefficients, λ, used in this study was 1.0, and the learning coefficient, which updates the weights in iterative calculation, was set at 0.01. The level of error tolerance was 1.0, and the threshold for determining the binary level for the activation function was 0.5. Network pruning was not required during the training process in this study. Details of using MLP can be found in the related bibliographies [17, 18, 26, 27].

2.4 Fractal dimension

Fractal dimension, D, was measured on location of specimens in individual rearing. The points recorded in every 0.25 sec in 1-hour segment of the movement tracks were used for calculation based on the Box-Counting method (MATLAB® 5.3.), eqn. (5):

$$N(r) = (1/r)^D, \qquad D = \frac{\log N}{\log(1/r)}$$

(5)

where $N(r)$ is the number of points observed within the box sized as r^2. The two values, $N(r)$ and r are presented as a linear form by the double logarithmic graph.

In each 1-hour segment of the movement tracks, overlapping was allowed for 30 min.

Fractal dimension was also measured for the data from group rearing. The data points (0.25 s interval) of four specimens in the 15-minute segments were used for calculating fractal dimension. In each segment of the movement tracks overlapping was allowed for 30 min and 7.5 min for individual and group rearing respectively.

3 Results

3.1 Characterization of behavioral patterns

Behaviors of fish have been reported to show typical patterns, including stationary movement, up-down swimming with circular motion, eating, agonistic behavior, hiding, etc [28]. In this research we also observed some clear movement patterns of medakas under the experimental conditions. Figure 2 shows the typical movement patterns of the tested specimens.

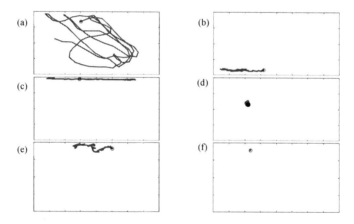

Figure 2:   The movement tracks (side view) showing the behavioral patterns of medaka specimens in 30 s segments (a); Swimming, (b); Feeding, (c); Surface movement, (d); Slow movement, (e); Frequent stop, (f); No movement).

'Swimming' presented the active state of specimens (Fig. 2(a)), being characterized by the highest speed (65.82 mm/s) and wide circling in the observation aquarium (Table 1). 'Feeding' showed the horizontal movement along the bottom of the observation aquarium in the limited range (Fig. 2(b)). The Y-position (180.02 mm) of 'Feeding' is higher than any other patterns (Table 1). In the observation system, the position on Y-coordinate movement is higher as it is closer to the bottom of the aquarium. 'Surface movement' showed horizontal activity near the top area of the aquarium (Fig. 2(c)). In contrast to the

'Feeding', Y-position (15.56 mm) of 'Surface movement' was lower than any other patterns (Table 1). 'Slow movement' showed the lower phase of activity: the stop time (20.92 s) was longer than any other patterns (Table 1). 'Stop' is defined as the specimens maintaining the same position for the duration of 0.25 s in this study. Stop time is calculated as summation of the time duration for each stop. 'Frequent stop' is another pattern showing slow phase of activity. The specimens repeated the 'stop' and 'short advancement'. Stop number in 'Frequent stop' was observed as frequently as in 'Slow movement', however stop time was distinctively shorter in 'Frequent stop' (6.2 s) than in 'Slow movement' (20.92 s) (Table 1). Overall, 'Frequent stop' presented somewhat more active states compared with 'Slow movement'.

Table 1:   Parameters characterizing the different movement patterns of medaka specimens before and after the treatments of copper (n=30 for each parameter for each pattern, (a); Swimming, (b); Feeding, (c); Surface movement, (d); Slow movement, (e); Frequent stop).

Parameters	Speed ($mm\ s$)		Y-position (mm)		Stop time (sec)		Stop number		Turning rate ($rad\ s$)		Meander ($rad\ mm$)	
Patterns	Mean	SD	Mean	SD	Mean	SD	Mean	SD	Mean	SD	Mean	SD
a	65.8	12.4	90.7	16.4	0.3	0.5	0.7	1.2	1.7	0.3	0.0	0.0
b	11.2	2.4	180.2	3.4	7.7	4.0	11.7	2.7	4.4	0.8	0.3	0.1
c	17.5	9.8	15.6	7.2	7.2	5.0	11.2	6.0	2.9	1.2	0.2	0.1
d	3.6	1.1	97.9	82.1	20.9	3.3	13.1	4.8	5.9	1.0	0.5	0.1
e	10.0	2.0	107.6	39.3	6.2	3.5	14.2	4.8	5.0	1.4	0.4	0.1

Table 2:   Analysis of variance (ANOVA) and Tukey test for multiple comparisons of parameters characterizing the different movement patterns of medaka specimens before and after the treatments (n=30 for each parameter for each pattern, (a); Swimming, (b); Feeding, (c); Surface movement, (d); Slow movement, (e); Frequent stop).

Parameters	F[1]	P	Comparison of parameters[2] (Tukey test, α=0.05)
Speed ($mm\ s$)	368.01	<0.001	a ≠ c ≠ b = e ≠ d
Y-position (mm)	59.38	<0.001	b ≠ e = d = a ≠ c
Stop time (sec)	133.75	<0.001	d ≠ b = c = e ≠ a
Stop number	48.99	<0.001	e = d = b = c ≠ a
Turning rate ($rad\ s$)	80.11	<0.001	d ≠ e = b ≠ a = c
Meander ($rad\ mm$)	112.23	<0.001	d ≠ e = b ≠ c = a

[1] $F_{0.05(2),\ 5,\ 150} = 2.66$

[2] Patterns were listed in the increasing order from left to right.

The parameters mostly appeared to be statistically different in different movement patterns according to the Tukey test [29] (Table 2). Regarding speed, 'Swimming', 'Surface movement' and 'Slow movement' were different among the patterns, but 'Frequent stop' and 'Feeding' were in the same range. 'Feeding' and 'Frequent stop' were also similar in the other parameters including stop time, stop number, turning rate and meander. Y-position, however, was different between the two patterns. The other movement patterns were uniquely distinguished each other and were statistically different (Table 2).

3.2 MLP applied to individual data

The parameters characterizing the movement patterns were effectively learned by MLP with the training rates mostly over 92%. As stated before the whole data set was used for testing. The pattern of input segment (30 s) was recognized by the trained MLP. Detection rates were calculated as the number of correct recognition divided the number of the total recognition for each individual. The detection rate for each specimen was subsequently averaged with 10 specimens. Table 3 shows changes in detection rates (%) for each pattern in averages for 10 specimens before and after the treatments.

Table 3:   Detection rate (%) of different movement patterns of medaka specimens before and after the treatment of copper (a; Swimming, b; Feeding, c; Surface movement, d; Slow movement, e; Frequent stop, f; No movement). (n=10).

Treatments	Patterns	Day		Night		All	
		Mean	SD	Mean	SD	Mean	SD
Before Treatment	a	31.48%	30.21	10.79%	17.93	21.13%	26.41
	b	13.10%	11.24	7.91%	6.84	10.50%	9.44
	c	6.56%	10.21	7.74%	5.07	7.34%	7.73
	d	5.80%	5.25	10.70%	5.68	8.25%	5.89
	e	4.24%	3.75	9.12%	9.12	6.68%	7.23
	f	6.19%	7.89	20.26%	19.55	13.22%	16.21
After Treatment	a	12.51%	13.25	3.51%	3.49	8.01%	10.5
	b	8.21%	5.93	3.23%	2.72	5.72%	5.17
	c	3.56%	3.94	2.45%	1.72	3.01%	3.01
	d	14.91%	7.33	19.45%	5.18	17.18%	6.6
	d	9.21%	5.87	8.77%	5.81	8.99%	5.69
	f	21.88%	11.37	37.93%	9.41	29.21%	13.07

Before the treatments, detection rate of 'Swimming' pattern was high with 21.13%, but the rate decreased to 8.01% after the treatments (Table 3). The 'Feeding' (from 10. 50% to 5.72%) and 'Surface movement' (from 7.34% to

3.01%) patterns also accordingly decreased. In contrast, detection rates for 'Slow movement' (from 8.25% to 17.18%), 'Frequent stop' (form 6.68% to 8.99%) and 'No movement' (from 13.22% to 29.21%) increased after the treatments of copper (Table 3). In general, detection rates for the patterns representing high activity (e.g., 'Swimming', 'Feeding' etc) were decreased after the treatments.

Higher variation was observed in the detection rates obtained by the trained network. Standard deviations in detection rates were in the higher range, and all the parameters were not distinguished with statistical tests. Table 4 shows the comparison of detection rates (%) before and after the treatments based on the t-test ($n=10$ for each pattern) in different light phases. The patterns of 'Slow movement' and 'No movement' were statistically significant accordingly in photo- and scoto-phase, and the total periods (Table 4). Although the average values showed differences, 'Swimming', 'Feeding' and 'Frequent stop' were not statistically different before and after the treatments. The trends of behavioral changes were similar in scoto- and photo-phases. 'Surface movement' was only different at night before and after the treatments (Table 4).

Table 4: Comparison of detection rates (%) in different movement patters before and after the treatments based on the t-test ($n= 10$ for each pattern, a; Swimming, b; Feeding, c; Surface movement, d; Slow movement, e; Frequent stop, f; No movement).

	Photophase		Scotophase		Total period	
	t	P	t	P	T	P
a	1.818	n.s.	1.259	n.s.	1.666	n.s.
b	1.215	n.s.	2.012	n.s.	1.646	n.s.
c	0.992	n.s.	3.123	$0.01<P<0.02$	1.815	n.s.
d	3.193	$0.01<P<0.02$	3.599	$0.005<P<0.01$	3.836	$0.002<P<0.005$
e	2.255	n.s.	0.100	n.s.	0.912	n.s.
f	3.587	$0.005<P<0.01$	2.576	$0.05<P<0.02$	3.293	$0.005<P<0.01$

$t_{0.05(2),9} = 2.262$

3.3 Fractal dimension applied to individual rearing

In contrast to the results from MLP, fractal dimension showed more consistency in revealing changes in behavioral states of medaka specimens after the treatments of copper (Fig. 3). Although there were individual variations, decrease in fractal dimension appeared consistently for all the tested specimens. The average in fractal dimension was 1.62 ± 0.10 before the treatments, but decreased to 1.42 ± 0.16 after the treatments. The Nested ANOVA indicated that the values of fractal dimension were statistically different between 'before' and 'after' the treatments (df = (1, 18), F = 6.2, $0.02<P<0.05$). The sub-group of individual specimens, however, was different (df = (18, 600), F = 3.07, $P<0.001$). This indicated that individual variation existed in the values of fractal dimension.

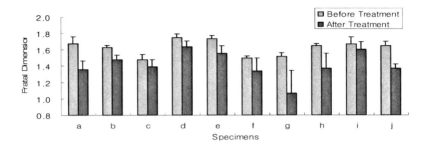

Figure 3:   Fractal dimension of the movement points in different specimens of medaka obtained from individual rearing before and after the treatments of copper, 1 mg/L.

3.4 Fractal dimension applied to group rearing

We further analyzed fractal dimension of the movement points when the specimens were reared in groups of 4 specimens (Fig. 4). Fractal dimension consistently decreased after the treatments of copper (1 mg/L) in different groups, being similar to the case of individual rearing. The average of fractal dimension was 1.63 ± 0.02 before the treatments and 1.46 ± 0.07 after the treatments. The values of fractal dimension from group rearing were more consistent compared with individual rearing. The Nested ANOVA showed that the values of fractal dimension were significantly different between the treatments (df = (1, 18), F = 23.35, $P<0.001$). In contrast to the case of individual rearing, however, the values of fractal dimension were also in the similar range between the tested groups: the sub-group difference was not significant (df = (18, 820), F = 0.05, $P>0.5$). This indicated that individual variation in fractal dimension could be minimized through group rearing of fishes.

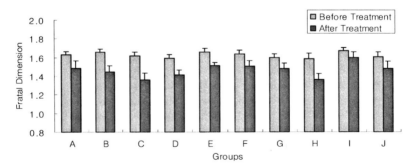

Figure 4:   Fractal dimension of the movement points in different groups of the 4 medaka specimens before and after the treatments of copper, 1.0 mg/L.

4   Discussion and conclusions

A computational system was developed for automatically detecting the movement states of medaka specimens in this study. Although individual variation occurred, MLP was useful for detecting movement patterns explicitly. The specific patterns such as 'Slow movement' and 'No movement' were statistically different before and after the treatments (Table 3). These patterns could be used as indicator patterns of medakas for detecting presence of toxic substance in environment.

We further showed that the higher variation in individuals could be decreased by using fractal dimension. The values of fractal dimension appeared to consistently decrease for all the tested specimens after the treatments (Fig. 3). The group testing, consisting of 4 medaka fishes, further minimized the variation of fractal dimension by showing no statistical difference among different groups (Fig. 4). Consistency in the measurement of fractal dimension was revealed in comparing Coefficient of Variation (CV: standard deviation divided by mean) (Fig. 5). CVs for group rearing were lower for both 'before' and 'after' the treatments. The difference between individual and group rearing was more clearly observed after the treatments with the statistical significance (df= (1, 18), $F = 6.01$, $0.02<P<0.05$). The statistical difference was not observed for CVs between individual and group rearing before the treatments (df= (1, 18), $F = 0.66$, $P>0.5$), however the average value was lower for group rearing (Fig. 5).

This study indicated that fractal dimension based on group rearing could be used as a reliable parameter to indicate behavioral changes of medakas after the treatments of copper. Another advantage of fractal dimension is the flexibility in recording data points in group rearing. In the image processing system, it is in general difficult to trace the movement tracks for each specimen in group rearing especially if the specimens are small in size. Fractal dimension, however, was measured from the positions of the specimens collectively, and tracing each individual movement was not necessary in this case.

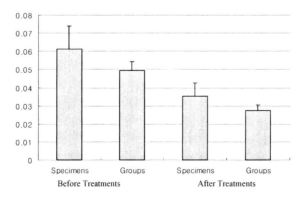

Figure 5:   Comparison of CVs of fractal dimension in individual and group rearing before and after the treatments of copper, 1.0 mg/L.

In real situation, MLP and fractal dimension could be used in combination for providing more practical information for warning system in risk assessment. While fractal dimension would provide information on the global change in behavioral states more consistently in a compressed form, MLP would reveal differences in the specific patterns, thus providing more detailed information on explicit response behaviors. The two methods could be combined to produce an efficient monitoring system for *in-situ* risk assessment in aquatic systems in the future.

In this study we used copper as toxic substances. Copper plays an essential role in mitochondrial function, detoxification of free radicals, neurotransmitter synthesis, cross-linking of connective tissue, and cellular iron metabolism. Copper causes mutations in genes encoding "P-type" transport ATPase and induces neurotic disease such as Lou Gehrig's and Wilson's disease [30]. The toxicological impact would consequently produce stressful responding behaviors of the organisms. Toxic responses to copper have been reported on some indicator species. Activity accordingly decreased in *Daphnia magna* and *Gammarus* [31, 32]. Not much quantitative research, however, has been conducted on behavioral changes especially on vertebrates such as fish. In this study, we demonstrated that computational methods such as MLP and fractal dimension could be efficiently used for monitoring contamination of copper by using fish as indicator specimens.

In conclusion, MLP could accommodate local information on response behaviors and would be useful for detecting changes in specific patterns. The consistency in behavioral detection was achieved by fractal dimension especially through group rearing, and the parameter could be useful source of a reliable indicator in determining behavioral states of specimens exposed to toxic chemicals. MLP and fractal dimension could be used in combination as an efficient means of *in-situ* monitoring by providing both 'local and more specific' (i.e., MLP), and 'global and more consistent' (i.e., fractal dimension) information concurrently.

Acknowledgements

This work was supported by grant No. R01-2004-000-11036-0 from the Basic Research Program of the Korea Science & Engineering Foundation. We appreciate provision of the original stock for test specimens from the Bioscience Center, Nagoya University, Japan.

References

[1]   Lemly, A.D. & Smith, R.J., A behavioral assay for assessing effects of pollutants of fish chemoreception. Ecotoxicology and Environmental Safety 11, pp. 210-218, 1986.
[2]   Dutta, H., Marcelino, J. & Richmonds, Ch., Brain acetylcholinesterase activity and optomotor behavior in bluegills, Lepomis macrochirus

exposed to different concentrations of diazinon. Arch. Intern. Physiol. Biochim. Biophys. 100, pp. 331-334, 1992.

[3]   Abgrall, P., Rangeley, R.W., Burridge, L.E. & Lawton, P., Sublethal effects of azamethiphos on shelter use by juvenile lobsters (*Homarus americanus*). Aquaculture 181, pp. 1-10, 2000.

[4]   Roast, S.D., Widdows, J. & Jones, M.B., Disruption of swimming in the hyperbenthic mysid Neomysis integer (Peracarida: Mysidacea) by the organophosphate pesticide chlorpyrifos. Aquatic Toxicology 47, pp. 227-241, 2000.

[5]   Ibrahim, W.L.F., Furu, P., Ibrahim, A.M. & Christensen, N.Ø., Effect of the organophosphorous insecticide, chlorpyrifos (Dursban), on growth, fecundity and mortality of Biomphalaria alexandrina and on the production of Schistosoma mansoni cercariae in the snail. Journal of Helminthology 66, pp. 79-88, 1992

[6]   Gray, M.A., Teather, K.L. & Metcalfe, C.D., Reproductive success and behavior of Japanese Medaka (*Oryzias latipes*) exposed to 4-tert-octylphenol. Environmental Toxicology and Chemistry 18, pp. 2587-2594, 1999.

[7]   Chon, T.-S., Park, Y.S. & Ross, M.H., Activity of German cockroach, *Blattella germanica* (L.) (Orthoptera: Blattellidae), at different microhabitats in semi-natural conditions when treated with sublethal doses of pesticides. Journal of Asia-Pacific Entomology 1, pp. 99-107, 1998a.

[8]   Chon, T.-S., Park, Y.S. & Ross, M.H., Temporal pattern of local activity at harborage in two strains of the German cockroach, *Blattella germanica* (L.), in semi-natural conditions after treated with sublethal doses of insecticides. Korean Journal of Entomology 28, pp. 77-83, 1998b.

[9]   Oshima, Y., Kang, I. J., Kobayashi, M., Nakayama, K., Imada, N. & Honjo, T., Suppression of sexual behavior in male Japanese medaka (*Oryzias latipes*) exposed to 17β-estradiol. Chemosphere 50, pp. 429-436, 2003.

[10]   Alt, W., Correlation analysis of two-dimensional locomotion paths. In: Alt, W., Hoffmann, G. (Eds.), Biological Motion. Lecture Notes in Biomathematics 89, pp. 254-268, 1989.

[11]   Berg, H. C., Random Walks in Biology. Trends in Neurosciences. 1983.

[12]   Rybaczuk, M. & Zieliński, W., The concept of physical and fractal dimension I. The projective dimensions. Chaos, Solitons, & Fractals 12, pp. 2517-2535, 2001.

[13]   Johnson, A.R., Milne, B.T. & Wiens, J.A., Diffusion in fractal landscapes: simulations and experimental studies of tenebrionid beetle movements. Ecology 73, pp. 1968-1983, 1992.

[14]   Wiens, J.A., Crist, T.O. & Milne, B.T., On quantifying insect movement. Environmental Entomology 22, pp. 709-715, 1993.

[15]   Weins, J.A., Crist, T.O., With, K.A. & Milne, B.T., Fractal patterns of insect movement in microlandscape mosaics. Ecology 79, pp. 663-666, 1995.

[16]   Alados C. L., Escos J. M. & Emlen J. M., Fractal structure of sequential behaviour patterns: an indicator of stress. Animal Behaviour, 51, pp. 437-443, 1996.

[17]   Lippmann, R.P., An Introduction to Computing with Neural Nets. IEEE ASSP Magazine, April. pp. 4-22, 1987.

[18]   Zurada, J.M., Introduction to Artificial Neural Systems. West Publishing Company, New York. 1992.

[19]   Chon, T.-S., Park, Y.S., Moon, K.H. & Cha, E.Y., Patternizing communities by using an artificial neural network. Ecological Modeling 90, pp. 69-78, 1996.

[20]   Lek, S., Delacoste, M., Baran, P., Dimopoulos, I., Lauga, J. & Aulagnier, S., Application of Neural Networks to modelling nonlinear relationships in Ecology. Ecological Modelling 90, pp. 39-52, 1996.

[21]   Chon, T.-S., Park, Y.-S., Park, J. Y., Choi, S.-Y., Kim, K. T. & Cho, E. C., Implementation of computational methods to pattern recognition of movement behavior of *Blattella germanica* (Blattaria : Blattellidae) treated with Ca2+ signal inducing chemicals. Appl. Entomol. Zool. 39, pp. 79-96, 2004.

[22]   Park, Y.-S., Chung, N.-I., Choi, K.-H. Cha, E. Y., Lee, S.-K. & Chon, T.-S., Computational characterization of behavioral response of medaka (*Oryzias latipes*) treated with diazinon. Aquatic Toxicology 71, pp. 215-228, 2005.

[23]   Kwak, I.-S., Chon, T. S., Kang, H. M., Chung, N. I., Kim, J. S., Koh, S. C., Lee, S. K. & Kim, Y. S., Pattern recognition of the movement tracks of medaka (*Oryzias latipes*) in response to sub-lethal treatments of an insecticide by using artificial neural networks. Environmental Pollution 120, pp. 671-681, 2002.

[24]   Kim, C.-K. & Kwak, I.-S., Implementation of wavelets and artificial neural networks to detection of toxic response behavior of chironomids (Chironomidae: Diptera) for water quality monitoring. Ecological Modelling 195, pp. 61-71, 2006.

[25]   Environmental Protection Agency., Ambient aquatic life water quality for copper. http://www.epa.gov/ost/pc/ambientwqc/copper1984.pdf., 1984.

[26]   Rumelhart, D.E., Hinton, G.E. & Williams, R.J., Learning internal representations by error propagation, in Parallel distributed processing: explorations in the microstructure of cognition. In: Rumelhart, D.E., McClelland, J.L. (Eds.), MIT Press, Cambridge MA. pp. 318 – 362, 1986.

[27]   Haykin, S., Neural Networks. Macmillan College Publishing Company, New York. 1994.

[28]   Hosn, W. A., Quantitative analysis and modelling of the behavioural dynamics of *Salvelinus fontinalis* (brook trout), Behavioural Processes. 6, pp. 105-120, 1999.

[29]   Zar, J.H., Biostatistical Analysis. Prentice-Hall International, Englewood Cliffs. 1984.

[30]   Wessling-Resnick, M., Understanding copper uptake at the molecular level. Nutrition Reviews. 60, pp. 177-179, 2002.

[31]   Untersteiner, H., Kahapka, J. & Kaiser, H., Behavioural response of the cladoceran Daphnia magna to sublethal Copper stress—validation by image analysis. Applied Toxicology 65, pp. 435-442, 2003.

[32]   Mills, C. L., Shukla, D. H. & Compton, G. J., Development of a new low cost high sensitivity system for behavioural ecotoxicity testing. Aquatic Toxicology 77, pp. 197-201, 2006.

Section 3
Effluent toxicity

Studies of produced water toxicity using luminescent marine bacteria

S. Grigson, C. Cheong & E. Way
Heriot-Watt University, Scotland, UK

Abstract

The main aqueous discharge from oil production platforms is produced water (PW). Produced water is contaminated with a range of pollutants including crude oil, inorganic salts, trace metals, dissolved gases, produced solids and oilfield chemical residues. Concern has been expressed on the impact these discharges, and particularly the dissolved oil component, may be having on the marine environment. In this investigation the toxicity of synthetic produced waters contaminated with petroleum hydrocarbons was compared to PW samples received from the field using the luminescent marine bacterium *Vibrio fisheri*. The objective was to correlate toxicity to specific PW components. Initial studies of individual oil components showed that both aromatic and aliphatic compounds exhibited toxicity. Naphthalene was the most toxic aromatic compound measured and cycloheptane the most toxic aliphatic. For benzenes, toxicity increased with alkyl substitution. Synthetic PW samples, based on the composition of those obtained offshore, had lower toxicities than the field PW samples. The addition of oilfield chemicals at dosage levels used offshore increased the toxicity of the synthetic PW mixtures, but not to the original values. Removal of the oil components by solid-phase extraction reduced PW toxicity in both synthetic and real samples. The results suggest that a range of hydrocarbons, both aliphatic and aromatic, along with heavy metals and oilfield chemical residues, contribute to the toxicity of produced water. Removal of petroleum hydrocarbons significantly reduces the acute toxicity of produced water. However, differences in toxicity between real and synthetic PW samples suggest that components other than hydrocarbons, heavy metals and oilfield chemical residues, are also influencing the toxicity of the effluent.
Keywords: toxicity testing, luminescent marine bacteria, produced water, hydrocarbons, oilfield chemicals.

WIT Transactions on Biomedicine and Health, Vol 10, © 2006 WIT Press
www.witpress.com, ISSN 1743-3525 (on-line)
doi:10.2495/ETOX060111

1   Introduction

Produced water is the main aqueous discharge from offshore oil production. The composition of produced water varies greatly between production fields. It consists mainly of water, with minor amounts of inorganic and organic constituents derived from the source geologic formation, associated hydrocarbon resource, and chemicals added during processing CAPP [1].

Detailed studies on the precise chemical composition of produced water have been carried out in the North Sea (Stephenson [2]; Utvik, [3]), Gulf of Mexico (Neff and Sauer [4]), and Canadian waters [1]. The organic compounds in produced water include mainly hydrocarbons, in the form of dissolved oil, phenols, organic acids, and naphthenic acids (Frost et al. [5]). The dissolved oil component contains mainly lower molecular weight aliphatic compounds (C5 and C6 normal, branched and cycloalkanes) and 1-3 ring aromatic hydrocarbons. The organic acids, the major organic component, are dominated by C1-C6 acids and phenols, alkylated up to C9. The inorganic component consists of salts (the salinity of produced water can range from a few ppt to 300 ppt [1]), heavy metals, for example zinc, lead, copper, cadmium, chromium and nickel (Garland et al. [6]), and in some produced waters, radionuclides [1]. In addition, produced water contains residues of oilfield chemicals used on the platform to prevent corrosion and scale formation, in the separation of oil and water (demulsifiers), and to mitigate a range of other operational problems.

1.1 Produced water toxicity studies

A wide range of species have been used to investigate the toxicity of produced water (Somerville et al. [7]; Stromgren et al. [8]; Flynn et al. [9]) including the marine bacterium *Vibrio fisheri* ([7, 9], Johnsen et al. [10]; Whale [11]). The objectives have been to elucidate those components posing the greatest threat to the marine environment through an assessment of their toxicity and to compare the sensitivities of different test species. In these studies whole produced waters were utilised for toxicity testing and attempts made to correlate toxicity to chemical composition.

In this investigation the toxicity of individual produced water components were assessed. Synthetic produced waters were then made up based on the composition of real produced waters obtained from the field and the toxicity of the two compared. Finally produced waters (both real and synthetic) were subjected to chromatography to remove the dissolved oil component and the subsequent toxicity measured. The objective was to obtain a better understanding of the toxicity contribution of individual components and to see whether synthetic produced waters could account for all of the toxicity of the field samples upon which they were based.

2   Methods

Aliphatic and aromatic solvents for toxicity testing were obtained from Rathburn Chemicals Limited, Walkerburn, Scotland. All solvents were HPLC Grade. All other chemicals were purchased from Sigma-Aldrich Co. Ltd, Gillingham, England. Toxicity testing was performed on a Microtox Model 500 analyser. Materials and reagents for Microtox testing were purchased from SDI Europe Limited, Hampshire, England.

Chemicals for testing were dissolved in brine (35 g/l sodium chloride in distilled water) or in 1% ethanol in brine. Poorly water soluble chemicals and synthetic produced water samples were made up in ethanol prior to dilution. Produced water samples were received from 14 North Sea production platforms. On arrival at the laboratory they were stored in the dark at 4°C until required for analysis. Prior to testing, the salinity and pH of each sample was checked to ensure it was in the optimal range for the bacteria. Toxicity testing was carried out using standard test protocols: Basic Test eight dilutions in duplicate or four dilutions, duplicate samples (Microbics Corporation 1992).

Solid phase extraction was performed using Strata C_{18}-E 6 ml 1 g solid-phase extraction (SPE) cartridges (Phenomenex UK, Cheshire, England). The presence of oil components was determined using a Thermoquest Trace gas chromatograph fitted with a 30 m, 0.25 mm i.d. HP-5MS capillary column (Agilent Technologies, UK) and flame ionisation detector.

2.1 Toxicity of individual produced water components

Toxicity values for a range of components typically found in produced water are shown in Figure 1. The components have been selected as those representing the most abundant members of their type reported in the literature. The values given are the mean of duplicate analyses. The toxicity of different components varies with exposure time. The exposure times reported represent maximum toxicity values. For the aliphatic and aromatic hydrocarbons this was 5 minutes and the organic acids 15 minutes. For the heavy metals maximum toxicity occurred after an exposure time of 30 minutes. The quality of the data was based on an assessment of the confidence range calculated for each EC50 value and the coefficient of determination (R^2), an expression of the quality of the estimating equation from which the EC50 is obtained. Ideally 95% confidence range values for each replicate should not exceed 30% of the EC50 value. Greater confidence range values were observed for some of the replicates, in particular the less water soluble components. This is detailed in the description of their toxicity.

2.1.1 The aliphatic hydrocarbons
Crude oil derived aliphatic hydrocarbons are present in produced water. They tend to be poorly water soluble and a solvent (1% ethanol) was used to aid solubility. The EC50 values of the normal alkanes (pentane C5, hexane C6 and heptane C7) were significantly higher than those of the other components tested (i.e. these compounds were less toxic to the bacteria). Mean EC50 values were 3668, 1403 and 231.3 mg/l respectively. The EC50 of trimethylpentane (TMP) C8 was 124.3 mg/l. There appears to be a trend of increasing toxicity with increasing carbon number (correlation coefficient r = -0.9785, p = 0.05).

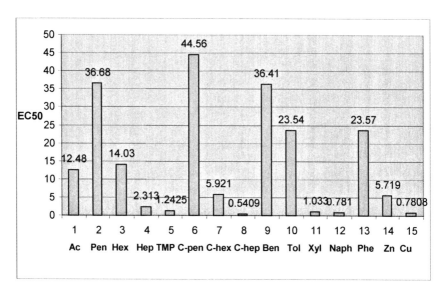

Figure 1:   EC50 values (mg/l) for produced water components. The EC50 values for Components 1-5 have been divided by 100 to enable easier comparison of data.

Caution must be observed as the 95% confidence range values exceeded 30% of the EC50 for some of the replicates. However, for all of the alkanes, either the two EC50 values were within 20% of one another or the 95% confidence range values of at least one of the replicates were less than 30%. Further, the coefficient of determination (R^2) for all replicates was >0.91, suggesting the estimating equation for calculating the EC50 was of reasonable quality for all replicates.

The C5 – C7 cycloalkanes were significantly more toxic than their linear counterparts (range 44.56 – 0.5409 mg/l). As with the normal alkanes, there appears to be a trend of increasing toxicity with increasing carbon number. For the cycloalkanes, EC_{50} values fell by an order of magnitude for each additional carbon atom in the ring.

2.1.2 The aromatic hydrocarbons

Benzenes (single ring aromatics) are relatively water soluble and are the most abundant aromatic hydrocarbons present in produced water. As with the aliphatic hydrocarbons, they are predominantly derived from crude oil. The toxicity of the benzene ring was observed to increase with increasing alkylation (r = -0.9657). This was particularly noticeable for xylene (dimethylbenzene), which was significantly more toxic than either benzene itself or toluene (methylbenzene). Naphthalene (a two ring aromatic hydrocarbon) had a similar toxicity to xylene. Significant levels of phenols have been reported in produced water. The toxicity of phenol was similar to toluene.

As expected, the aromatics were more toxic than the linear and branched alkanes. Interestingly, the cycloalkanes were of similar toxicity to the aromatics.

2.1.3 The organic acids and heavy metals

Organic acids are the most abundant organic material present in produced water. The most abundant acid reported, acetic, was relatively non-toxic (EC50 1123 mg/l). The toxicity of zinc (the most abundant heavy metal reported in a majority of the field samples) was 5.719 mg/l. The EC50 for copper (also present in some of the produced waters tested) was 0.7808 mg/l.

2.2 Toxicity of real versus synthetic produced waters

The toxicity of 17 produced water samples from 14 different North Sea oil platforms was measured. The values ranged from 3.74 – 37.34%, the majority (14) having values between 3 and 10%. Synthetic produced waters were made up based on the chemical compositions of six of the field samples. Their compositions are shown in Table 1. Their EC50 values compared to the field samples upon which the compositions were based are shown in Figure 2.

Table 1:   Synthetic produced water compositions.

Chemical Group	Component	Platform					
		A	B	C	D	E	F
Oil in water (OIW) content (mg/l)		39	36	25	12	22	16
Oil in water aliphatics (mg/l)	Pentane	5.5	5.5	1.5	2	5.5	5
	Cyclopentane	5.5	5.5	1.5	2	5.5	5
BTEX (mg/l)	Benzene	1.9	8.2	7.2	3	5	2.9
	Toluene	1.3	3	3.2	1.8	3.2	3.2
	Xylene	0.7	0.7	0.9	0.5	1	1.1
NPD (ug/l)	Naphthalene	861	352	328	473	880	795
	Phenanthrene	74	39	6	24	98	21
	Dibenzothiophene	50	17	3	19	27	19
PAH (ug/l)	Fluorene	2	5	2	4	4	8
	Chrysene	0	0	0	0	2	0
Organic acids (mg/l)	Acetic Acid	260	86	231.7	116.2	377.4	566.9
	Formic Acid	0.3	0	1	7	0.4	3.3
Heavy metals (mg/l)	Copper	0	0	0	0	0.42	0
	Zinc	0.36	0.1	0.05	0	0.24	0

The toxicity of the synthetic produced waters was significantly less than the field samples upon which they were based. One reason may be the presence of oilfield chemical residues in the samples collected from the field. In order to investigate this, the synthetic produced waters were dosed with production chemicals at concentrations used at the respective platforms. A list of the chemical types and dosage levels is given in Table 2, the resultant toxicity values are shown in Figure 2.

Table 2: Types and concentrations of oilfield production chemicals dosed at the different platforms.

PLATFORM	CHEMICAL AND DOSING CONCENTRATION (MG/L)
A	demulsfier (3) corrosion inhibitor/scale inhibitor (10)
B	demulsfier (4) scale inhibitor (6) antifoamer (2)
C	demulsfier (4) corrosion inhibitor (10) scale inhibitor (6) antifoamer (2)
D	demulsfier (7) corrosion inhibitor/scale inhibitor (30) antifoamer (10)
E	demulsifier (10)
F	demulsfier (2) scale inhibitor (30) water clarifier (15)

The addition of oilfield chemicals significantly increased the synthetic produced water toxicity values for all samples. However, with the exception of platforms A and D, the values were less than the field samples. This suggests that toxicity may be a function of compounds other than those measured. For platform D the toxicity of the synthetic PW + oilfield chemicals was significantly greater than the field sample. It should be noted that the addition of oilfield chemicals in the laboratory does not take into account partitioning in the topside process stream nor degradation/loss of components following dosing offshore. The toxicity measured may therefore be an overestimate of the field situation.

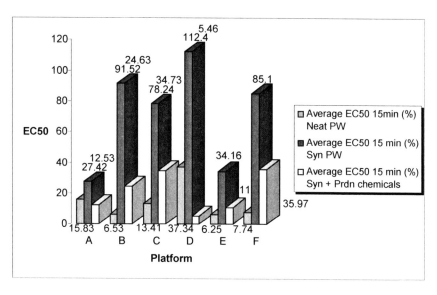

Figure 2: Microtox EC50 15 minute determinations of the field PW samples, synthetic PW and synthetic PW + production chemicals.

2.2.1 Statistical analysis of produced water sample data

PCA analysis was performed on the compositional data from the 17 field PW samples. Components not common to all samples were removed from the analysis. PCA suggested that toxicity did not appear to have a strong dependence on a single variable or group of variables – all had an effect to a certain extent.

2.3 The effect of dissolved oil on PW toxicity

It has been suggested that produced water toxicity is primarily dependent on dissolved oil components [12]. This was examined in this study. Brent blend crude oil (50 ml) was partitioned with brine (200 ml, 35 g/l NaCl) by shaking in a separating funnel and allowing to equilibrate for 24hrs. The oil concentration measured by infrared analysis in the brine following equilibration was 19.89 mg/l. The 15 minute Microtox EC50 value for the oily brine was 9.14% (confidence range 8.58 – 9.72%) indicating that dissolved oil may contribute significant toxicity to produced water. In order to assess the impact of removing these components, the 17 field produced water samples were eluted through C18-E Strata SPE cartridges. An initial study had shown that these cartridges removed 40 mg/l crude oil in brine (as determined by gas chromatography). Elution through SPE cartridges reduced the toxicity of the field PW samples significantly. In seven of the samples hormesis (stimulation of the bacteria) was observed. In the remaining samples toxicity was reduced by a factor ranging from 2 to 8 times. This data suggests that although dissolved oil is a major contributor to acute toxicity at some fields other components also influence the toxicity of the produced water.

3 Discussion

The Microtox EC50 values for the produced waters tested lay within the range of North Sea production platforms reported in the literature [7, 9, 10]. There is a general assumption that aromatic hydrocarbons contribute significantly to produced water toxicity [1]. However, studies have implicated a range of different components that contribute toxicity to produced water. These include: zinc and hydrocarbons [8]; the aromatic and phenolic fraction [9, 10]; and naphthenic acids [5]. Oilfield chemicals have also been reported to increase the "risk" (Roe et al. [13]). Little attention has been paid to the "non-toxic" aliphatic component of crude oil.

This study has shown that aliphatic hydrocarbons do indeed contribute toxicity to produced water. Their toxicity contribution appears to be related to their structure, molecular weight and their octanol-water partition coefficient (K_{ow}). These structure activity relationships (SARs) have been used for many years to predict the impact of chemicals on the environment (Clements [14]). QSARS (Quantitative Structure Activity Relationships) have been used by Sverdup and Kelley [15] to investigate the toxicity of n-pentane, n-hexane, benzene and toluene to *Daphnia magna*. Their results showed good agreement between measured toxicity and theoretical predictions based on K_{ow}. Indeed, the normal alkanes, cycloalkanes and benzenes all gave increasing toxicity with increasing molecular weight and K_{ow}. However, our studies indicate that toxicity of these compounds to *Vibrio fisheri* depends on both K_{ow} and chemical structure, with toxicity varying between each hydrocarbon class. Both the cycloalkanes and benzenes were significantly more toxic than the normal alkanes, with the cycloalkanes producing similar EC50s to the benzenes. This may be significant when assessing the toxicity of produced waters as cycloalkanes are the most common molecular structures in crude oil (Hunt [16]). Further, they are significantly more water soluble, and hence higher produced water concentrations would be expected, than for their linear counterparts. They have also been reported as "not readily biodegradable" (ARKEMA [17]).

Principal component analysis suggested the levels of soluble, oil-derived organics strongly influence the produced water toxicity. Removal of these components with SPE significantly reduced PW toxicity. However, at some platforms the reduction was only a factor of two. This suggests that at these platforms other components were having a significant influence. Frost [5] has suggested that naphthenic acids may contribute significant toxicity to produced water. These compounds have not been previously studied and are not routinely measured in produced water. Phenols have also been reported to contribute significant toxicity to produced water [9, 10] although the toxicity of phenol itself is less than a number of other components measured, including cyclohexane. Phenols were not included in the analytical reports received for the field samples and were therefore not included in the synthetic PW samples.

The toxicity of synthetic produced waters was also significantly less than the field samples upon which they were based. One possible reason was the omission of oilfield chemicals from these samples. The addition of oilfield chemicals to the synthetic produced waters significantly increased their toxicity, in one case to close to and, in another case, exceeding that of the field samples. The effect of oilfield chemicals on the acute toxicity of produced waters depends on their structure, oil:water partitioning characteristics (K_{ow}) and their dose rate. Johnsen et al. [10] concluded that, at normal dosing concentrations, process chemicals did not alter the toxicity of produced water significantly. At high concentrations there was evidence that a minor contribution to measured toxicity was observed. Studies in our own laboratory have also indicated that, for the majority of chemicals tested, the acute toxicity of the aqueous phase (following oil partitioning) was not significantly altered by process chemicals at normal dosing levels, although their was some evidence that certain chemicals may increase the partitioning of oil components into the aqueous phase by an order of magnitude when applied at high dosage rates [12]. The significant increases in toxicity observed in the synthetic produced water samples probably reflects a lack of partitioning of the more oil-soluble, and often toxic, components. Oilfield chemicals almost certainly increase the environmental risk associated with produced water discharges. Roe et al. [13] used modelling to show that this was the case with produced water from the Statfjord and Gullfaks North Sea oil fields. Frost [5] calculated that oilfield chemicals contributed 41% of the Environmental Impact Factor (EIF) at an unnamed Statoil operated field.

Microtox data for whole effluents (produced water) from oil platforms in the North Sea have been reported to be similar to toxicity data from a number of other marine toxicity tests [11]. However, differences can be observed for individual components. The EC50 values for the normal and branched alkanes measured in this study were all >100 mg/l, which would place them in the least toxic categories (D/E) of the Revised OCNS for offshore chemicals (CEFAS [18]). However, wide variation in EC50 values for these compounds is reported in the literature. This may reflect difficulties in maintaining these compounds in aqueous solution over the duration of the test or differences in species sensitivity. Sverdup and Kelley [15] reported significantly lower pentane and hexane EC50 values for *Daphnia magna* than were measured in this study using *Vibrio fisheri*. Indeed the *Daphnia* EC50 values would place these compounds in OCNS category B (>1 – 10mg/l), the second most toxic category. The mechanism of action of the alkanes is thought to be non-polar narcosis. The endpoint of the *Daphnia* test is immobilisation and the different sensitivity may reflect different modes of action in the two test species.

The reported sensitivity of different species to other produced water components can also vary greatly. In the case of cyclopentane, EC50 values ranging from 11.1 µg/l for brown shrimp to 116 mg/l for algae, have been reported [17]. By contrast, the toxicity to *Vibrio* of more water soluble compounds, such as benzene, was similar to that reported for *Daphnia*, as was the EC50 value of naphthalene.

The results of this study suggest that a number of components, including those not normally regarded as toxic, contribute to the acute toxicity of produced water. However, reducing the levels of oil-in-water may not have a concomitant effect on PW toxicity as other components may also be making a contribution.

Once produced water has been discharged, dilution will have the most immediate effect in mitigating potential environmental impacts of produced water [7]. Volatilisation and biodegradation are also important mechanisms. For example, benzene is a known carcinogen but is relatively water soluble and will disperse in the water column very rapidly. It is also reported to be readily biodegradable [9].

References

[1]   Canadian Association of Petroleum Producers (CAPP). Produced Water Waste Management technical report August 2001.
[2]   Stephenson, M.T., Components of produced water: a compilation of industry studies. *Journal of Petroleum Technology*, **5**, pp. 548-603, 1992.
[3]   Utvik, T.I.R., Chemical characterisation of produced water from four offshore oil production platforms. *Chemosphere*, **39(15)**, pp. 2593-2606, 1999.
[4]   Neff, J.M. & Sauer T.C., Aromatic hydrocarbons in produced water: bioaccumulation and trophic transfer in marine food webs. *Produced Water 2: Environmental Issues and Mitigation Technologies,* ed. M. Reed, S. Johnsen, Plenum Press: New York, pp. 163-175, 1996.
[5]   Frost, T.K., Hustad, B.M. & Vindstad, J.E., The relevance of naphthenic acids on produced water management at a Statoil-operated field. *Proc. Of*

the 14th International Oil Field Chemistry Symposium, 23-26 March, Geilo, Norway. The Norwegian society of Petroleum Engineers 2003.

[6]   Garland, E.M., Aquitaine, E., & E&P Forum, 1998. Produced water in the North Sea: a threat for the environment or a threat for the industry? *SPE International Conference on Health, safety and Environment in Oil and Gas Exploration and Production, Caracas, Venezuela, 7-10 June, 1998.* Society of Petroleum Engineers, 46706.

[7]   Somerville, H.J., Bennett, D., Davenport, J.N., Holt, M.S., Lynes, A., Mahieu, A., McCourt, B., Parker, J.G., Stephenson, R.R., Watkinson, R.J. and Wilkinson, T.G., Environmental effects of produced water from North Sea oil operations. *Marine Pollution Bulletin*, **18(10)**, pp. 549-558, 1987.

[8]   Stromgren, T., Sorstrom, S.E., Schou, L., Kaarstad, I., Aunaas, T., Brakstad, O.G. & Johansen, O., Acute toxic effects of produced water in relation to chemical composition and dispersion. *Marine Environmental Research*, **40(2)**, pp. 147-169, 1995.

[9]   Flynn, S.A., Butler, E.J. & Vance, I., Produced water composition, toxicity and fate. *Produced Water2: Environmental Issues and Mitigation Technologies,* ed. M. Reed, S. Johnsen, Plenum Press: New York, pp. 69-80, 1996.

[10]   Johnsen, C.S., Smith. A.T. & Brendenhaug, J., Identification of acute toxicity sources in produced water. *Second International Conference on Health, safety and Environment in Oil and Gas Exploration and Production, Jakarta, Indonesia 25-27 January*, pp. 383-390, Society of Petroleum Engineers, 27138, 1994.

[11]   Whale, G.F., Potential applications of the Microtox toxicity test system within the offshore oil and gas industry. *Second International Conference on Health, safety and Environment in Oil and Gas Exploration and Production, Jakarta, Indonesia 25-27 January*, pp. 687-698, Society of Petroleum Engineers, 27176, 1994.

[12]   Henderson, S.B., Grigson S.J.W., Johnson P. & Roddie, B.D., Potential impact of production chemicals on the toxicity of produced water discharges from North Sea oil platforms. *Marine Pollution Bulletin*, **38(12)**, pp. 1141-1151, 1999.

[13]   Roe, T.I., Johnsen, S. & The Norwegian Oil Industry association (OLF), Discharges of Produced Water to the North Sea. *Produced Water 2: Environmental Issues and Mitigation Technologies,* ed. M. Reed, S. Johnsen, Plenum Press: New York, pp. 13-25, 1996.

[14]   Clements, G.C., (ed). *Estimating toxicity of industrial chemicals to aquatic organisms using structure-activity relationships.* Environmental Effects Branch, Health and Environmental Review Division, Office of Pollution Prevention and Toxics, U.S. Environmental Protection Agency, Washington, DC 20460. August 30, 1996.

[15]   Sverdrup, L & Kelley, A., Environmental classification of petroleum substances: evaluation of test principles and CONCAWEs classification. Aquateam – Norwegian Water Technology Centre A/S. Report no: 00-004 January 13.2000.

[16]   Hunt, J.M., *Petroleum Geochemistry and Geology,* W. H. Freeman & Company, San Francisco, 1979.
[17]   ARKEMA Inc. Cyclopentane Material Safety Data Sheet. Issued: 11 Oct 2004.
[18]   Guidelines for the UK revised Offshore Chemical Notification Scheme. CEFAS, Burnham-on-Crouch, Essex, England. March 2000.

In situ sludge reduction by microorganisms

S. Banerjee & U. Hooda
Institute of Paper Science and Technology and School of Chemical and Biomolecular Engineering, Georgia Institute of Technology, USA

Abstract

A mixture of facultative microorganisms added to pulp mill aeration stabilization basins is able to reduce sludge *in situ* over several months. Laboratory work shows that pulp fibers exposed to a cell-free extract prepared from the microorganisms are degraded through consumption of their hemicellulosic fraction. However, the process is quite slow; a controlled field study run for seven months did not lead to mineralization of the sludge.
Keywords: sludge, microorganisms, hemicellulose, treatment systems, fiber, ASB.

1   Introduction

Wastewater treatment in the paper industry is done with activated sludge systems or with aerated stabilization basins, ASBs [1, 2]. ASBs need to be dredged periodically and the associated sludge handling and disposal costs are substantial [3, 4]. A recent innovation introduced by Remediation Resources Inc (RRI) of Pembroke, GA, controls the sludge depth *in situ* by introducing a suite of *Pseudonomads* (isolated from oil well corings) into the inlet of the lagoon. The organisms are grown onsite in a fermenter fed by an influent sidestream and are released continuously into the lagoon.   The microorganism formulation combines the desiccated *Pseudonomads* with nutrients and is supported on carbon beads, which are continuously dispensed into the fermenter. A number of mills have confirmed that releasing the RRI mcroorganisms into the lagoon leads to a gradual reduction of the sludge bed. In one mill, a sludge inventory of 600,000 m^3 decreased by 90,000 m^3 over a year [5]. In another, the hydraulic retention time increased from 2.5 to 3.4 days over two years [6]. In this paper we review laboratory and field work in light of the mechanism [7] of the process.

WIT Transactions on Biomedicine and Health, Vol 10, © 2006 WIT Press
www.witpress.com, ISSN 1743-3525 (on-line)
doi:10.2495/ETOX060121

2 Results and discussion

The ability of the organism to grow in the influent, which usually contains components toxic to microorganisms, was confirmed in laboratory work. Clarifier overflow was obtained from a mill in Georgia and attempts were made to grow organisms obtained from activated sludge as well as the RRI organisms in this medium. No growth was observed for the activated sludge samples, whereas growth (rod-shaped) was observed with the RRI organisms. The growth rate, measured in a chemostat, was about fivefold lower than that typical of organisms in the back end of an ASB. Hence, the RRI organisms are resistant to influent toxicity, which could be one reason that they are effective.

It seemed likely that the organisms degraded fiber through extracellular enzymes and this viewpoint was tested by exposing pulp fiber (in place of primary sludge) to cell-free extracts prepared from the organisms. The changes in fiber properties were followed with a Fiber Quality Analyzer (FQA). Unbleached and bleached softwood kraft pulps were screened separately through a Bauer-McNett fiber fractionater using 14 and 48 mesh screens, which correspond to openings of 1.6 and 0.4 mm, respectively. The fibers retained by the two screens were designated as long and short fibers, respectively. The fibers were also hydrolyzed in sulfuric acid [8] and the solution analyzed for carbohydrates.

The results, shown in Figures 1 and 2, demonstrate that the fiber length progressively decreases for both brown and bleached fibers, respectively. The results are comparable for the two different fiber lengths used. The curl index also decreases, especially for the long fibers. The results for the brown fibers are illustrated in Figure 3; values for the bleached fibers were very similar. The mixture of fiber and cell-free extract was also sealed between microscope slides. A micrograph illustrating the breakdown of a fiber exposed to the cell-free extract is illustrated in Figure 4. This level of degradation was not observed for all the fibers; some showed extensive damage, while others remained relatively unaffected. Control fibers kept in autoclaved media showed no change whatsoever.

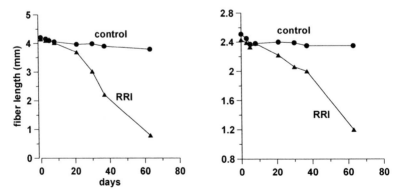

Figure 1:   Fiber length changes in long (left) and short (right) brown fiber.

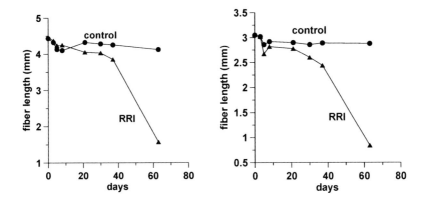

Figure 2:   Fiber length changes in long (left) and short (right) bleached fiber.

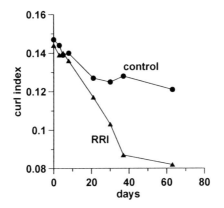

Figure 3:   Changes in curl for long brown fiber.

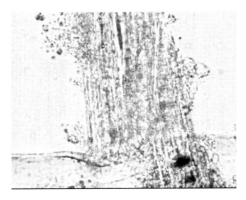

Figure 4:   Degradation of a stained fiber in cell-free extract (100X).

Lignin cannot be the principal component involved in the degradation; otherwise the bleached and brown fibers would have behaved differently. Hence, the cellulosic components must be implicated in the disintegration of the fiber. The changes in these components with time are listed in Table 1. The levels of arabinan and xylan in the fiber decrease progressively as illustrated in Figure 5. A similar decrease also occurs with galactan, although there is more scatter. No significant change is seen for either glucan or mannan.

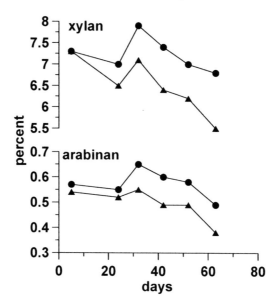

Figure 5: Profiles of arabinan and xylan with (triangles) and without (circles) added organisms.

Table 1: Degradation of cellulose and hemicellulose in the cell-free extract.

day	arabinan (%)		galactan (%)		glucan (%)		xylan (%)		mannan (%)	
	control	RRI	control	RRI	control	RRI	control	RRI	control	RRI
5	0.57	0.54	0.44	0.39	77	79	7.3	7.3		7.3
24	0.55	0.52	0.40	0.41	71	77	7.0	6.5	6.9	7.2
32	0.65	0.55	0.47	0.41	85	84	7.9	7.2	7.9	7.9
42	0.61	0.49	0.44	0.40	75	78	7.4	6.4	7.1	7.5
52	0.58	0.49	0.39	0.40	73	79	7.0	6.2	6.7	7.3
63	0.49	0.38	0.28	0.21	70	66	6.8	5.5	6.3	6.0

It appears that the organisms settle in the sludge bed and release enzymes that gradually degrade the sludge. Given that the organisms are formulated on carbon beads that are denser than water it seemed reasonable to seed the lagoon with the beads themselves.   The beds would sink to the sludge bed and the microorganisms would be delivered directly to the sludge where the enzymes would be more effective. Accordingly, an area of a lagoon in a Georgia mill was seeded with the beads. The sludge bed depth was measured by echosounding as well as with a sludge-light.   Depth comparisons made at several locations ranging from 1–5 m showed the correspondence to be within 1.3%. The sludge light went off at a TSS of 3.2% so the sludge bed was operationally defined to begin at this depth.   However, surveys taken over three months showed no improvement in the treated region over the others. It is possible that growth of the organisms is too slow under the anoxic conditions prevalent in the sludge bed.

An attempt was made to verify some of the laboratory findings in a controlled field study in a mill in Georgia. The technique used has been described earlier [9, 10] and consists of placing a mixture of sludge and microorganism in a vial capped with semi-permeable membranes and placing it in the lagoon. The membranes allow water to enter the vials but prevent the solids from leaving them.   The intent was to determine the rate at which the organisms would mineralize sludge under field conditions.

Sludge was collected from a side of the lagoon just after the inlet and was dosed with the RRI formulation. The vials were then inserted into holders constructed from PVC tubing. Each tube supported five vials and was long enough to ensure that the vials were submerged. Controls were run with sludge but without the added microorganisms. Surprisingly, no change in the sludge mass occurred over 200 days. Evidently, the added organisms did not reduce sludge mass even though they were applied as an overdose. In order to confirm that the RRI organisms were still viable after exposure, a heterotrophic plate count of the 149-day sample was taken. The control contained an average of 2.2×10^6 CFU/mL, whereas the sludge with the added organism showed a higher count of 2.4×10^7 CFU/mL, demonstrating that the added organisms were still alive. Clearly, the process is slow. It is possible that the fibers held in the membrane-capped vials in the field trial were degraded by the RRI microorganisms, but not to the point where they were able to leave the vials. In other words, the fiber length decreased but the fragments were large enough to be held in place by the 0.01 μm membrane.

The sludge bed contains fiber only in the front end of the lagoon; the sludge in the rest of the basin mainly consists of settled microorganisms. The RRI organism is also likely to be effective on this material; otherwise the sludge bed would not have decreased uniformly in the mill applications. While our work does not address this situation, we note that the cell walls of gram-negative bacteria contain polysaccharides, which should be susceptible to degradation.

In conclusion, laboratory trials and full-scale field experience demonstrate that a suite of organisms isolated from oil well corings degrade sludge *in situ* in aerated stabilization basins.   Enzymes released from the sludge attack hemicellulose entities in the sludge. The process is effective, but slow, and several months to a year is required before the effect can be quantified in the field.

Acknowledgements

We thank Georgia-Pacific, International Paper, Bowater, and Stora Enso for financial support and Remediation Resources Incorporated for a gift of the microorganism formulation and for conducting the field trial.

References

[1]   Fels, M., Pinter, J. & Lycon, D.S., Optimized design of wastewater treatment systems: Application to the mechanical pulp and paper industry. *Can. J. Chem. Eng.*, **75(2)**, pp 437-451, 1997.

[2]   Stuthridge, T.R., Campin, D.N., Langdon, A.G., Mackie, K.L., McFarlane, P.N. & Wilkins, A.L., Treatability of bleached kraft pulp and paper-mill wastewaters in a New-Zealand aerated lagoon treatment system. *Wat. Sci. Technol.* **24(3-4)**, pp 309-317, 1991.

[3]   Dorica, J.G., Harland, R.C. & Kovacs, T.G., Sludge dewatering practices at Canadian pulp and paper mills. *Pulp and Paper Canada* **100(5)**, 19-22, 1999.

[4]   Pickell, J. & Wunderlich, R., Sludge disposal - current practices and future-options - innovative methods are discussed. *Pulp & Paper Canada* **96(9)**, 41-47, 1995.

[5]   Massengill D., Pazderski R. & MacDonald S. *Tappi Env. Conf.,* Montreal Canada, 2002.

[6]   Bryant C.W., Kloeker J.W., Hege J., Meech M. & Nilsson K. *Tappi Env. Conf.*, Charlotte, NC (2001).

[7]   Banerjee, S. & Hooda, U., In-situ reduction of fibrous sludge in a pulp mill aerated stabilization basin. *Tappi J.*, **4 (5)**, pp 3-7 (2005).

[8]   TAPPI standard methods T 249 cm-00 Tappi Press, Atlanta, GA, 2000.

[9]   Williams, C.L., Banerjee, S., Kemeny, T.E., Sackellares, R., Juszynski, M.D. & Degyansky M.E., In-situ measurement of local biodegradation during secondary treatment: Application to bleached pulp mill chloroorganics. *Environ. Sci. Technol.* **31**, pp 710-716, 1997.

[10]   Williams, C.L., Mahmood, T., Corcoran, H., Zaltzmann, M.E. & Banerjee, S., Tracing the efficiency of secondary treatment systems. *Environ. Sci. Technol.*, **31**, pp 3288-3292, 1997.

Section 4
Pharmaceuticals
in the environment

Pharmaceuticals and personal care product residues in the environment: identification and remediation

E. Heath[1], T. Kosjek[1], P. Cuderman[1] & B. Kompare[2]
[1]Jožef Stefan Institute, Department of Environmental Sciences, Slovenia
[2]University of Ljubljana, Faculty of Civil and Geodetic Engineering, Institute of Sanitary Engineering, Slovenia

Abstract

In this paper we investigate the possible impact on the ecosystems of the active ingredients in pharmaceutical and personal care products (PPCPs). These molecules are in their nature biologically active substances whose effects, once released into the environment, remain unclear. Commonly used drugs and cosmetics have been detected in drinking and surface waters suggesting possible consequences for environment and human health. Considering the possible bioaccumulation in the food chain and chronic toxicity due to a combination effect it is necessary to better understand their environmental fate.

Pharmacologically and cosmetic active substances involved in our study were chosen according to their wide application in central Europe and to their suspected toxicity and bioaccumulation (nonsteroidal anti-inflammatory drugs - NSAIDs, sunscreen agents, antiseptics). For isolation and identification of the selected compounds GC-MS procedure has been developed in our laboratory. To determine the presence of PPCP in the Slovene aquatic environment, different water samples (river, potable, well, lake, sea, swimming pool and waste water) were analyzed. The results show that the average PPCP contamination in Slovenian waters is comparable to published results for Central and Western Europe. To understand the environmental fate of these substances we studied the biodegradation paths in controlled conditions using laboratory scale bioreactor facilities. Degradation experiments were made with spiked water samples of different concentrations of pollutants (1µg/L-1ng/L) starting with activated biomass from the active wastewater treatment plant. After 6 months of continuous operation, a steady removal of all observed compounds (NSAID representatives) was achieved (up to 90%). In the future other PPCP representatives will be studied and the reactor configuration will be optimized.
Keywords: pharmaceutical, personal care products, NSAIDs, UV filters, antimicrobial, analytical procedure, environmental samples.

WIT Transactions on Biomedicine and Health, Vol 10, © 2006 WIT Press
www.witpress.com, ISSN 1743-3525 (on-line)
doi:10.2495/ETOX060131

1 Introduction

The focus of environmental research is extending beyond classic environmental pollutants, such as PCBs, dioxins, and pesticides, to a group of compounds with a common name "new emerging contaminants" such as pharmaceuticals and personal care products (PPCPs). Thousands of tons of pharmacologically active substances are used annually to treat or prevent illnesses, or to help people to face up to the stresses of modern life. As well as pharmaceuticals, the amount of chemicals used in personal care products is in the thousands of tons per years. The world's population consumes enormous quantities of skin care products, dental care products, sunscreen agents, to name just a few (Ternes et al. [1]) In the future, their application is predicted to increase due to the growth in the world's population, longer life expectancy, development of pharmaceutical industry and an increased awareness of the danger that modern life poses over our civilization, between which we will mention only thinning of ozone layer, DNA radical damage and unphysiological aging.

The main sources of pharmaceutical residues in the environment are human and veterinary medicine and as part of the waste produced by the pharmaceutical industry. Unlike pharmaceuticals, personal care products may enter surface waters directly, for example, from the skin during swimming and bathing or indirectly via wastewaters. The discharge of therapeutic and cosmetic agents from production facilities, hospitals and private household effluents as well as improper disposal of unused or expired products, direct discharge of sunscreens and veterinary medicines lead to surface water, ground water and drinking water where they pose a potential risk to humans. During and after treatment, humans and animals excrete a combination of intact and metabolized pharmaceuticals. Consequently, many bioactive compounds enter wastewaters and the receiving water bodies without any tests for specific environmental effects and therefore it is both the parent compound and active metabolites that should be the subject for a risk assessment.

The acute toxic effects of pharmaceutical compounds are well investigated, since they are required for a marketing authorization permit. On the other hand the toxicity of cosmetic compounds does not need to be assessed by detailed clinical trials even though some cosmetics are known to cause local irritations on sensitive skin and there is no reliable data regarding their toxicity after ingestion. The effects of continuous low dosage ingestion are not adequately recognized in both, pharmaceutical and cosmetic compounds, moreover, synergistic or additive toxic effects are possible between different parent compounds and/or metabolites. Also, it is not known how different the toxicity of complex mixtures are from that of individual compounds and although the amounts of pharmaceutical and personal care product (PPCP) residues, when released to the environment, are low, their continuous input may lead to chronic low level exposures and accumulation with potential negative effects on the environment. Considering the possible bioaccumulation in the food web and chronic toxicity due to a synergistic or additive effect, it is necessary therefore to understand better their environmental fate.

To date commonly used pharmaceuticals have been detected in waste, surface and even drinking waters (Kosjek et al. [2], Kümmerer [3]), cosmetic residues in surface waters (Giokas et al. [4]) and antimicrobial agents in waste and surface waters (Lindström et al. [5]). PPCPs are often subjected to waste water treatment and therefore, it is essential to know what is the WWTP efficiency to eliminate them from wastewaters to be aware of the amounts of these compounds that can enter environmental cycle this way.

This study reports on the development of analytical procedures for determination of selected pharmaceutical and personal care product active ingredients in environmental and wastewater samples and determination of selected PPCPs in aqueous samples from Slovenia. Further, elimination of the pharmaceutical compounds was studied in a small-scale pilot wastewater treatment plant (PWWTP).

2   Experimental

2.1 Chemicals

Pharmacologically and PCP active substances involved in our study were chosen according to their wide application in Slovenia and Central Europe and to their suspected toxicity or bioaccumulation. The pharmaceutical substances belong to a group of NonSteroidal AntiInflammatory Drugs (NSAIDs): Ibuprofen, Ketoprofen, Diclofenac and Naproxen. UV filters with commercial name Eusolex (Homosalate, 4-Methylbenzylidene Camphor, Benzophenone-3, Octocrylene, Butyl Methoxydibenzoylmethane, Ethylhexyl Methoxycinnamate) and anti-microbial agents (Clorophene and Triclosan) were chosen for our study.

2.2 Analytical method

In order to determine the presence of PPCP residues in various water samples (waste, surface, ground, tap and drinking water), we developed an analytical procedure which involves solid phase extraction (SPE) and gas chromatography with mass spectrometric detection (GC-MSD) [2]. Prior to extraction, water samples were filtered (0.45 μm filter, Sartorius, Goettingen, Germany) and acidified to pH 2 to ensure successful trapping of the acidic compounds onto the SPE sorbent. For SPE commercially available 3 mL SPE cartridges with 60 mg of Strata™ X (surface modified styrene-divinylbenzene polymer) sorbent (Phenomenex®, Torrance, ZDA) were applied. For each group of compounds (pharmaceuticals and UV filters with antimicrobial agents), different SPE extraction and elution procedures were applied. For determination of NSAIDs the cartridges were first conditioned with methanol, which was removed with acidified deionised water having the same pH as the sample (2,6). Compounds of interest were retained on the sorbent and afterwards eluted with 1,5 mL of methanol. The methanol from the extract was evaporated under a stream of N_2 and replaced with toluene. For the determination of UV filters and antimicrobial agents, the cartridges were first conditioned with 1,5 mL ethyl acetate/ dichloromethane (1:1, v/v), followed by 1,5 mL of methanol, which was then removed with acidified deionised water having the same pH as sample loaded afterwards (pH =2,6). Selected compounds were eluted from SPE sorbent with 1,5 mL of ethyl acetate/ dichloromethane (1:1, v/v). The extract solvent was evaporated and replaced by toluene. The compounds were derivatized to their trimethylsilyl esters using N-methyl-N-trimethylsilyl trifluoracetamide (MSTFA) to convert the polar compounds to volatile esters. GC-MSD analysis was performed on a HP 6890 gas chromatograph (Hewlett-Packard, Waldbronn, Germany), fitted with a HP-5MS capillary column and mass spectrometric detection.

2.3 Elimination of pharmaceuticals on pilot wastewater treatment plant (PWWTP)

To study the elimination of compounds of interest in WWTP, we designed a PWWTP that mimics the classical treatment in a municipal WWTP. The PWWTP uses active sludge obtained from Slovenian municipal wastewater treatment plant. The design of the PWWTP consists of three parallel reactors. Two reactors R1 and R2 operate under continuous input of higher ($R1 = 0,05$ mg L^{-1}) and lower ($R2 = 0,005$ mg L^{-1}) concentrations of selected NSAIDs, while R0 is a control reactor. We designed the reactors to be economical in terms of the operating costs while containing a sufficient volume to allow installation of equipment (aerators, pumps) and maintenance (cleaning, repair). The reactor design was optimized at 4,0 L and configured to allow the reactor to be operated under aerobic and anaerobic, or anoxic conditions. Figures 1 and 2 show the layout of the reactor. Each reactor consists of a selector, anaerobic, and aerobic chamber and a sedimentation basin.

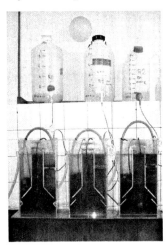
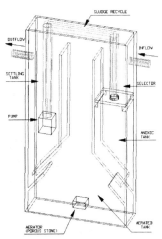

Figure 1: The experimental setup of the three pilot WWTP reactors (RO = control, R1 = 0,05 and R2 = 0,005 mg L^{-1} of pharmaceutical mixture).

Figure 2: The 3D scheme of the pilot reactor showing inflow, selector, anoxic anaerobic tank, aerated tank, settler with sludge return to aerated tank and sludge recycle.

All three reactors are fed with an aqueous nutrient and mineral solution, while NSAIDs are added in R_1 (0,05 mg L^{-1}) and R_2 (0,005 mg L^{-1}) only. The flow rate of the nutrient-mineral solution was set at approximately 2,0 L per day, with hydraulic residence time of 48 hours. We collect both influent and effluent samples. During first six months of operation, samples were collected on a weekly basis and then on monthly basis. Each sample is acidified, filtered and stored at 4°C prior to analysis.

3   Results and discussion

To estimate the extent of NSAID pollution in the Slovene environment, analytical procedure was developed, validated and described in details elsewhere (Kosjek et al. [2]). Using developed analytical procedure 16 river, 2 well and 11 tap water samples were collected. In all of the tap and well water samples NSAID residues were under the detection limit, while naproxen was determined in 9 and diclofenac in 6 out of 16 river water samples collected in 2004 (Table 1).

Table 1:   Determined concentrations (ng/L) of the selected NSAIDs in river water samples.

River samples	Date of sampling	Concentration (ng L^{-1})		
		Naproxen	**Diclofenac**	**Ketoprofen**
KRKA before pharm. factory	Jul-04	<LOQ	<LOD	<LOD
	Sep-04	<LOD	<LOD	<LOQ
KRKA after pharm. factory	Jul-04	313	282	<LOQ
	Sep-04	60	49	<LOQ
LJUBLJANICA1	Jul-04	<LOD	<LOD	<LOD
	Sep-04	<LOD	<LOD	<LOD
LJUBLJANICA2	Jul-04	73	<LOQ	<LOD
	Sep-04	<LOD	<LOD	<LOQ
SAVA	Jul-04	80	9	<LOQ
	Sep-04	<LOQ	<LOQ	<LOD
MURA	Jul-04	49	41	<LOQ
DRAVA1	Jul-04	46	26	<LOQ
	Sep-04	<LOD	<LOD	<LOD
DRAVA2	Jul-04	24	32	<LOQ
	Sep-04	42	<LOD	<LOD
PŠATA	Jul-04	17	<LOQ	<LOD

The analytical procedure and its validation for isolation of UV filters and antimicrobial agents from aqueous samples is described in full elsewhere (Cuderman [6]. To estimate the pollution of the Slovene aquatic environment with UV filters and antimicrobial agents were analysed in 4 recreational and 4 waste waters. The results revealed the concentrations of the selected compounds to be above the limit of detection in half of the samples. Clorophene was not detected. The most frequently detected UV filter was Benzophenone-3 (11-400 ng/L), while of the antimicrobial agents we could only find traces of Triclosan in the river Kolpa (68 ng/L) and in one hospital effluent (122 ng/L). Table 2 gives the concentrations of compounds of interest in ng/L range.

Table 2:   Determined concentrations (ng/L) of the selected UV filters and antimicrobial agents in sea, pool, lake, river and wastewater.

Matrix	sampling site	Eusolex HMS	Eusolex 6300	Eusolex 4360	Eusolex 6007	Eusolex 2292	Eusolex OCR	Tri-closan	Cloro-phene
River	Nadiža-Soča	345	< LOD	< LOD	< LOD	< LOD	35	< LOD	< LOD
	Kolpa	165	< LOD	114	47	88	34	68	< LOD
Lake	Rakitna	< LOD	< LOD	85	34	92	< LOD	< LOD	< LOD
	Lake Bohinj	< LOD	< LOD	32	< LOD	< LOD	< LOD	< LOD	< LOD
	Šobec	< LOD	< LOD	58	< LOD	< LOD	< LOD	< LOD	< LOD
	Lake Bled	< LOD	< LOD	66	< LOD	< LOD	31	< LOD	< LOD
	Bakovci	< LOD	< LOD	< LOD	24	< LOD	< LOD	< LOD	< LOD
swim-ming pool	Kodeljevo	< LOD.	< LOD	103	< LOD	< LOD	15	< LOD	< LOD
	Portorož	< LOD	330	400	17	< LOD	< LOD	< LOD	< LOD
hospital effluent	hospital 1	< LOD	< LOD	< LOD	< LOD	< LOD	< LOD	122	< LOD

3.1 Removal of pharmaceutical substances in a PWWTP.

To calculate the elimination efficiency for each of the NSAIDs understudy we compared the difference between the influent and effluent concentrations. The results (Table 3) show high and constant elimination of Ibuprofen, Ketoprofen and Naproxen (> 89%), whereas the elimination of Diclofenac was less efficient deviating significantly from the average removal rate (48 to 62%). Our results are in agreement with the data from the literature (Zwiener and Frimmel [7]) and show Diclofenac to be more persistent to elimination in WWTP biodegradation.

In our future studies, we intend to optimise the reactor operating conditions in terms of higher and constant elimination of NSAIDs and study pharmaceutical elimination mechanism. The elimination of selected NSAID in PWWTP might be a consequence of physical, chemical or biochemical transformation paths, e.g. abiotic degradation, biological or photodegradation or adsorption to the active sludge or reactor wall surface. Our future research will focus upon understanding NSAID removal mechanisms in a PWWTP and isolation of stable degradation products. In our preliminary investigations we compared the total ion chromatograms (TIC) of the influent and observed additional chromatographic peaks possibly originating from NSAIDs degradation. Figure 3 shows chemical structures of the parent compound (diclofenac) and a compound which is, according to the TIC, mass spectra and its NIST library comparison a diclofenac's degradation product. However, the proposed structure will need to be confirmed using standard addition and high-resolution mass spectrometry.

Table 3:   Elimination efficiency of the selected pharmaceuticals in PWWTP.

% REMOVAL RATE		median	average	standard deviation
IBUPROFEN	R1	98,9	94,6	9,2
	R2	90,2	89,3	5,5
NAPROXEN	R1	98,7	96,3	5,3
	R2	89,7	90,7	4,8
KETOPROFEN	R1	98,2	93,5	9,8
	R2	90,5	90,8	6,2
DICLOFENAC	R1	60,4	62,6	27,7
	R2	59,5	48,1	27,5

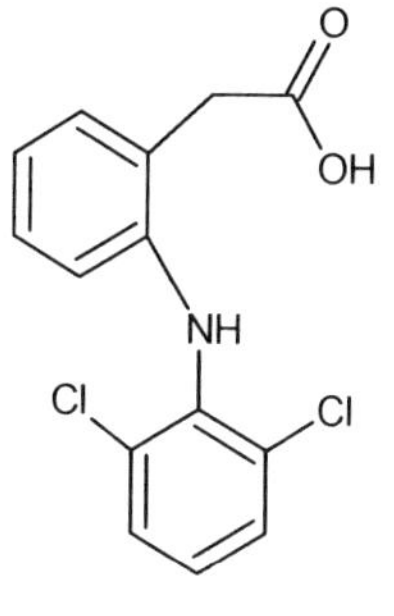

Diclofenac: 2-[(2,6-dichlorophenyl)amino] benzeneacetic acid

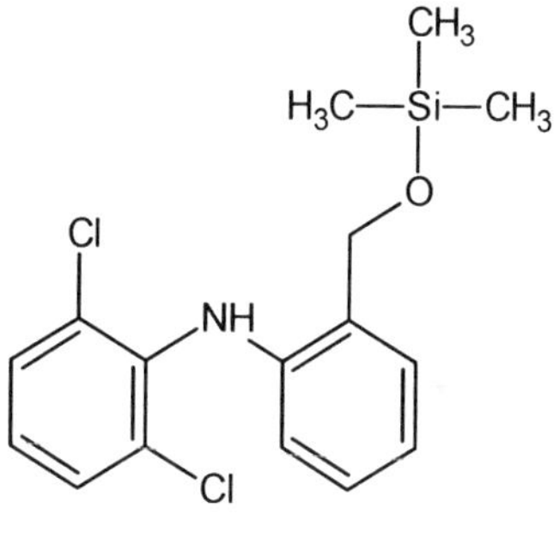

Degradation product (silylised): trimethylsilyl ether of 2-[(2,6-dichlorophenyl)amino]benzyl alcohol

Figure 3:   Chemical structure of Diclofenac and its proposed degradation product.

4   Conclusions

In our study analytical procedures for determination of selected PPCPs were developed and include solid phase extraction with Strata™ X, derivatisation to trimethylsilyl esters and determination with GC-MSD. To estimate the PPCP pollution of Slovene aqueous environment, surface (river, tap, well and recreational waters) and wastewaters were sampled. In the case of NSAIDs (diclofenac, ibuprofen, naproxen and ketoprofen), the content of NSAIDs in all tap and well waters were under the detection limit, while naproxen was determined in most of river water samples and ketoprofen in every fourth (as a percentage) tested river samples. Our results also revealed concentrations of UV filters and antimicrobial agents to be under the limit of detection in most samples tested.

We also studied removal of selected NSAIDs in PWWTP and showed that with an exception of Diclofenac (up to 60%) the other tested compounds were eliminated with over 90% efficiency. In the future, we will improve the elimination efficiency of NSAIDs and attempt to understand the mechanism of their elimination and identify their stable degradation products. Also, the spectra of PPCPs studied in PPWTP will be expanded to psychiatric drugs, lipid-lowering agents etc.

To evaluate toxicity of complex environmental samples, we took samples from this study and subjected them to cytotoxic and genotoxic testings (bacterial SOS/umu test and mammalian MTT and comet assay). Results of this combined study will be presented at the same conference (Environmental Toxicology 2006).

References

[1] Ternes T.A. et al.: Environmental Science & Technology 2004, October 15
[2] Kosjek T., Heath E., Krbavčič A., Environment International, 31(5), 679-685, 2005.
[3] Kümmerer K., Pharmaceuticals in the Environment; Sources, Fate, Effects and Risks; 1st ed., Springer – Verlag Berlin Heidelberg New York, 1–7, 2001.
[4] Giokas D. L. et al.: Determination of residues of UV filters in natural waters by solid-phase extraction coupled to liquid chromatography-photodiode array detection and gas chromatography-mass spectrometry, J. Chromatogr. A, 1026, 289-293, 2004.
[5] Lindström A. et al.: Occurrence and Environmental Behavior of the Bactericide Triclosan and Its Methyl Derivative in Surface Waters and in Wastewater, Environ. Sci. Technol., 36, 2322-2329, 2002.
[6] Cuderman P., Heath E., Zupančič-Kralj L.: UV Filters and Antimicrobial Agents: Occurence and Remediation in Slovene Environment, snet for publication in Enviroment International (June 2006).
[7] Zwiener C., Frimmel F.H., The Science of the Total Environment 309, 201-211.

The presence and effect of endocrine disruptors on reproductive organs of fish in a reservoir used for aquiculture

L. S. Shore[1] & M. Gophen[2]
[1]Kimron Veterinary Institute, Israel
[2]Migal Technologies, Israel

Abstract

A reservoir receiving some sewage water from an urban area, in addition to fresh water sources, was used for raising three species of fish. Prior to the introduction of the fish, the reservoir water contained appreciable amounts of testosterone, estrogen, ethinylestradiol and medroxyprogesterone. After introduction of the fish the level of hormones and drugs were reduced to nearly non-detectable levels while in a control reservoir without fish, the levels of the compounds remained constant. Of the three species studied, after eight months only the female carp developed gonads, which were small for body size. Observations suggest that the fish absorb the hormones and drugs in the early growing season and this has a negative effect on their reproductive development.
Keywords: endocrine disrupters, ethinylestradiol, gonadal atrophy, carp, tilapia.

1 Introduction

Household, industrial and agricultural (recycled effluents) water sources are liable to contain Endocrine Disrupting Agents (EDAs). The EDAs are a heterologous group which can influence the endocrine system of various organisms [5] and to cause endocrine disruption in the parent organism or its offspring. In particular, chemicals that bind to the receptor of the natural hormone can mimic or depress the actions of the natural ligand. These compounds have been demonstrated to affect wildlife and domestic animals but their effects in humans are not well documented [5, 8, 9]. In the last two decades, there has been an increased awareness of the effects of these compounds on animals and man and this has been reflected in scientific publications as well as

WIT Transactions on Biomedicine and Health, Vol 10, © 2006 WIT Press
www.witpress.com, ISSN 1743-3525 (on-line)
doi:10.2495/ETOX060141

the media in general. This heterogenic group of EDAs contains many compounds with estrogenic actions (1) steroidal hormones (e.g. estrone, estradiol, testosterone) from animal (wildlife or domestic) or human sources. (2) Anthropogenic compounds (industrial effluent, drugs). (3) Mycoestrogens from *Fusarium sp.* and (4) Phytoestrogens, a group of non-steroidal compounds with estrogenic activity found naturally in plants.

1.1 Source of environmental hormones in the Jordan River catchment area

We have previously reported on the presence of natural and synthetic steroids (testosterone, estrone, estradiol, ethinylestradiol) in the catchments of the Jordan River [1, 10]. The major sources of the hormones were shown to be runoff from cattle pasture and effluent from aquaculture or sewage water treatment (SWT) ponds. The hormonal profile was characteristic of each source, e.g. the presence of ethinylestradiol was indicative of SWT pond effluent. In SWT ponds, the concentrations of estrogen and testosterone can reach levels of hundreds of nanograms/liter [9].

1.2 Impact of EDAs on fish

EDAs are well documented to affect reproduction in fish [4]. In particular the effects of sewage water plant effluent on intersex and sexual organs are well documented [12]. The most likely EDC causing these effects is ethinylestradiol, a synthetic steroid found in birth control pills, although other EDAs have been suggested [4, 7].

2 Materials and methods

2.1 Site description

The reservoir held 5.6 mcu^3 over area of 100 acres and a depth of 15 m. The reservoir received water from three sources: sewage effluent from an urban area of about 8000, excess water flow from a nearby artificial lake and Jordan River water from the western canal. The water was use to raise about 400 tons of fish. The limonological properties of the reservoir have been described in detail [6].

Samples were taken on 16.02.05 before fish were introduced and 04.05.05, 04.08.05 and 19.11.05 representing 1, 5 and 8 months after the introduction. Samples were taken at depths of 1, 3 and 6 m except on the final sampling before harvesting when the depth was only 0.8 m. Samples were taken from a control reservoir without fish on the same dates.

2.2 Measurements of steroids and pharmaceuticals

One liter aliquots of water were extracted on C-18 maxi-extraction columns (Mega BE-C18 1 gm, 6 ml PN 12256001 Varian, Middleburg, The Netherlands) as previously described [1, 10]. Testosterone and estrogen were measured by radioimmunoassay [1, 10]. Medroxyprogesterone, ethinylestradiol, androstenedione and estriol were measured using commercial ELISA kits (DRG GmbH, Marburg, Germany). The ELISA kits for determination of benzodiazapines and barbiturates were only qualitative.

2.3 Histology

Gonads were placed in 10% buffered formalin and slides prepared by the Dept. of Pathology at the Kimron Veterinary Institute. Gonads were evaluated according to Degani et al. [2, 3].

3 Results

3.1 Natural steroids

The natural steroids, testosterone, estrogen (estradiol and estrone), androstenedione and estriol were present in the reservoir prior to introduction of the fish (Table 1). After one month there was a decline in the concentrations of testosterone, estrogen and androstenedione, which then rose towards maturation. Estriol, which only comes from human sources (pregnant women), was only present before the fish were introduced.

Table 1:   Natural hormone concentrations in ng/l in reservoir water.

	Testo-sterone	Estrogen	Androstene-dione	Estriol
Before introduction of fish	4.13	4.21	3.00	1.43
One month after	2.37	0.97	1.62	<0.1
Five months after	3.75	2.20	4.67	<0.1
Eight months after	5.73	7.07	18.23	<0.1

3.2 Pharmaceuticals

Prior to introduction of the fish, the reservoir was found to contain appreciable amounts of ethinylestradiol, medroxyprogesterone (both components of contraceptive pills); and benzodiazepines but was negative for barbiturates (Table 2). After three months in the presence of the fish the levels of these compounds were essential non-detectable (<0.5 ng/l). In contrast, a control reservoir without fish, the level of medroxyprogesterone and ethinylestradiol remained between 1.3 and 1.7 ng/l during the same time period.

These results indicate that there was a decrease in the pharmaceutical compounds, apparently due to absorption by the fish. On the other hand the natural steroids increased as the fish produced the gonadal steroids.

Table 2:   Pharmaceutical concentrations in ng/l in reservoir water.

	Ethinyl-estradiol	Medroxy-progesterone	Benzo-diazepenes
Before introduction of fish	1.43	0.54	Weak
One month after	1.20	0.50	Negative
Three months after	<0.5	<0.5	Negative
Eight months after	0.57	<0.5	Negative

3.3 Effect on fish gonadal development

There were three species of fish in the reservoir. The Tilapia had been treated with methyltestosterone so no ovaries were expected. However the fish should have had masculine gonads, which were absent. The mullet usually take two years or three years to mature so no gonads were expected in the mullet. However the carp should have had gonads weighing about 800 grams at eight months and the ratio of the GSI should have been in the order of 5%. We found that the GSI in the females was between 0.4 to 3% and in the males, the testes were very small, in the order of 200 mg. Histological examination of the ovaries indicated the ovaries were fully matured and there was no evidence of intersex. The male testes were too small for histological analysis.

4   Discussion

The effects of ethinylestradiol on fish reproductive organs in controlled experiments have been well documented and it is the probable agonist present in SWTP effluent that is responsible for the widespread intersex in fish in English rivers [12]. However, the present report would the first documenting gonadal atrophy by exposure to SWTP effluent in a commercial aquaculture pond. Interesting, the brief exposure to EDAs for only one month shortly after hatching affected the gonads at maturity.

Acknowledgement

This work was supported by a grant from the Israel Ministry of the Environment.

References

[1]   Barel-Cohen, K., Shore, L.S., Shemesh, M., Wenzel, A., Mueller, J. & Kronfeld-Schor, N., Monitoring of natural and synthetic hormones in a polluted river. *Journal of Environmental Management*, **78**, 16-23, 2005.
[2]   Degani, G, Boker. R & Jackson. K., Growth hormone, gonad development, and steroid levels in female carp. *Comparative Biochemistry and Physiology C: Pharmacology, Toxicology & Endocrinology*, **115(2)**, pp. 433-430, 1998.

[3] Degani, G., Boker, R. & Jackson, K., Growth hormone, sexual maturity and steroids in male carp (*Cyprinus carpio*). *Comparative Biochemistry & Physiology C: Pharmacology, Toxicology & Endocrino*logy, **120**, pp. 433-430, 1998.

[4] Lange, R., Hutchinson, T.H., Croudace, C.P. & Siegmund, F., Effects of the synthetic estrogen 17 alpha-ethinylestradiol on the life-cycle of the fathead minnow (*Pimephales promelas*). *Environmental Toxicology & Chemistry*, **20**, pp.1216-1227. 2001.

[5] Lintelmann, L., Katayama, A., Kurihara, N., Shore, L. & Wenzel, A., Endocrine disruptors in the environment (IUPAC Technical Report). *Pure & Applied Chemistry*, **75**, pp. 631-681. 2003.

[6] Milstein, A., Zoran, M. & Krembeck, H.J., Seasonal stratification in fish culture and irrigation reservoirs: potential dangers for fish culture. *Aquaculture International*, **3**, pp. 1-7. 1995.

[7] Panter, G.H., Thompson R.S. & Sumpter J.P., Adverse reproductive effects in male fathead minnows (*Pimephales promelas*) exposed to environmentally relevant concentrations of the natural oestrogens, oestradiol and oestrone. *Aquatic Toxico*logy, **42**, pp. 243-253, 1998.

[8] Shemesh, M. & Shore, L.S., Non-steroidal oestrogens of dietary origin: Activity, distribution and mechanism of action. *Israel Journal of Veterinary Medicine*, **43**, pp. 192-197, 1987.

[9] Shore, L.S., Gurevich, M. & Shemesh, M., Estrogen as an environmental pollutant. *Bulletin of Environmental Contamination & Toxicology*, **51**, pp.361-366. 1993.

[10] Shore, L.S. Reichman, O., Shemesh, M., Wenzel, A. & Litaor, M., Washout of accumulated testosterone in a watershed. *Science of the Total Environment*, **332**, pp. 193-202, 2004.

[11] Shore, L.S. & Shemesh, M., Naturally Produced Steroid Hormones and their Release into the Environment. *Pure & Applied Chemistry,* **75 (11–12)**, pp.1859–1871, 2003.

[12] Tyler, C.R. & Routledge E.J., Oestrogenic effects in fish in English rivers with evidence of their causation. *Pure & Applied Chemistry,* **70(8)**, pp. 1795-1804, 1998.

Ocular manifestations of occupational exposure to corticosteroids in the pharmaceutical industry

F. M. Metwally[1], S. El-Assal[2] & E. A. Kamel[3]
[1]Department of Occupational Health and Industrial Medicine,
National Research Center, Dokki, Cairo, Egypt
[2]Memorial Institute of Ophthalmology, MOHP, Giza, Egypt
[3]Department of Community Medicine, Mansoura University,
Mansoura, Egypt

Abstract

The therapeutic use of corticosteroids has long been associated with ocular side effects such as the development of cataracts, glaucoma, and retinal and choroidal emboli. This study was conducted to assess ocular disorders, which may occur in workers engaged in manufacturing corticosteroids in the pharmaceutical industry. The study population included 58 workers involved in the manufacturing of corticosteroid preparations. They were subdivided according to duration of exposure into 2 groups: Gr. I of <15 years duration of employment and Gr. II of 15 + years duration of employment. Methods included answering a structured questionnaire enquiring about occupational and medical histories, full ocular and clinical examination, and measurement of certain biochemical parameters, such as plasma cortisol, serum total cholesterol, serum triglycerides, fasting blood sugar and serum calcium. Results showed that 25 cataracts were diagnosed in the studied group with a rate of 21.55 per one hundred eyes. Of Gr.II, 43.75% of eyes had cataracts compared to 5.88% of Gr.I. Odds ratio to develop cataracts was 12.44 for Gr. II compared to Gr.I. Ocular hypertension and open-angle glaucoma were present in 25.00% of the eyes of the studied groups with odds of 5.83 times more when being employed for 15 or more years. Significant +ve correlation was found between intraocular pressure and relevant biochemical parameters. In the present study, a clear relationship was detected between the length of the period of exposure to corticosteroids as environmental pollutants and the development of ocular medical disorders such as cataracts, ocular hypertension and glaucoma. Further research work is needed to study the effects of exposure to corticosteroids as pollutants in the work environment, gaining access into body via unusual routes of entry, and for periods of time not reported in the therapeutic literature.
Keywords: corticosteroid, ocular manifestations, cataracts, glaucoma, ocular hypertension, occupational, hypertension, diabetes mellitus, lipid profile, pharmaceutical.

WIT Transactions on Biomedicine and Health, Vol 10, © 2006 WIT Press
www.witpress.com, ISSN 1743-3525 (on-line)
doi:10.2495/ETOX060151

1 Introduction

Corticosteroids are widely prescribed in everyday medical practice for the therapy of a vast array of systemic and organ disorders. The risk of occurrence of side effects due to the therapeutic use of corticosteroids is proportional to the dosage and duration of therapy. Topical, oral, and intravenous corticosteroids have long been associated with ocular side effects as the development of cataracts, glaucoma, and retinal and choroidal emboli. Data reported by Carnahan and Goldstein [1] suggested that inhaled corticosteroids were associated with the development of cataracts and increased intraocular pressure (IOP). Haimovici et al. [6] claimed that central serous chorioretinopathy may be associated with inhaled or intranasal corticosteroids. Mitchel et al. [9] found an association between ever use of inhaled corticosteroids and a finding of elevated IOP or glaucoma in subjects with glaucoma family history.

The widespread use of corticosteroids as therapeutics necessitates that the pharmaceutical industries manufacture them in quantities and in preparations that parallel the increasing market demands. Workers involved in the manufacturing processes are inevitably exposed to corticosteroids for periods of time and in concentrations which may mimic long term therapeutic use. Routes of entry of corticosteroids as environmental pollutants may include inhalation of respirable dust suspended in the work atmosphere and absorption through respiratory mucus membrane into blood, direct ocular absorption of soluble dust deposited on conjunctival mucosae, ingestion of dust while eating or smoking with contaminated hands at the workplace, and skin absorption of traces of ointments or creams which may be deposited on the uncovered skin. Cumulative exposure to such an environmental pollutant may lead to long-term consequences on the health of the susceptible individuals. Few, if any studies, discussed ocular and other complications resulting from daily exposure to corticosteroids as environmental contaminants in industry. So, the objective of the present work was to examine the association between the period of occupational exposure to corticosteroids as pollutants in a manufacturing plant and the possible development of ocular and other complications among the exposed workers.

2 Subjects and methods

2.1 The study population

The study population included 58 workers involved in the manufacturing of corticosteroid preparations in the pharmaceutical industries. Both male and female workers were included in the study. No age or duration of employment was exempt. None of the workers reported any of the systemic diseases or local conditions necessitating prolonged use of corticosteroids as therapeutics. The studied workers were grouped into 2 groups: Gr.I of (<15) years' duration of employment and Gr.II of (15 +) years' duration of employment.

2.2 Job and workplace description

The flowchart of the industrial process included the following sections:
A- Storage of raw material which took place in a separate location away from the manufacturing process.   Raw materials were composed of powder of fluocortolone and difluocortolone. After weighing the desired amount, the powder was carried manually to the following sections.
B- The preparation unit which included preparation of the following:
I- Ointments and creams:- by using the 'wet granulation method' by mixing about 10 kg of water with 100 gm of powder in open stainless steel pans 'mills' where mixing was done manually till a colloidal suspension was formed. Then the suspension was transferred into a stainless steel mixer which was operated for one hour. The resulting mixture was then emptied into tanks. Vaseline or paraffin was then added to get the desired texture. Mixing was done manually. The mixture was then emptied into 'hoover tubes' for automatic filling .
II- Ampoules:- which involved heating of a mixture of water and powder in open containers followed by filling of ampoules and sealing using heat. Sterilization in autoclave then followed.
C- The packaging and labeling of products then the despatching of final product.
    The whole process was open with minimal enclosure, if any. The preparation unit was a   room of 4 m X 6 m leading to a lengthy rectangular hall with partitions, of 15 m X 5 m in dimension. No exhaust ventilation was present. No protective equipment was used.   Inhalation of dust and contact of dust with eyes or skin took place during the transport of raw material and the preparation of the products.
    Workers rotated through various sections of production. They were equally exposed to the same pollutants in the work atmosphere.

3   Methods

Each participant in the study was subjected to:
- Full medical and occupational history taking with emphasis on probable side effects of corticosteroids.
- Ocular examination that included:
- Full ophthalmological examination.
- IOP measurement in mm Hg by applanation tonometry.
- Assessment of visual fields with automated perimetry (Humphrey 76 point suprathreshold screening test). Glaucoma was diagnosed by the finding of matching optic disc cupping and disc rim thinning (cup-disc ratio $\geq$ 0.7 or cup-disc asymmetry $\geq$ 0.3) plus characteristic visual field defects on automated perimetry.
- Biochemical parameters measurement which included: Plasma cortisol measured by radioimmunoassay, Serum total cholesterol, Serum triglycerides, Fasting blood sugar and Serum calcium.
- Statistical analysis of data was performed using SPSS package version 10.00 and EPI info package version 6.00.

4 Results

The mean age of the whole group was found to be 38.00 ± 11.97 years. Females had a mean age of 42.62 ± 11.65 years, while males had a mean age 32.30 ± 9.86 years. The age difference between females and males was statistically significant. All age groups were represented in the studied group with age ranging from 19 years up to 58 years.

Table 1:   Age distribution of the exposed workers.

Age Group in Ys	Females (n=32)		Males(n=26)		All (n=58)	
	No.	%	No.	%	No.	%
< 25	2	6.25	4	15.38	6	10.34
25 - <30	6	18.75	12	46.15	18	31.03
30 - <40	2	6.25	4	15.38	6	10.34
40 - <50	14	43.75	4	15.38	18	31.03
50 +	8	25.00	2	7.69	10	17.24
Mean (SD)	42.62 (11.65)		32.30 (9.86)		38.00 (11.97)	
Median	46.50		29.00		34.00	
Min. – Max.	21 - 58		19 - 53		19 - 58	
t-test, p-value	3.58 (df 56), P < 0.01					

Table 2:   Distribution of the exposed workers by duration of employment.

Duration of Employment Group in Years	Females (n= 32)		Males (n= 26)		All Group (n= 58)	
	No.	%	No.	%	No.	%
< 5	8	25.00	4	15.38	12	20.68
5 - <15	6	26.08	16	61.53	22	37.93
15 - <25	0	0.00	4	15.38	4	6.89
25 +	18	56.25	2	7.69	20	34.48
Mean (SD)	20.43 (14.60)		9.76 (7.70)		15.65 (13.06)	
Median	26.50		8.00		10.00	
Min. - Max.	1 - 38		0.5 - 28		0.5 - 38	
t-test, p-value	3.36 (df 56), P < 0.01					

Mean duration of employment for the whole group was 15.65 ± 13.06 years, ranging from 0.5 years up to 38 years. Females had a mean duration of employment of 20.43 ± 14.60 years. While for males, the mean was 9.76 ± 7.70 years. The difference was statistically significant. Fifty six point twenty five per cent of females were employed for 25 years or more, compared to 7.69 per cent of males. All duration of employment groups were represented in the studied group.

Table 3:   Distribution of the ocular manifestations among the exposed groups (n= 58 workers, 116 eyes).

Ocular Manifestation	Group I (< 15 Ys) (n=68 eyes) No. %		Group II (15 + Ys) (n= 48 eyes) No. %		All Group (n= 116 eyes) No. %	
Cataract						
-Diagnosed	4	5.88	12	25.00	16	13.79
- Extracted	0	0.00	9	18.75	9	7.75
• Total	4	5.88	21	43.75	25	21.55
Odds ratio	12.44 (3.56 < OR < 47.0) , P < 0.001					
Ocular Hypert.	8	11.76	12	25.00	20	17.24
Open-angle Glaucoma	0	0.00	9	18.74	9	7.75
Total	8	11.76	21	43.75	29	25.00
Odds ratio	5.83 (2.11 < OR < 16.56) , P < 0.001					

Twenty five cataracts were diagnosed in the studied group with a rate of 21.55 per one hundred eyes. Of Gr.II, 43.75% of eyes had cataracts compared to 5.88% of Gr.I. The Odds Ratio which is an estimate of the relative risk for eyes to acquire cataracts when being employed for 15 years or more in the manufacturing of corticosteroids was 12.44 times more than the risk when being employed for less than 15 years, i.e., the probability to develop cataracts when being employed for 15 years or more was 12 times more than the probability to develop it when being employed for <15 years. Ocular hypertension and open angle glaucoma were present in 25.00% of eyes of the studied groups with odds of 5.83 times more when being employed for 15 + years.

Table 4:   Distribution of IOP among the exposed groups.

IOP in mm Hg	Group I (n= 68 eyes) No. %		Group II (n= 48 eyes) No. %		All Group (n= 116 eyes) No. %	
10 - < 15	34	50.00	12	25.00	46	39.65
15 - < 20	28	41.17	10	20.83	38	32.75
20 - < 25	6	8.82	14	29.16	20	17.24
25 +	0	0.00	12	25.00	12	10.34
Mean (SD)	14.91 (3.31)		19.66 (5.16)		16.87 (4.78)	
Median					16	
Min. – Max.					10 - 27	
t- test (df 114), P-value	- 6.03, P < 0.001					

Mean IOP differed significantly between Gr.I and Gr.II exposed groups with Gr.II showing higher IOP of 19.66 ± 5.16 mm Hg. The mean IOP for the whole group was 16.87 ± 4.78 mm Hg.

Table 5:   Mean values of biochemical parameters among the exposed groups.

Parameter	Group I (n= 34) Mean ± SD	Group II (n= 24) Mean ± SD	All Group (n= 58) Mean ± SD
Plasma Cortisol: (Normal: 8.5- 13.5 ug/dL)	13.48 ± 1.75	15.54 ± 0.67	14.33 ± 1.73
t-test, P-value	- 5.45, < 0.001		
Serum total Cholesterol : (Normal : 140- 270 mg/dL)	206.70 ± 23.29	262.16 ± 11.35	229.65 ± 33.54
t-test, P-value	- 10.77, < 0.001		
Serum triglycerides: (Normal : 40- 140 mg/dL)	138.41 ± 10.72	155.91 ± 4.24	145.65 ± 12.22
t-test, P-value	- 7.57, < 0.001		
Fasting blood sugar: (Normal : 70- 100 mg/dL)	119.64 ± 17.02	172.16 ± 19.00	141.37 ± 31.53
t-test, P-value	- 11.03, < 0.001		
Serum calcium Ca^{++} (Normal: 9- 11 mg/dL)	8.90 ± 1.29	6.87 ± 1.30	8.06 ± 1.63
t-test, P-value	5.87, < 0.001		

The two groups differed significantly in the mean values of the biochemical parameters with Gr.II showing higher values than Gr.I. The mean plasma cortisol for the whole group was 14.33 ± 1.73 ug/dL which is higher than the upper limit value of the normal range for this parameter.

Table 6:   Relation between biochemical parameters and IOP of the exposed groups.

Biochemical Parameter	IOP (n= 116 eyes) r
Plasma cortisol	0.491 **
Fasting blood sugar	0.346 *
Serum triglycerides	0.391 *
Serum total cholesterol	0.392 *
Serum calcium, ca $^{++}$	- 0.47
Age (Ys)	0.493 **
Duration of exposure (Ys)	0.491 **

**   Correlation is significant at P-value < 0.01 (2-tailed).
*   Correlation is significant at P-value < 0.05 (2-tailed).

Significant positive correlation was found between IOP of eyes of the exposed groups and relevant biochemical parameters, especially with plasma cortisol i.e., IOP increases with increase in plasma cortisol.

Table 7:   Distribution of medical disorders among the exposed groups.

Medical disorder	Group I (n= 34) No.	%	Group II(n= 24) No.	%	Chi-square, P-value
Endocrinological: Weight gain	22	64.70	18	75.00	0.67, P > 0.05
Diabetes M. under treatment	0	0.00	7	29.16	
Acne	16	47.05	6	25.00	2.90, P > 0.05
Immunological: Recurrent infections: Chest infections	14	41.17	6	25.00	1.63, P > 0.05
Herpes zoster	0	0.00	0	0.00	
Fungal infections	0	0.00	0	0.00	
Gastrointestinal: Hyperacidity	22	64.70	16	66.66	0.02, P > 0.05
Increased appetite	10	29.41	4	16.66	1.24, P > 0.05
Musculoskeletal: Bone aches	8	23.52	20	83.33	20.15,P<0.001
Easy fatiguability	0	0.00	4	16.66	
Reproductive: Menstrual disturbances:	0	0.00	2	8.33	
Amenorrhea	4	11.76	0	0.00	
Irregular menses Early menopause	0	0.00	8	33.33	
Cardiovascular: Hypertension under treatment	2	5.88	10	41.66	10.97, P < 0.01
Hypertension (currently Diagnosed on Clinical ex.)	2	5.88	10	41.66	
Total Hypertension	**4**	**11.76**	**20**	**83.33**	**29.7, P < 0.001**
Thrombo-embolic phenomenon	0	0.00	0	0.00	
Arrhythmia	0	0.00	4	16.66	
Neurological sym.: Recurrent headaches	18	52.94	8	33.33	2.18, P > 0.05
Migrainous attacks	4	11.76	0	0.00	

Hypertension was prevalent in 41.37% of the studied groups, in 83.33% of Gr.II subjects and in 11.76% of Gr.I subjects with the difference being statistically significant. Diabetes Mellitus was present in 29.16% of Gr.II subjects. Hyperacidity was present in about 60% of either groups. Bone aches were reported by 83.33% of Gr.II, with a statistically significant difference compared to Gr.I. Recurrent headaches were complained of by half of Gr.I and one third of Gr.II.

Table 8(A): Comparison of cases of ocular hypertension and open-angle glaucoma with subjects with normal IOP.

Item	Cases (n= 15) Mean ± SD	Subjects with normal IOP (n= 43 Mean ± SD	t- test (df=56), P- value
Age	48.12 ± 6.43	34.14 ± 11.35	6.07, P <0.001
Duration of job	24.00 ± 10.54	12.47 ± 12.61	5.19, P <0.001
Plasma cortisol	15.72 ± 0.59	13.80 ± 1.73	4.28, P <0.001
Fasting bl. sugar	161.62 ± 21.87	133.66 ± 31.40	3.91, P <0.001

Table 8(B): Age distribution of cases of ocular hypertension and open-angle glaucoma (n= 15).

Age	34 Ys	45 Ys	48 Ys	52 Ys	53 Ys	55 Ys	56 Ys	58 Ys
No.	1	2	2	2	2	2	2	2
%	6.66	13.33	13.33	13.33	13.33	13.33	13.33	13.33

Table 8(A) shows that cases of ocular hypertension and glaucoma among the exposed groups were older, with more years of employment, higher plasma cortisol and fasting blood sugar. Table 8(B) shows that the distribution of age of cases included ages as young as 34 years, 45 years and 48 years. Meanwhile, no family history of glaucoma was reported by participants in the study.

Table 9(A): Comparison of cases of cataracts with subjects not suffering from cataracts.

Item	Cataract cases (n=16) Mean ± SD	Other Subjects (n=42) Mean ± SD	t-test (df=56), P-value
Age	48.12 ± 6.43	34.14 ± 11.35	4.63, P <0.001
Duration of job	24.00 ± 10.54	12.47 ± 12.61	3.24, P < 0.01
Plasma cortisol	15.72 ± 0.59	13.80 ± 1.73	4.30, P < 0.001
Fasting bl. Sugar	161.62 ± 21.87	133.66 ± 31.40	3.26, P <0.01

Table 9(B): Age distribution of cases of cataracts.

Age	34 Ys	45 Ys	48 Ys	49 Ys	52 Ys	53 Ys	56 Ys
No.	2	2	4	2	2	2	2
%	12.5	12.5	25.00	12.5	12.5	12.5	12.5

Table 9(A) shows that cases of cataracts differed from other subjects in age, duration of employment, plasma cortisol and fasting blood sugar. Age of cataract cases started from 34 years onwards which is considered unusually young for cataracts to develop.

5   Discussion

In the present study, a clear relationship was detected between the length of the period of exposure to corticosteroids as environmental pollutants and the development of ocular medical disorders such as cataracts, ocular hypertension and open-angle glaucoma.   The prevalence of these ocular disorders among the exposed groups exceeded that among the general population and the patients using corticosteroids as therapeutics (Khan et al. [7], Leske et al. [8] and Carnahan and Goldstein [1]).

The occupational exposure to corticosteroids as air pollutants may mimic the therapeutic use of inhalants and nasal preparations, with limitations. Also, direct ocular exposure to dust containing corticosteroids may mimic the therapeutic use of ophthalmic preparations. The prevalence of cataracts in the present study was 21.55% and it affected workers as young as 34 years.   True prevalence of cataracts among the general population is obscure.   The Framingham Eye study reported a prevalence of cataracts of 18% among persons older than 65 years of age.   The prevalence of cataracts among the adult population of both sexes of the outpatient clinics and the inpatients of the Memorial ophthalmology Institute of MOHP at Giza was calculated to be 4.74% in a 6-month period. The prevalence of cataracts among the studied groups can be attributed to the excess exposure to corticosteroids as environmental pollutants in their work atmosphere as evidenced by the higher plasma cortisol in their sera. The use of corticosteroids by nasal spray and inhalation has been associated with the development of posterior subcapsular cataracts as reported by Fraunfelder and Meyer [4]. Garbe et al. [5] similarly concluded that prolonged administration of high doses of inhaled corticosteroids increases the likelihood of undergoing cataract extraction in elderly patients.   They found out that the odds ratio for cataract extraction in patients with prolonged cumulative exposure to inhaled corticosteroids compared with non-users was 3.40.   Other risk factors may have been operating as well, like exposure to ultraviolet rays and diabetes mellitus. Risk factors together with the occupational exposure to corticosteroids pollutants were mostly responsible for the development of cataracts in the susceptible 20% of workers and at an earlier age than noticed among the general population as observed by Khan et al. [7], West [13], and Leske et al. [8].

The ocular hypertensive response to intravenous, oral, topical dermatologic, topical ocular, and periocular corticosteroids has been well established by investigators since the 1960s (Dryer [3] and Opatowsky et al. [11]).   Increased IOP and odds of developing glaucoma in the studied groups signify that corticosteroids as pollutants in the work atmosphere reach ocular structures in levels sufficient to provoke an ocular hypertensive response in susceptible individuals.

Systemic absorption of 50% or more of steroids administered by nasal spray has been documented by the United States Pharmacopeial Convention [12]. Studies have shown that 90% of aerosol deposition is present in the oropharyngeal, mucosa where the drug could be systemically absorbed as reported by Davies [2] and Newman et al. [10]. Other possible routes of entry and absorption include droplet deposition on the ocular surface, ingestion with absorption from the gastrointestinal tract or inhalation of the nasal spray. The same pattern of routes of entry can be extrapolated to the occupational exposure to corticosteroid dust as an environmental pollutant in the work atmosphere. In therapy, the side effects of absorbed corticosteroids may depend on the type of the drug administered. At the workplace, corticosteroid dust may represent a combination of more than one type of corticosteroids which may produce more intense adverse effects than those expected to result from therapy. Of course, environmental industrial hygiene measures, if applied, can moderate such effects.

The prevalence of ocular hypertension and glaucoma in the present study mounted to 25.00% of the studied groups. Compared to rates from a study by Mitchell et al. [9] to investigate the association between ever use of inhaled corticosteroids and elevated IOP or glaucoma in subjects with a glaucoma family history, our present rates are much higher. Open-angle glaucoma is expected to affect 1 in 200 of the general population. In Mitchell's study, ocular hypertension affected 4.37% of patients while glaucoma affected 2.95% of them, with overall rate of 7.33%. So, the occupational exposure clearly resulted in prevalence rates of ocular hypertension and glaucoma which were much higher than rates in the general population or in patients on corticosteroids therapy.

Increased IOP induced by corticosteroids may occur due to diminished outflow facility as a result of increased hyaluronic acid content of endothelial lining cells due to the steroid induced change in lysosomal membrane of goniocytes in the meshwork (Opatowsky et al. [11]). No family history was reported by any of the glaucoma cases in the present study. Failure to elicit positive family history may be due to lack of knowledge of the morbid condition among other family members or the presence of co-morbid medical disorders which masked the morbid condition leading to missed diagnosis and underreporting. However, lack of family history – if true – may signify that the occupational exposure to corticosteroids pollutants was intense and consistent enough to induce ocular hypertension and glaucoma irrespective of family history as a potent risk factor for this medical disorder as denoted by Mitchell et al. [9]. Other medical disorders which may be attributed to excess exposure to cortisone were more prevalent in the Group II of exposed workers as hypertension, diabetes mellitus, hyperacidity and headaches with an evident exposure gradient. These disorders, specifically diabetes mellitus, might have attributed to the occurrence of ocular manifestations among the exposed groups. Further research work is needed to study the effects of exposure to corticosteroids as pollutants in the work atmosphere, gaining access into body via multiple routes of entry, and for periods of time not reported in the therapeutic literature. Intervention strategies should be evaluated to minimize hazards and risks. Periodic full ocular examination of susceptible workers should be carried out for the early detection of medical disorders before irreversible complications set in.

References

[1]   Carnahan MC, and Goldstein DA. Ocular complications of topical, peri-ocular, & systemic corticosteroids. Curr Opin Ophthalmol 2000; 11(6):478- 483.
[2]   Davies DS. Pharmacokinetics of inhaled substances. Post-grad Med j 1975; 51 (Suppl 7): 69- 75.
[3]   Dryer EB. Inhaled steroid use and glaucoma 'letter'. N Engl J Med 1933; 329: 1822.
[4]   Fraunfelder FT, and Meyer SM. Posterior subcapsular cataracts associated with nasal or inhalation corticosteroids. Am J Ophthalmol 1990; 109: 489- 490.
[5]   Garbe E, Suissa S, and LeLorier J. Association of inhaled corticosteroids with cataract extraction in elderly patients. JAMA 1998; 280: 539- 543.
[6]   Haimovici R, Gragoudas ES, Duker JS, Sjaarda RN, and Eliott D. Central serous chorioretinopathy associated with inhaled or intranasal corticosteroids. Ophthalmology 1997; 104 (10): 1653- 1660.
[7]   Khan HA, Leibowitz HM, and Ganley JP. The Framingham Eye Study: Outline and major prevalence findings. Am J Epidemiol 1977; 106: 17- 27.
[8]   Leske MC, Chylack LT, and Wu SY. The lens opacities – case control study. Risk factors for cataract. Arch Ophthalmol 1991; 109: 244.
[9]   Mitchell P, Cumming RG, and Mackey DA. Inhaled corticosteroids, family history, and risk of glaucoma. Ophthalmology 1999; 106 (12): 2301- 2306.
[10]   Newman SP, Moren F, & Pavia D. Deposition of pressure suspension aerosols inhaled through extension devices. Am Rev Respir Dis 1981;124: 317- 320.
[11]   Opatowsky I, Feldman RM, Gross R, and Feldman ST. Introcular pressure elevation associated with inhalation and nasal corticosteroids. Ophthalmology 1995; 102 (20):177- 179.
[12]   United States Pharmacopeial Convention. Drug information for the health care professional,12[th]ed. Rockville, MD: US Pharmacopeial Convention 1992; 56- 62.
[13]   West SK. Editorial: who develops cataracts? Arch Ophthalmol 1991; 109: 196.

Section 5
Genotoxicity

Toxicity and genotoxicity studies of surface and waste water samples using a bacterial SOS/*umu* test and mammalian MTT and comet assay

B. Žegura[1], E. Heath[2], A. Černoša[3] & M. Filipič[1]
[1]*National Institute of Biology Department of Genetic Toxicology and Cancer Biology, Ljubljana, Slovenia*
[2]*Jožef Stefan Institute, Department of Environmental Sciences, Ljubljana, Slovenia*
[3]*University of Ljubljana, Faculty of Pharmacy, Ljubljana, Slovenia*

Abstract

Surface and waste waters are complex mixtures that may contain thousands of different pollutants of different origins (industrial, agricultural and domestic). Many of them show toxic and/or genotoxic effects and are therefore potentially hazardous for humans and the environment.

It is extremely difficult to quantify the risk associated with xenobiotics in environmental samples because they usually occur in concentrations too low to allow chemical analytical determination. Additionally, single and combined biological effects of most of the micropollutants are not known. The best approach to evaluate potential toxic/genotoxic risks of such mixtures is to use biological test systems with living cells or organisms that give a global response to the pool of micropollutants present in the sample.

In this study we evaluated the cytotoxic/genotoxic potential of 51 different water samples (river, potable, well, lake, and waste waters) potentially contaminated with pharmaceutical and personal care products (PPCP). The samples were evaluated for their genotoxic potential with the bacterial SOS/*umu* test with *Salmonella typhimurium* TA1535/pSK1002 and for their cytotoxic potential with mammalian cell based MTT assay with human hepatoma (HepG2) cells. Genotoxicity of seven selected samples was further tested with the comet assay with metabolically competent HepG2 cells.

The results from the present study confirmed that biological tests are indispensable for the reliable assessment of cytotoxic and genotoxic potential of surface and waste waters. There is also a need for chemical analytical characterisation of cytotoxic/genotoxic samples in order to identify and quantify the compounds responsible for the cytotoxicity/genotoxicity.
Keywords: SOS/umuC, Salmonella typhimurium, MTT, comet assay, HepG2, water samples, cytotoxicity, genotoxicity.

WIT Transactions on Biomedicine and Health, Vol 10, © 2006 WIT Press
www.witpress.com, ISSN 1743-3525 (on-line)
doi:10.2495/ETOX060161

1 Introduction

Water pollution by toxic and genotoxic micropollutants represents one of the most critical problems concerning public health and protection of aquatic ecosystem. Major sources of surface water pollution by a variety of substances are industry, agriculture and domestic households, including municipal wastewaters. Numerous toxic and genotoxic contaminants such as trace metals, polycyclic aromatic hydrocarbons, heterocyclic amines, polychlorinated biphenyls, pesticides, dyes, pharmaceutical and personal care products and many more, can be identified as the components of complex aquatic environmental mixtures [1–3]. These chemicals are released either deliberately or unintentionally into rivers, lakes and seas and can accumulate in sediments. Many genotoxins can also be produced by chemical and biological transformation after the emition into the environment.

Because of the complexity of pollution, standard targeted chemical analyses are limited in their ability to give adequate information regarding toxic or genotoxic potential. The two reasons are: a) toxicological properties of many pollutants are not known and b) chemical analysis cannot predict potential synergistic effects. On the other hand bioassay assessment of polluted samples provides means to evaluate and compare toxic or genotoxic potential of different samples without detailed knowledge of their chemical composition. In this context monitoring for potential genotoxicity is of particular importance as exposure to genotoxic contaminates may be associated with the risk for cancer development.

In the present study genotoxicity and cytotoxicity of environmental water samples was studied by the combination of a bacterial assay SOS/umu test with *Salmonella typhimurium* TA1535/pSK1002 and with mammalian assays: MTT and comet assay with human hepatoma cell line, HepG2.

2 Materials and methods

2.1 Water samples

Samples were collected in Slovenia from lakes, rivers, hospital and chemical industry effluents, waste water treatment plant influents and effluents, wells and potable water during the 2004 and 2005. The samples were stored at -20°C and filtered (0,22 µm pore size) before genotoxicity testing.

2.2 SOS/umu assay

The cytotoxic and genotoxic effects of surface and waste water samples were evaluated using SOS/umuC test performed according to Reifferscheid et al. (1991) [4], with minor modifications as described in ISO standards [5]. The tester strain *Salmonella typhimurium* TA1535/pSK1002 carries the plasmid pSK1002 with *umuC* operon fused with the *lacZ* gene for ß-galactosidase activity. This allows monitoring *umuC* induction by measuring ß-galactosidase activity. Water testing was performed with and without S9 metabolic activation.

Briefly, the overnight culture was diluted 10 times with fresh TGA medium and incubated at 37°C for 1.5 hour with shaking until the bacteria reached exponential growth phase. The test was carried out in triplicate on the microtiter plate. The incubation mixture consisted of 180 µl un-concentrated water sample, 20 µl 10x TGA and 70 µl bacterial culture or in the case of metabolic activation 180 µl un-concentrated water sample, 20 µl 10x TGA with cofactors and 70 µl S9 bacterial culture mixture prepared as described in ISO standard [5]. The microtiter plate was incubated at 37°C for 2 hours with shaking. Afterwards the incubation mixture was diluted 10 times with fresh TGA medium. After 2 hour incubation at 37°C the growth rate of the biomass was measured (600 nm) and the induction of the umuC gene was determined by measuring ß-galactosidase activity, using ONPG as a substrate (420 nm). Biomass was calculated by the formula: $G = $ (sample OD_{600} – blank OD_{600}/ control OD_{600} – blank OD_{600}). Growth inhibition of biomass for more than 25% was considered to be indicative of water samples cytotoxicity. ß-galactosidase in relative units was calculated by the formula: $U = $ (sample OD_{420} – blank OD_{420}/ sample OD_{600} – blank OD_{600}) and induction ratio (IR) by the formula: $(1/G)$ x (sample OD_{420} – blank OD_{420}/ control OD_{420} – blank OD_{420}). The induction ratio with a threshold 1,5 was used as a measure of genotoxic potency of water samples. 1-Methyl-3-nitro-1-nitrosoguanidine (MNNG; 6 µM) was used as a positive control without and aflatoxin B1 (AFB1; 2 µg/ml) with metabolic activation.

2.3 Cell culture

HepG2 cells (a gift from Dr. Firouz Darroudi, Department of Radiation Genetics and Chemical Mutagenesis, University of Leiden, Netherlands) were grown in William's medium E (Sigma, St. Louis, USA) containing 15% FBS, 2 mM L-glutamine and 100 U/ml pen/strep at 37°C in 5% CO_2. In each experiment the growth medium control and distilled water control were included in order to exclude possible effects of medium dilution. In all the experiments the results of water sample treated cells were compared to the results of distilled water control.

2.4 Cytotoxicity assay (MTT)

Cytotoxicity of water samples was determined with 3-(4,5-dimethylthiazol-2-yl)-2,5-diphenyltetrazolium bromide (MTT) according to Mosmann (1983) [6], with minor modifications. This assay measures the conversion of MTT to insoluble formazan by dehydrogenase enzymes of the intact mitochondria of living cells. HepG2 cells were seeded at a density of 1×10^4 cells/well into 96-well microtiter plates in five replicates. After 4 hours incubation at 37°C, growth medium was replaced with fresh medium containing 30 vol % of water samples and the cells were incubated for 20 hours. After the treatment MTT was added at a final concentration of 0.5 mg/ml and the cells were further incubated for 3h at 37°C. The medium was removed and the formazan crystals were dissolved in DMSO. The optical density (OD) was measured at 570 nm (reference filter 690 nm) using a spectrofluorimeter (Tecan, Genios). Cell survival (viability) was determined by comparing the OD of the wells containing cells treated with water samples to cells exposed to 30 vol % distilled water in growth medium.

2.5 Cell treatment and alkaline comet assay

HepG2 cells were seeded in 12 well cell culture cluster plates (Corning Costar Corporation, New York, USA) and allowed to attach to the plate. After 4 hours the cells were treated with water samples (5, 10, 20 or 30 vol %) in growth medium for 20 hours. BaP (40 µM) was used as a positive control. After the treatment the cells were trypsinized and centrifuged at 800 rpm for 5 min.

The Comet assay was performed according to Sing et al. (1988) [8], with minor modifications. Briefly, 30µl of cell suspension was mixed with 70µl 1% LMP agarose and added to fully frosted slides precoated with 80µl of 1% NMP agarose. Subsequently, the slides were lysed (2,5M NaCl, 100mM EDTA, 10mM Tris, 1% Triton X-100, pH10) for 1h at 4°C, placed into alkaline solution (300mM NaOH, 1mM EDTANa$_2$, pH13) for 20 min at 4°C to allow DNA unwinding and electrophoresed for 20 min at 25V (300mA). Finally, the slides were neutralized in 0,4 M Tris buffer (pH 7,5) for 15 minutes, stained with EtBr (5µg/ml) and analysed using a fluorescence microscope (Nikon, Eclipse 800). Images of 50 randomly selected nuclei per experimental point were analysed with the image analysis software (Comet Assay IV, Perceptive Instruments, UK). The results from three independent experiments are expressed as % of tail DNA and are shown as Figures in column scatter. The median value is shown as a solid line through the column. One-way analysis of variance (ANOVA, Kruskal-Wallis) was used to analyze the differences between treatments within each experiment. Dunnett's test was used for multiple comparison versus the control; p<0,05 was considered as statistically significant (*).

3   Results

Table 1 summarizes the results of genotoxic potential of water samples determined with the SOS/umuC test and results of toxic potential determined with MTT assay with human hepatoma HepG2 cells.

The viability of *S.typhimurium* TA1535pSK1002 was not affected by water samples, except by one (CIS1) of the two samples from chemical industry. The sample reduced bacterial viability bellow 70% in the presence of S9 metabolic activation (Table 1). The samples of the effluents from chemical industry (CIS1 and CIS2) were sampled at the same place, but at different time periods.

Cytotoxicity was determined also with metabolically competent human hepatoma cell line (HepG2) using the MTT assay (Table 1). The effluents from hospitals (HS1-5) were not cytotoxic for HepG2 cells, while both effluents from chemical industry (CIS1 and CIS2) reduced cell viability for more than 50%. Influents to wastewater treatment plants (WWI1-4) were not cytotoxic, however, two (WWE1 and WWE2) of five effluent samples reduced cell viability. Only two (RS1 and RS21) out of twenty-two river samples decreased cell survival, while samples of lake (LS1-6), potable (PS1-6) and well (WS1 and WS2) water were not cytotoxic.

Table 1:   The induction of SOS response in *Salmonella typhimurium* TA 1535/pSK1002 and the effect on HepG2 cells survival (MTT assay) after treatment with water samples.

| SAMPLE | | *Salmonella typhimurium* TA1535/pSK1002 | | | | | HepG2 |
| | | Without S9 | | | With S9 | | MTT |
		Viability (%)	U±SD	IR	Viability (%)	U±SD	IR	Viability (%)
HS	1	112,0	0,30 ± 0,04	1,37	121,5	0,24 ± 0,02	1,02	88,10
	2	90,2	0,48 ± 0,03	**2,17**	128,3	0,38 ± 0,06	**1,61**	130,61
	3	100,6	0,27 ± 0,03	0,88	116,9	0,25 ± 0,02	1,03	90,90
	4	103,1	0,47 ± 0,07	**1,51**	121,2	0,35 ± 0,02	1,43	101,14
	5	115,6	0,40 ± 0,04	1,27	169,1	0,30 ± 0,03	1,32	91,90
CIS	1	79,8	1,18 ± 0,27	**5,34**	**66,7**	2,44 ± 0,54	**10,25**	**46,11**
	2	82,0	0,54 ± 0,13	1,69	141,1	0,49 ± 0,03	**1,96**	**52,62**
WWI	1	126,9	0,28 ± 0,02	1,28	149,2	0,27 ± 0,03	1,13	98,05
	2	114,3	0,33 ± 0,05	1,04	147,6	0,24 ± 0,04	0,99	95,27
	3	111,4	0,30 ± 0,02	0,95	158,5	0,26 ± 0,03	1,07	82,28
	4	116,9	0,25 ± 0,01	1,14	119,4	0,25 ± 0,01	1,04	81,35
WWE	1	110,8	0,33 ± 0,04	1,05	150,0	0,28 ± 0,01	1,15	**71,21**
	2	112,4	0,29 ± 0,02	1,28	120,4	0,33 ± 0,05	1,38	**66,60**
	3	109,7	0,37 ± 0,04	1,19	149,8	0,27 ± 0,01	1,11	105,14
	4	114,7	0,52 ± 0,26	**2,35**	125,9	0,26 ± 0,02	1,10	84,27
RS	1	108,4	0,30 ± 0,03	0,96	112,1	0,26 ± 0,04	1,08	**70,72**
	2	101,6	0,30 ± 0,08	1,37	120,0	0,24 ± 0,02	1,02	82,28
	3	106,0	0,35 ± 0,05	1,12	110,5	0,27 ± 0,04	1,09	97,92
	4	106,0	0,25 ± 0,05	0,80	144,8	0,33 ± 0,01	1,45	89,36
	5	113,3	0,35 ± 0,12	**1,59**	110,8	0,27 ± 0,03	1,15	106,82
	6	104,9	0,31 ± 0,05	1,01	143,7	0,35 ± 0,02	**1,52**	91,83
	7	114,5	0,36 ± 0,00	**1,61**	134,7	0,31 ± 0,07	1,33	102,89
	8	100,0	0,30 ± 0,03	0,97	121,0	0,25 ± 0,02	1,03	108,73
	9	96,2	0,27 ± 0,03	0,85	113,5	0,31 ± 0,02	1,25	109,56

Table 1: Continued.

SAMPLE		Salmonella typhimurium TA1535/pSK1002						HepG2
		Without S9			With S9			MTT
		Viability (%)	U±SD	IR	Viability (%)	U±SD	IR	Viability (%)
	10	96,4	0,26 ± 0,05	0,85	119,0	0,21 ± 0,04	0,88	102,11
	11	100,0	0,32 ± 0,07	1,02	120,8	0,29 ± 0,05	1,19	111,44
	12	105,3	0,27 ± 0,01	1,21	120,1	0,27 ± 0,02	1,15	79,20
	13	105,6	0,33 ± 0,05	1,07	132,5	0,28 ± 0,00	1,15	117,02
	14	103,9	0,24 ± 0,02	1,08	115,4	0,24 ± 0,01	1,03	97,53
	15	103,9	0,34 ± 0,01	1,08	119,4	0,27 ± 0,04	1,11	82,79
	16	107,4	0,28 ± 0,11	0,92	113,9	0,27 ± 0,05	1,10	92,47
	17	105,7	0,30 ± 0,00	0,97	119,4	0,27 ± 0,02	1,19	96,96
	18	93,2	0,36 ± 0,03	1,44	114,1	0,23 ± 0,04	0,95	98,79
	19	115,0	0,34 ± 0,01	**1,52**	141,0	0,36 ± 0,04	1,55	97,60
	20	103,2	0,29 ± 0,07	0,94	127,3	0,28 ± 0,02	1,23	93,70
	21	106,7	0,38 ± 0,04	1,21	115,3	0,23 ± 0,03	0,95	**74,42**
	22	102,5	0,33 ± 0,01	1,06	140,3	0,30 ± 0,03	1,31	100,11
LS	1	110,4	0,30 ± 0,03	1,35	147,1	0,30 ± 0,02	1,33	85,23
	2	103,5	0,31 ± 0,03	1,38	124,1	0,38 ± 0,03	**1,64**	107,10
	3	101,4	0,31 ± 0,03	1,40	141,9	0,33 ± 0,05	1,44	93,01
	4	102,7	0,28 ± 0,03	1,27	143,9	0,33 ± 0,03	1,43	102,17
	5	103,3	0,45 ± 0,19	**2,00**	151,4	0,29 ± 0,03	1,28	99,06
	6	101,2	0,34 ± 0,04	**1,51**	125,9	0,35 ± 0,04	**1,53**	95,62
PS	1	88,1	0,30 ± 0,07	1,21	129,8	0,25 ± 0,02	1,05	98,80
	2	96,2	0,27 ± 0,02	1,08	120,2	0,21 ± 0,06	0,84	103,79
	3	107,3	0,39 ± 0,08	1,23	116,7	0,29 ± 0,04	1,19	94,79
	4	110,0	0,28 ± 0,07	0,91	115,7	0,27 ± 0,07	1,09	107,25
	5	108,4	0,25 ± 0,04	0,82	116,5	0,28 ± 0,04	1,13	109,76
	6	89,6	0,39 ± 0,05	**1,57**	123,2	0,23 ± 0,01	0,95	105,24

Table 1: Continued.

| SAMPLE | | Salmonella typhimurium TA1535/pSK1002 | | | | | | HepG2 |
| | | Without S9 | | | With S9 | | | MTT |
		Viability (%)	U±SD	IR	Viability (%)	U±SD	IR	Viability (%)
WS	1	81,5	0,28 ± 0,03	1,11	119,4	0,25 ± 0,08	1,05	105,21
	2	91,4	0,28 ± 0,02	1,12	119,6	0,25 ± 0,03	1,02	85,59
MNNG		92,9		**2,48**	/	/		/
AFB1		/	/	/	100,9		**2,93**	/

HS-hospital sample; CIS-chemical industry sample; WWI-waste water influent; WWE- waste water effluent; RS-river sample; LS-lake sample; PS-potable water sample; WS-well sample. Positive results are shown in bold type.

The genotoxicity testing results showed for most of the samples that induction ratio (IR) of β-galactosidase activity between exposed and control bacteria was around 1 indicating that these samples did not induce umuC system. However, the two hospital effluents (HS2 and HS4) and both samples from chemical industry (CIS1 and CIS2) induced genotoxic responses. The IR in bacteria exposed to sample CIS2 was 5 and 10 in the absence and presence of S9, respectively. Samples of wastewater influents were not genotoxic, while one sample of wastewater effluent (WWE4) induced IR of 2.3. Four (RS5, 6, 7 and 19) out of twenty-two river samples and three (LS2, 5 and 6) out of six lake samples induced IR above 1,5. Interestingly also one potable water sample (PS6) showed genotoxic effect, while no well water induced genotoxic response.

In order to confirm the genotoxic activity of water samples in mammalian cells seven samples were selected and tested for their genotoxic potential using the comet assay. The HepG2 cells were exposed to different concentrations of water samples (0, 10, 20 and 30 vol %) for 20 hours.

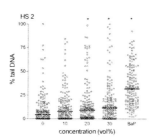

Figure 1:   The level of DNA damage induced by effluent from hospital (HS2).

The comet assay results showed that hospital sample (HS 2) which was positive in the SOS/umuC assay also increased the amount of single strand breaks in HepG2 cells dose dependently (Figure 1).

DNA damage of HepG2 cells was detected after exposure of cells to 20 vol % of wastewater influent (WWI 4) sample, while the effluent from wastewater treatment plant (WWE 4) did not induce significant increase of DNA damage, although this sample was positive in SOS/umuC test (Figure 2).

Both samples of effluents from chemical industry (CIS1 and CIS2) (Figure 3) were tested at lower concentrations than other samples (0, 5, 10 and 20 vol %) due to their cytotoxicity for HepG2 cells. Both samples were shown to be highly genotoxic in SOS/umuC assay and in mammalian cell test system.

We also determined the genotoxic potential of two surface water samples (Figure 4). Lake water sample LS6, which was positive in SOS/umu test showed slight genotoxic activity only at the highest concentration used (30 vol %), while river water sample (RS14), which was negative in the SOS/umu test did not induce DNA strand breaks.

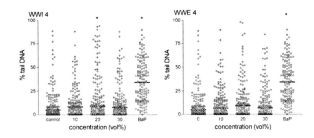

Figure 2:   The level of DNA damage induced by wastewater treatment plant influent (WWI 4) and effluent (WWE 4).

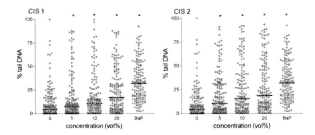

Figure 3:   The level of DNA damage induced by chemical industry effluent (CIS 1 and CIS 2).

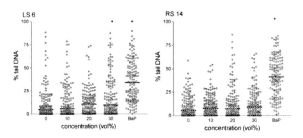

Figure 4:   The level of DNA damage induced by lake (LS6) and river (RS14) water samples.

4   Discussion

Surface waters such as lakes and rivers can be used as a source of drinking water, in agricultural and recreational activities as well as for the industrial purposes. As these waters can be potentially polluted with unknown compounds, this can have adverse effects on humans and indigenous biota. Indeed, in the past years a variety of bioassays were used to show that industrial effluents and wastes as well as environmental surface waters can have mutagenic activity (reviewed in [2]).

In the present study genotoxicity and cytotoxicity of surface and waste water samples collected in Slovenia were studied by a combination of SOS/*umuC* on *Salmonella typhimurium* TA1535/pSK1002 with and without metabolic activation and MTT and the comet assay with human hepatoma HepG2 cells. Our results showed that in the SOS/umuC test out of 51 water samples 11 (21%) were genotoxic without and 7 (13%) with metabolic activation. One sample was toxic for prokaryotic organisms, while 7 samples reduced viability of eukaryotic cells. As already described in other studies [1, 9], our experiments confirmed that the use of metabolic activation did not increase the sensitivity of *Salmonella* tester strain. This indicates that putative genotoxins in both surface and waste waters are primarily direct-acting.

Genotoxic compounds present in mixed municipal wastewaters can include sanitary wastes, pesticides used in agriculture, runoffs from roadways, commercial and industrial areas, hospitals and research institutions [1].

Hospitals represent an incontestable release source of many pharmaceutical drugs, DNA damaging agents and their metabolic products. Some of the substances found in hospital wastewaters are genotoxic and are suspected to be a possible risk factor for increased occurrence of the cancers observed during the last decades [8]. Our study showed that hospital effluents showed genotoxic effects in both assays used in this study. The data confirmed that effluents from hospitals represent a serious threat for environment and human health.

White and Rasmussen [1] presented that municipal wastewaters rank low in genotoxic potency, however, they can achieve loading values that are several orders of magnitude greater that wastes from industries. For this reason domestic wastewaters can constitute great genotoxic hazard to aquatic environment and biota. Wastewater treatment plant influent and effluent samples tested in the present study showed little or no cytotoxic/genotoxic activity. On the contrary, the chemical industry effluent water samples had high genotoxic potency in both test systems. However, we should bear in mind that these effluents are prior to release into the environment treated in the municipal wastewater treatment plant.

We showed that one third of lake and one fifth of river samples increased genotoxic response in bacteria. The data obtained with the comet assay confirmed the positive result in lake and the negative result in river sample. Apart from domestic and sanitary wastes, urban runoffs, pharmaceutical and chemical compounds, natural aquatic environment can be polluted also with compounds such as organic UV filters and antimicrobial agents, which are used in personal care products and cosmetics. These substances can enter water indirectly through wastewater treatment plants or directly from recreational activities such as swimming and bathing in lakes and rivers. Once released in the aquatic environment they can be transformed by photooxidation to potentially more harmful compounds. Chemical analysis of lake and river water samples revealed the presence of compounds originating from personal care products (UV filters and antimicrobial agents). However, personal care products active ingredients represent only one group of possible contaminants in water samples. Further chemical characterisation of samples is needed before coming to conclusions.

With the SOS/*umu* assay we detected genotoxicity of one sample of potable water. Genotoxicity of potable water is often the consequence of the formation of disinfection by-products of which many are known to be genotoxic and also carcinogenic. In the case of chronic exposure of potentially genotoxic drinking water this can pose a serious health risk of unknown magnitude.

The combination of SOS/umuC test with prokaryotic organism and MTT and comet assay with mammalian cells appeared to be appropriate and sufficiently sensitive assays to monitor genotoxicity of natural surface and waste waters. This study again confirmed that bioassays can be an integral tool for identification of toxicity/genotoxicity of the complex mixtures and to monitor effluents and their products that are formed during degradation as well as to provide data for comparative risk assessment.

References

[1]   White, P.A. & Rasmussen, J.B. The genotoxic hazards of domestic wastes in surface waters. Mutation Research 410, 223–236, 1998.

[2]   Ohe, T., Watanabe, T. & Wakabayashi, K. Mutagens in surface waters: a review. Mutation Research 567, 109-149, 2004.

[3]   Kosjek, T, Heath, E. & Krbavčič, A. Determination of non-steroidal anti-inflammatory drug (NSAIDs) residues in water samples. Environment International, 31, 5, 679-685, 2005.

[4]   Reifferscheid, G., Heil, J., Oda, Y. & Zahn, R.K. A microplate version of the SOS/umu-test for rapid detection of genotoxins and genotoxic potentials of environmental samples. Mutation Research 253, 215-222, 1991.

[5]   ISO/CD 13829: Water quality- Determination of the genotoxicity of water and waste water using umu-test. 2000.

[6]   Mosmann, T. Rapid colorimetric assay for cellular growth and survival: application to proliferation and cytotoxicity assays. Journal of Immunological Methods, 65, 16, 55-63, 1983.

[7]   Singh, N.P., McCoy, M.T., Tice, R.R. & Schneider, E.L. A simple technique for quantitation of low levels of DNA damage in individual cells. Experimental Cell Research, 175, 1, 184-191, 1988.

[8]   Jolibois, B. & Guerbet, M. Hospital wastewater genotoxicity. Annals of Occupational Hygiene 50(2), 189-196, 2006.

Cytogenetical and histochemical studies on curcumin in male rats

A. El-Makawy[1] & H. A. Sharaf[2]
[1]Department of Cell Biology, National Research Center, Giza, Egypt
[2]Department of Pathology, National Research Center, Giza, Egypt

Abstract

Curcumin is a major component of the curcuma species, commonly used as a yellow coloring and flavoring agent in foods. Curcumin has been demonstrated to have potent antioxidant, anti-inflammatory activity and has shown anticancer properties in many rodent models. There is the perception that since compounds are natural, they are devoid of toxicity and safe to use. Some of the active compounds in supplements have inherent toxicity. In this study, five doses (0.5, 5, 10, 25 and 50 mg/kg/daily) of curcumin spice were orally administered to male rats daily for four weeks. The effect of curcumin was studied genetically by evaluation of chromosomal aberrations and micronucleus formation in bone marrow cells of male rats. Histopathological and histochemical investigations were studied in different tissues (liver, kidney) of males. The cytogenetical results showed that curcumin caused a statistically significant dose-dependent increase in the number of micronucleated polychromatic erythrocytes (MNPCEs) and in the frequencies of total chromosomal aberrations over the control. Also, results showed that there were significant differences between positive control and all curcumin-tested doses, except the dose of 50 mg/kg showed no significant difference. Histopathological results showed different degrees of alterations, as manifested by vacuolar degeneration in hepatocytes, tubular degeneration in renal tissues. We conclude that over use of curcumin spice may cause genotoxical and histopathological effects.
Keywords: curcumin spice, ADI, bone marrow, chromosome aberrations, micronucleus polychromatic erythrocytes (MNPCEs), liver, kidney.

1 Introduction

Tumeric is a spice that comes from the root Curcuma longa, a member of the ginger family, Zingaberaceae. Curcuminoids are components of tumeric, which include mainly curcumin (diferuloyl methane), demethoxycurcumin and bisdemethoxycurcmin [1]. Curcumin, bis(4-hydroxy-3-methoxyphenyl)-1,6-diene-3,5-dione, is a yellow-orange dye derived from the rhizome of the plant Curcuma longa. Curcumin is widely used as a spice and coloring agent in several foods, such as curry, mustard, bean cake, cassava paste and potato chips, as well as in cosmetics and drugs [2]. It exhibits chemopreventive and growth inhibitory activity of the initiation and promotion of many cancers [3, 4]. The dietary photochemical curcumin possesses anti-inflammatory, antioxidant, and cytostatic properties. Curcumin acts as a potent anticarcinogenic compound. Curcumin induces apoptotic cell death by DNA-damage and preventing cancerous cell growth [5, 6]. In spite of curcumin is a natural antioxidant known to possess therapeutic properties and has been reported to scavenge free radicals and to inhibit clastogenesis in mammalian cells [7]. Curcumin has been reported to induce a significant increase in the frequency of chromosomal aberrations in Chinese hamster ovary (CHO) cells [8]. So it was necessary to evaluate the genotoxic effects and histochemical changes induced in male rat cells as a result to oral administration of different doses of curcumin for four weeks.

2 Materials and methods

An experiment was carried out on male Wistar rats (Rattus norvegicus) weighing 150-200gm, receiving standard food and water ad libitum. Each experimental group consisted of 10 animals. The first group (negative control) received distilled water orally by gastric intubation. The second group (positive control) injected IP with single dose of 25mg/kg cyclophosphamide. Other five groups received curcumin spice (0.5, 5, 10, 25 and 50 mg/kg body weight). Curcumin was dissolved in distilled water and 1ml/animal of suspension given orally to male rat daily for four weeks. The dose 0.5 is the curcumin acceptance daily intake (ADI) and the other doses were chosen according to [9]. Twenty-four hours after the last administration, Animals were injected IP with colchicine solution Two hours later animals were killed by cervical dislocation. The femur bones were quickly separated one femur bone was used for preparation of micronucleus polychromatic erythrocytes according to method of [10]. While the other one was used for chromosomal preparation according to [11]. For histopathological and histochemical studies, paraffin sections of liver and kidneys were stained with Hx and E, Feulgen stain for DNA and periodic acid Schiff technique for mucopolysaccharide. DNA and PAS were evaluated as optical density values of their specific color using computer- assisted image analyzer. The analysis of variance test was used for the chromosomal aberrations data analysis. While the t- test was used to analyse the data of MNPCEs and histochemical studies.

Table 1:   The frequencies of MNPCEs in male albino rate bone marrow cells of all experimental groups.

	Total Counted PCEs/ animal	MNPCEs		Mean ± SD
		No.	%	
Negative control	2000	94	1.57	5.67 ± 2.16
Positive control	2000	604	10.07	100.67 ± 6.77 **
0.5 mg/kg bw curcumin	2000	91	1.52	NS 5.17 ± 1.47 ♦
5mg/kg bw curcumin	2000	142	2 .37	* 23.67 ± 4.59 ♦♦
10 mg/kg bw curcumin	2000	222	3 .70	** 37.00 ± 4.52 ♦♦
25 mg/kg bw curcumin	2000	381	6.35	** 63.50 ± 6.83 ♦♦
50 mg/kg bw curcumin	2000	587	9.78	** 97.83 ± 3.31 NS

*significant comparison to −ve control at ($p < 0.05$) **significant compared to −ve control at (0.01)
♦significant comparison to +ve control at ($p < 0.05$) ♦♦significant compared to +ve control at (0.01)

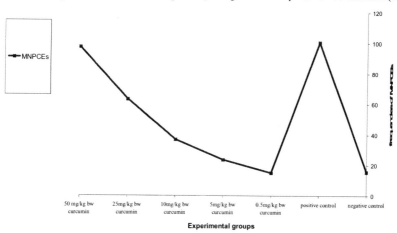

Figure 1:   Frequencies of MNPCUs in male albino rat bone marrow cells of all experimental groups.

3 Results

3.1 Cytogenetical results

Results of micronucleus test in bone marrow cells of male rats of all experimental groups are summarized in table 1 and presented in figure 1. Data indicated that curcumin exhibited a dose dependent increase in the frequency of MNPCEs when compared with negative control. Since the dose of 5mg/kg bw induced significant increase at ($p < 0.05$), whereas, the 10, 25 and 50 mg/kg bw doses of curcumin induced significantly increase in the frequencies of MNPCEs at ($p < 0.01$) than control. At the comparison to positive control, results showed that there were statistically significant differences between the frequencies of MNPCEs of curcumin treated animals with doses (0.5, 5 and 25 mg/kg bw) and positive control. In contrast, no statistically significant difference between the dose of 50mg/kg bw and the positive control was observed.

Table 2: Frequency of different chromosomal aberration male rat bone marrow cells of different experimental groups.

	Experimental groups						
	Negative control	Positive control	0.5mg/kg curcumin	5mg/kg curcumin	10mg/kg curcumin	25mg/kg curcumin	50mg/kg curcumin
Hypo-ploidy	0.17c ± 0.41	1.50b ± 1.05	0.50c ± 0.55	0.83c ± 0.75	1.50b ± 0.55	2.50a ± 0.55	3.00a ± 0.00
Hyper-ploidy	0.00d ± 0.00	0.67bc ± 0.88	0.00d ± 0.00	0.00d ± 0,00	0.50cd ± 0.55	1.17ab ± 0.41	1.33a ± 0.52
Polyploidy	0.00b ± 0.00	1.83a ± 0.75	0.00b ± 0.00	0.00b ± 0.00	0.00b ± 0.00	0.17b ± 0.41	1.67a ± 0.75
Chromatid gap	0.67d ± 0.82	4.83a ± 1.47	1.00d ± 0.63	2.17c ± 0.41	2.67bc ± 0.52	3.33b ± 0.82	5.17a ± 0.52
Chromatid break	0.33b ± 0.52	4.83a ± 1.03	0.17b ± 0.41	0.83b ± 0.41	1.50b ± 0.55	2.00a ± 0.89	2.67a ± 0.52
Chromosome break	0.00c ± 0.00	1.00a ± 0.89	0.00e ± 0.00	0.0de ± 0.00	0.17d ± 0.41	0.83bc ± 0.41	1.33b ± 0.75
Fragment	0.00d ± 0.00	3.00a ± 1.16	0.33cd ± 0.52	0.83bcd ± 0.75	1.17cd ± 0.41	1.67b ± 1.05	2.83a ± 0.75
Deletions	0.00b ± 0.00	0.50a ± 0.55	0.00b ± 0.00	0.00b ± 0.00	0.00b ± 0.00	0.33ab ± 0.52	0.67a ± 0.52
Centromeric attenuation	3.33bc ± 1.03	5.17a ± 1.47	2.33c ± 0.82	4.67ab ± 1.37	5.83a ± 1.49	5.17a ± 1.17	5.17a ± 1.17
Total chromosomal aberrations	4.50e ± 1.38	22.83a ± 2.71	4.33e ± 0.82	9.17d ± 0.75	13.33c ± 1.03	17.17b ± 0.75	23.83a ± 1.60

(Means with different letters within each column are significant at 5% level.)

Table 2 and figure 2 illustrate the mean values of structural and numerical chromosomal aberrations in bone marrow cells of all experimental groups. Results showed that curcumin accepted daily intake (ADI) dose not induced significant difference as compared to negative control. In bone marrow cells of animals administered with the other four doses, a dose dependant increase in the frequencies of all individuals and total chromosomal aberrations were observed. At the comparison between the frequencies of all individuals and total chromosomal aberrations induced in positive control and those of different curcumin doses treated animals, results showed that there were significant differences ($p<0.01$) between positive control and groups of 0.5, 5, 10, and 25 mg/kg curcumin. Whereas, there was no significant difference between positive control and the group of 50 mg/kg curcumin treated animals. These results indicated that the high dose of curcumin (50 mg/kg) is the more effective to induce significant increase in the frequencies of MNPCEs and chromosomal aberrations in rat bone marrow cells.

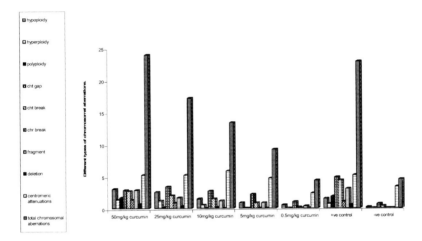

Figure 2:   Frequencies of different chromosomal aberrations induced in all experimental groups.

3.2 Histological and histochemical results

The microscopic observation in control and 0.5 mg/kg curcumin treated rats showed normal structure of hepatic lobules in which the hepatocytes arranged in cords radiating from the central vein. Liver sections of curcumin treated rats revealed cytoplasmic degeneration, necrosis, cytoplasmic vacuolar damage and dilated and congested with few inflammatory cells in the portal tract. These observations were highly pronounced in rats treated with 5,10 and 25 mg/kg curcumin and were prominent in rat treated with 50mg/kg curcumin (fig 3). As regard to kidney tissues, histological observation of sections of animals receiving 0.5, 5, and 10 mg/kg of curcumin showed that almost all the renal tubules were normal without histopathological changes. While, the kidney tissues of animals received the dose of 25 mg/kg of curcumin showed some picture of damage in tubular epithelial cells, thickening in blood vessels, mild cellular infiltration and interstitial hemorrhage. While, the renal tubules of animals received the dose of 50 mg/kg of curcumin revealed advanced tubular degeneration, increased in cellular infiltration, interstitial hemorrhage and some glomeruli were damage (fig 4). Histochemical evaluation to DNA and PAS were measured as mean value per nucleus by image analyzer. DNA was demonstrated by using Feulgen reaction technique. The Results showed that curcumin different doses induced significant decrease (p<0.01) in the mean value of DNA content of liver tissues when compared with control (table 3 and fig 5). In kidney cells, also results showed that the curcumin induced significant decrease (p<0.01) in the mean value of DNA content than in control (table 4 and fig 6). As regard to PAS + ve material, curcumin caused significant decrease (p<0.01) in the level of PAS in both liver and kidney tissues of all groups as compared to control as shown in (tables 5, 6 and fig 7, 8). while in renal tubule, all treated groups revealed moderate reaction in the basement membrane and brush border.

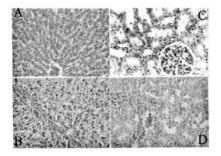 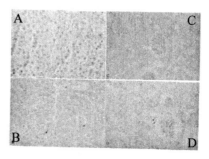

Figure 3:   Section of control and   Figure 4:   Section of control and
             treated liver (A and B)   treated liver (A and B)
             and kidneys (C and D)   and kidneys (C and D)
             H and E X300.   Feulgen reaction X300.

Table 3:   Mean values of DNA content in liver tissues of all treated groups.

Groups	Control	0.5mg/kg	5mg/kg	10mg/kg	25mg/kg	50mg/kg
DNA Content	0.35 ± 2.31E02	0.25 ± **3.01E-02**	0.22 ± 2.20E-02	0.20 ± 2.58E-02	0.18 ± **0.19E-02**	0.18 ± 1.49E-02

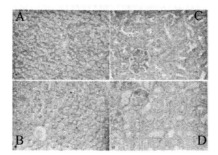

Figure 5:   Section of control and treated liver (A and B) and kidneys (C and D) PAS X300.

Table 4:   Mean values of DNA content in kidney tissues of all treated groups.

Groups	Control	0.5mg/kg	5mg/kg	10mg/kg	25mg/kg	50mg/kg
DNA Content	0.30 ± 2.87E-02	0.25 ± 3.71E-02	0.25 ± 3.25E-02	0.24 ± 2.75E-02	0.22 ± 1.87E-02	0.22 ± 2.46E-02

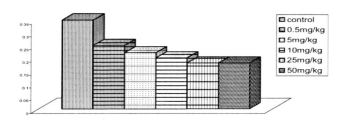

Figure 6:   Mean values of DNA content in hepotocytes of all treated groups.

Table 5:   Mean values of PAS content in hepatic cells of all treated groups.

Groups	Control	0.5mg/kg	5mg/kg	10mg/kg	25mg/kg	50mg/kg
DNA Content	0.42 ± 4.89E-02	0.18 ± 2.11E-02	0.13 ± 5.18E-02	0.27 ± 6.54E-02	0.13 ± 6.79E-02	0.28 ± 1.61E-02

Table 6:   Mean values of PAS content in renal cells of all treated groups.

Groups	Control	0.5mg/kg	5mg/kg	10mg/kg	25mg/kg	50mg/kg
DNA Content	0.37 ± 2.20E-02	0.31 ± 1.99E-02	0.25 ± 1.74E-02	0.28 ± 1.75E-02	0.28 ± 1.56E-02	0.28 ± 1.66E-02

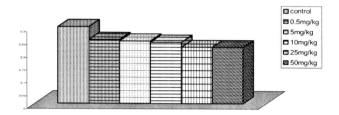

Figure 7:   Mean values of DNA content in kidney cells of all treated groups.

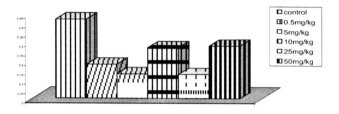

Figure 8:   Mean values of PAS content in hepatic cells of all groups.

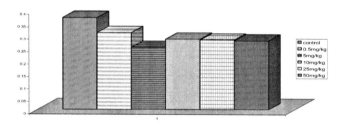

Figure 9:   Mean values of DNA control in hepotocytes of all treated groups.

4   Discussion

Curcumin has been used as a herbal medicine. Curcumin shows a variety of physiological and pharmacological effects. Several studies indicate curcumin to be anticarcinogenic [12] and anti-inflammatory [13] Curcumin further shows antioxidant properties, it acts as a superoxide radical scavenger [7, 14] and as a singlet oxygen guencher [15]. Contrary to the antioxidant nature of curcuminoids, much evidence for cytotoxic properties of curcumin was reported, and its cytotoxicity is suggested to be due to production of reactive oxygen species and causes oxidative DNA damage [16]. The cytogenetic results of the present study demonstrate that the curcumin administration induced a dose dependent significant increase in the frequencies of MNPCEs and total chromosomal aberrations in bone marrow cells of male rats. Histochemical studies confirmed these results. Since, the histochemical investigation of the present study showed that curcumin caused significant decrease in DNA content in both liver and kidney tissues of treated animals. Many studies have found that the antitumor- promoting effects of curcumin, could be contributed to its apoptosis-inducing activity. Since, curcumin induces a growth arrest in the G2/M–phase of the cell cycle, it exerts profound effects on mitotic spindle organization and leads to formation of monopolar spindles that are unable to segregate chromosomes normally. Cells with monopolar spindles arrest in M-phase for extended periods, and eventually leave mitosis and enter interphase with grossly aberrant nuclei consisting of numerous large micronuclei.

The production of cells with extensive micronucleation are due to its ability to disrupt normal mitosis, and raises the possibility that curcumin may promote genetic instability under some circumstances [17]. Histochemical examination of PAS +ve material indicated that curcumin caused significant decrease of PAS content in both liver and kidney tissues of all groups. The decrease in PAS +ve material observed in this study was interpreted due to the most probably consequent to the degenerative changes [18]. Previous literatures have demonstrated that antioxidants can act as prooxidants under some circumstances [19]. It is found that b-carotene [20], vitamin E [21], quercetin [22], N-acetylcysteine [23] , and caffeic acid [24] can act as potent DNA damaging agents. Also, a potential clastogenic effect of known antioxidant compounds has been reported by others. S-vanillin were enhanced the chromosome aberrations induced by alkylating agents in cultured Chinese hamster cells [25]. Bcarotene and ascorbic acid were enhanced the clastogenicity induced by BLM in CHO cells [26and27].Curcumin, like ascorbic acid, can become a pro-oxidant agent depending on the redox state of the biological environment [28]. Therefore, the mutagenic effects of curcumin found in the present work could be explained by the fact that curcumin would act as a pro-oxidant agent at the highest concentrations tested under the conditions of the present study. Clastogenic effects of curcumin were also observed by others. Curcumin showed a slight increase in the number of chromosomal aberrations in treated mice [29]. Curcumin acute and chronically treated induced a significant increase in SCE and a weak increase in the frequency of chromosomal aberrations in mice and rats [30]. The histological examination of the present study also showed that curcumin induced hepatotoxicity. Liver tissues of curcumin treated animals revealed highly pronounced cytoplasmic degeneration, necrosis and cytoplasmic vacuolar damage. In addition, the portal tract was dilated and congested with few inflammatory cells were present in animals treated with curcumin high dose. Other studies confirmed this result. It was reported that the administration of high curcumin dose for longer duration showed hepatotoxicity represented in focal necrosis or focal necrosis with regeneration both in mice and rats [31, 32]. Thus putative chemo-preventive antioxidants may have carcinogenic effect. Therefore, much consideration to safety should be required when curcumin is used for nutrition supplement.

References

[1]   Chainani-Wu N. (2003): Safety and anti-inflammatory activity of curcumin: a component of tumeric (Curcuma longa). J Altern Complement Med., 9(1):161-8. 2-

[2]   Lin J.K. & Shiau S.Y. (2001): Mechanism of cancer chemoprevention by curcumin. Proc. Natl. Sci. Counc.25 (2): 59-66.

[3]   Blasiak J., Trzeciak A. & Kowalik J. (1999a): Curcumin damages DNA in human gastric mucosa cells and lymphocytes. J Environ Pathol Toxicol Oncol. 18(4):271-6.

[4]   Deeb D., Xu Y.X. Jiang H., Gao X,. Janakiraman N., Chapman R.A. & Gautam S.C. (2003): Curcumin (diferuloyl-methane) enhances tumor necrosis factor-related apoptosis-inducing ligand-induced apoptosis in LNCaP prostate cancer cells. Mol Cancer Ther. 2 (1):95-103.

[5]   Chen, H. W. & Huang, H. C.(1998): Effect of curcumin on cell cycle progression and apoptosis in vascular smooth cells. Br. J. Pharmacol., 124, 1029–1040.

[6]   Martin-Cordero C., Lopez-Lazaro M., Galvez M. & Ayuso M. J.(2003): Curcumin as a DNA topoisomerase II poison. J. Enzyme Inhib. Med.Chem.18:505-509.

[7]   Ruby, A.J., Kuttan, G., Babu, K.D., Rajasekharan, K.N. and Kuttsn, R.(1995): Antitumor and antioxidant activity of natural curcuminoids. Cancer Lett. 94: 79-83.

[8]   Araujo MC, Antunes L.M., Takahashi CS. (2001): Protective effect of thiourea, a hydroxyl-radical scavenger, on curcumin-induced chromosomal aberrations in an in vitro mammalian cell system. Teratog Carcinog Mutagen. 21(2):175-80.

[9]   Choudhary D, Chandra D, Kale RK. (1999): Modulation of radioresponse of glyoxalase system by curcumin. J Ethnopharmacol. 64(1): 1-7.

[10] Salamone M.F., Heddle J.A. and Katz M. (1980) : The mutagenic activity of 41 compounds on the in vivo micronucleus assay, In : J Ashby and Fde Serres (F. DS.), proceedings of International program for THE Evaluation of Short- TerM Tests for Carcinogenicity. Elsever. Amsterdam.

[11] Yosida, T.H. and Amano K. (1965): Autosomal polymorphic in laboratory bred and Wild Norway rats -Rattus norvegicus, found in Misima.Chromosoma,16:658-66.

[12] Conney, A.H., Lysz, T., Ferraro, T., Abidi, T.F., Manchand, P.S., Laskin, J.D. and Huang, M.T. (1991): Inhibitory effect of curcumin and some related dietary compounds on tumor promotion and arachidonic acid metabolism in mouse skin. Adv. Enzyme Regul.31 385-396.

[13] Huang M.T., Lysz, T., Ferraro, T., Abidi, T.F., Laskin, J.D. and Conney, A.H (1991): Inhibitory effects of curcumin on in vitro lipoxygenase and cytooxygenase activities in mouse epidermis. Cancer Res. 51: 813-819

[14] Reddy, A.C. and Lokesh, B.R. (1994): Studies on the inhibitory effects of curcumin and gingerol on the formation of reactive oxygen species and oxidation of ferrous iron. Moll. Cell. Biochem. 137: 1-8

[15] Das, K.C. and Das, C.K.(2002) : Curcumin (diferuloylmethane), a singlet oxygen (o2) quencher. Biochem. Biophys. Res. Commun. 295:62-66.

[16] Sakano K. and Kawanishi S. (2002): Metal- mediated DNA damage induced by curcumin in the presence of human cytochrome P450 isozymes. Arch. Biochem. Biophys. 405(2) :223-230.

[17] Holy, J.M. (2002): Curcumin disrupts mitotic spindle structure and induces micronucleation in MCF-7 breast cancer cells., Mut. Res. 518(1): 71-84.

[18] Khadiga I.M.M.& Shadia, M. M. (1995) : Structural changes in rat induced by dietary addition of monosodium glutamate. Egypt. J. Histol.(2): 423-430.

[19] Yoshino, M., Haneda, M., Naruse, M., Htay, H.H., Tsubouchi, R., Qiao, S.L., Li, W.H., Murakami, K. and Yokochi, T. (2004): Prooxidant activity of curcumin: copper-dependent formation of 8-hydroxy-2-deoxyguanosine in DNA and induction of apoptotic cell death. Toxicology in vitro,18(6): 783-789.

[20] Murata M. and Kawanishi S. (2000): oxidative DNA damage by vitamin A and Its Derivative via Superoxide Generation. J Biol Chem,. 275(3,): 2003-2008

[21] Yamashita N, Murata M, Inoue S, Burkitt MJ, Milne L, Kawanishi S.(1998): Alpha-tocopherol induces oxidative damage to DNA in the presence of copper(II) ions. Chem Res Toxicol. 11(8):855-62.

[22] Yamashita N. and Kawanishi S.(2000) : Distinct mechanisms of DNA damage in apoptosis induced by quercetin and luteolin. Free Radic Res. 33(5):623-33.

[23] Oikawa S., Yamada K., Yamashita N., Tada-Oikawa S. and Kawanishi S. (1999): N-acetylcysteine, a cancer chemopreventive agent, causes oxidative damage to cellular and isolated DNA Carcinogenesis, 20(8): 1485-1490.

[24] Inoue S., Ito K, Yamamoto K. and Kawanishi S. (1992): Caffeic acid causes metal-dependent damage to cellular and isolated DNA through H2O2 formation Carcinogenesis, (13): 1497-1502.

[25] Matsumura, H., Watanabe, K. and Ohta, T. (1993). s-Vanillin enhances chromosome aberrations induced by alkylating agents in cultured Chinese hamster cells. Mutat. Res. 298: 163-168.

[26] Salvadori, D.M.F., Ribeiro, L.R. and Natarajan, A.T. (1994). Effect of b-carotene on clastogenic effects of mitomycin C, methyl methanesulphonate and bleomycin in Chinese hamster ovary cells. Mutagenesis 9: 53-57.

[27] Cozzi, R., Ricordy, R., Agliti, T., Gatta, V., Perticone, P. and De Salvia, R. (1997). Ascorbic acid and b-carotene as modulators of oxidative damage. Carcinogenesis 18: 223-228.

[28] Sahu, S.C. and Washington, M.C. (1992). Effect of ascorbic acid and curcumin on quercetin-induced nuclear DNA damage, lipid peroxidation and protein degradation. Cancer Lett. 63: 237-241.

[29] Mukhopadhyay, M.J., Saha, A. and Mukherjee, A. (1998): Studies on the anticlastogenic effect of turmeric and curcumin on cyclophosphamide and mitomycin C in vivo. Food Chem. Toxicol. 36: 73-76.

[30] Giri, A.K., Das, S.K., Talukder, G. and Sharma, A. (1990). Sister-chromatid exchange and chromosome aberrations induced by curcumin and tartrazine on mammalian cells in vivo. Cytobios 62: 111-118.

[31] Deshpandea, S.S., Lalithab, V.S., Inglec, A.D, Rasted, A.S., Gadred, S.G. and Marua G.B. (1998): Subchronic oral toxicity of tumeric and ethanolic tumeric extract in female mice and rats. Toxicol. Lett. 95(3): 183-193.

[32] Kandarkar S.V., Sawant S.S., Ingle A.D., Deshpande S.S. and Maru G.B. (1998): Subchronic oral hepatotoxicity of tumeric in mice histopathological and ultrastructural studies. Indian J Exp. Biol. 36(7): 675-679.

Section 6
Bioaccumulation
of chemicals

Total mercury and methylmercury in fish from a tropical estuary

A. P. Baêta[1], H. A. Kehrig[2], O. Malm[2] & I. Moreira[1]
[1]*Chemistry Department,*
Pontifícia Universidade Católica do Rio de Janeiro, Brazil
[2]*Biophysics Institute Carlos Chagas Filho,*
Federal University of Rio de Janeiro, Brazil

Abstract

Mercury content, as methylmercury, in aquatic biota varies greatly among species from the same location. Many parameters may affect its accumulation and concentration in fish tissues. The study assessed total mercury (T-Hg) and methylmercury (MeHg) in the muscle, liver and gonad of *Micropogonias furnieri* - carnivorous fish, *Bagre* spp. - omnivorous fish and *Mugil liza* - iliophagous fish from a polluted eutrophic estuary in the Brazilian Southeast coast, Guanabara Bay. Fish were collected during the years 1990, 1999 and 2003. T-Hg was determined by CV-AAS with sodium borohydride as a reducing agent. MeHg was identified and quantified in the toluene layer by GC-ECD. In all cases, the liver appears to be the preferential organ for mercury accumulation. T-Hg in muscle was higher and more variable in carnivorous than in omnivorous and iliophagous fishes. Carnivorous and omnivorous fishes presented a similar percentage of MeHg (99% and 97%) in the muscle. Iliophagous fish, which is at the lower level of this food chain, presented the lowest % MeHg in the muscle and liver, 52% and 9% respectively. However, the percentage of MeHg to T-Hg was around 25% in the liver of carnivorous and omnivorous fishes. In all cases, the gonad presented the lowest T-Hg, and the ratio of MeHg to T-Hg was around 1. In the year 1999, the samples of carnivorous and iliophagous fishes presented the highest T-Hg in the muscle. In this year, the fish specimens of both species showed sexual maturity and the highest total length. The sex of the specimens did not show any influence in the accumulation of mercury by the fish. The total length of the fish specimens presented a significant relationship with T-Hg in the muscle. The different feeding habits of the studied species are important for the accumulation of mercury and methylmercury by the organisms. Mercury, as methylmercury, is biomagnified through this food chain.
Keywords: tropical fish species, different feeding habits, biological parameters, methylmercury, total mercury, Brazilian estuary.

 WIT Transactions on Biomedicine and Health, Vol 10, © 2006 WIT Press
www.witpress.com, ISSN 1743-3525 (on-line)
doi:10.2495/ETOX060181

1   Introduction

Shallow estuarine and near-shore marine waters have become increasingly degraded over recent years. In spite of efforts to improve natural resource management, industrial plants continue to release toxic compounds into the environment, via liquid effluents and atmospheric emissions. Estuaries and coastal zones, particularly near high population density centres, are of special concern, as they receive the largest exposure to chemical contamination due to source proximity. Toxic compounds, such as mercury, can affect productivity, reproduction and survival of coastal and marine organisms, and can eventually be hazardous to human health (IPCS [1]).

The presence and behaviour of mercury in aquatic systems is of great interest and importance since it is the only heavy metal which bioaccumulates and biomagnifies through the aquatic food chain (Lindqvist et al [2]). Methylmercury, the most abundant organic form of mercury in the environment, has been recognised as a serious pollutant of aquatic ecosystems. However only limited information about the manner in which it spreads through the tropical estuarine and marine food chains is available. Methylmercury is largely responsible for the accumulation of mercury in organisms (bioaccumulation) and the transfer of mercury from one trophic level to another (biomagnification).

The trophic transfer of trace elements along marine food webs has been increasingly recognized as an important process influencing metal bioaccumulation and geochemical cycling (Fisher and Reinfelder [3]).

In the marine environment, almost all of the mercury in the muscle of fish is methylated (Joiris et al [4], Kehrig et al [5, 6]). However, the major part of mercury accumulated in the internal organs especially in the liver, exist as inorganic mercury, suggesting that demethylation of methylmercury is possible (Kehrig et al [6], Holsbeek et al [7]). The literature has proposed that the liver of the aquatic animals may act as an organ for mercury demethylation and/or the sequestration of both organic and inorganic forms of this element from the body (Endo et al [8]).

Mercury content in aquatic biota varies greatly among species from the same location. Many parameters may affect the accumulation and concentration of mercury in fish tissues. The concentrations of mercury and methylmercury accumulated by fishes are a function not only the water and sediment quality, but also of seasonal factors, temperature, salinity, diet, spawning and individual variation (Huchabee et al [9], Kehrig et al [10]).

Guanabara Bay (22°S, 43°W, 400 km^2), in Rio de Janeiro state, can be considered as one of the most important estuaries for fish production on the South-eastern Brazilian coast. The bay receives untreated domestic and industrial sewage from a densely populated area, and from the second largest industrialized region in the country, with around 10,000 industrial plants, two harbours, shipyards and oil terminals (FEEMA [11]). In some areas, the ecosystem is heavily impacted by organic matter, oil and heavy metals, including mercury. Consequently, elevated concentrations of toxic metals and hydrocarbons in sediments and changes in the pelagic and benthic communities can be detected

(FEEMA [11], Carreira *et al* [12]). An important point source of mercury for this estuary is a chlor-alkali plant located at the most polluted region of its watershed, on the North western side. The bay is among the most productive marine ecosystem in Rio de Janeiro State, presenting a high phytoplankton density and also high nutrients concentrations (C, N, P) that result in a high primary production waters, with an average net primary production (NPP) of 0.17 mol C m^{-2} day (Carreira *et al* [12], Rebello *et al* [13]). This estuary has been the object of numerous studies, but very few of them dealing with mercury and methylmercury in the aquatic biota (Kehrig *et al* [5, 6, 10], Costa *et al* [14]).

The present study assesses the accumulation of total mercury (THg) and methylmercury (MeHg) concentrations and the ratios of MeHg to THg in the muscle, liver and gonad of three species of fish, *Micropogonias furnieri* - carnivorous fish, *Bagre* spp. - omnivorous fish and *Mugil liza* - iliophagous fish from a polluted eutrophic estuary in the Brazilian Southeast coast, Guanabara Bay.

These fish species, which are present in high abundance and are widely distributed on the Southeast Brazilian coast, are those most frequently consumed by human populations. These organisms are characteristic of tropical areas in the Southern Atlantic.

2   Materials and methods

A total of 110 fish specimens with different feeding habits, *Micropogonias furnieri* – carnivorous fish (N=65), *Bagre* spp. – omnivorous fish (N=14), *Mugil liza* – iliophagous fish (N=31), were collected in different periods between 1990 and 2003 at Guanabara Bay. Fish samples were obtained from local fishermen who use a variety of fishing techniques. Following determination of the weight and total length of the specimens, a skinless cube of dorso-lateral muscle tissue was extracted. Cubes of tissue were stored in airtight plastic bags at below $-10^{\circ}C$ until analysis. Muscle, liver and gonad samples were analysed for total mercury (T-Hg) and methylmercury (MeHg) at Federal University of Rio de Janeiro laboratory.

For determination of total mercury, wet tissue samples were acid-digested and subjected to cold vapour atomic absorption spectrometry (FIMS system, Perkin-Elmer) with sodium borohydride as a reducing agent (Bastos *et al* [15]). For methylmercury, an analytical procedure developed at the National Institute for Minamata Disease (NIMD-Japan) laboratory and subsequently adapted at the UFRJ laboratory was used. The methylmercury analysis in the wet fish tissues was performed by digesting samples with an alcoholic potassium hydroxide solution followed by dithizone-toluene extraction and analysis by Gas Chromatography with Electron Capture Detector (GC-ECD) (Akagi and Nishimura [16]; Kehrig and Malm [17]). Precision and accuracy of the analytical methods were determined using certified standard materials from International Atomic Energy Agency (IAEA-350) and National Research Council Canada (DORM-2). Certified reference material IAEA-350 and DORM-2 was analysed in all sample batches. The overall reproducibility for the analysis period was determined from the results obtained using certified samples. On the basis of triplicate analyses, the coefficient of variation (a measure of random error in determination of mercury concentration) was less than 10%.

The results for total mercury DORM-2 (N=15) were 4.54 ± 0.13 μg.g^{-1}. The CRM has a certified T-Hg value of 4.64 ± 0.26 μg.g^{-1}. Our routine methylmercury results for the reference sample IAEA 350 (N=39) were 3.59 ± 0.38 μg.g^{-1}; the CRM MeHg value is 3.65 ± 0.35 μg.g^{-1}.

3 Results and discussion

In this study, the concentrations of total mercury (T-Hg) and methylmercury (MeHg) (on a fresh weight basis), and the percentage of methylmercury (% MeHg) in the tissues, muscle and liver, of the three fish species presented data that did not differ greatly to the measurements reported in the literature for fishes with different feeding habits from the South America coastal regions: Argentina (Marcovecchio [18]), Brazil (Niencheski *et al* [19], Sant'Anna *et al* [20], Kehrig *et al* [5, 6, 10], Pinho *et al* [21]), Suriname (Mol *et al* [22]) and Uruguay (Viana *et al* [23]).

The average total mercury concentration (T-Hg) and the percentage of methylmercury as T-Hg (% MeHg) in the tissues of the fish (muscle, liver, and gonad) from Guanabara Bay, and also the total length and weight of the fish specimens are summarised in table 1.

In all cases, T-Hg in the liver of the three fish species (*Micropogonias furnieri*, *Bagre* spp., *Mugil liza*) were higher than those found in their muscle tissue and gonad, presenting statistically significant difference (p<< 0.05). These high hepatic mercury concentrations are probably related to the role played by the liver in terms of pollutant bio-transformation (Frodello *et al* [24]).

The average ratio of liver T-Hg: muscle T-Hg was 2.4 (range 1.2-3.9, n=14) for the carnivorous, 4.2 (range 2.9-5.9, n=14) for the omnivorous and 94.5 (range 1.1-485.5, n=13) for iliophagous fishes.

Carnivorous and omnivorous fishes presented the average ratio of liver MeHg: muscle MeHg close to 1:1, 0.82 for carnivorous and 0.99 for omnivorous. However, the muscle and liver of the carnivorous and omnivorous fishes, both predators' species, presented similar MeHg concentrations (p>>0.15). Meanwhile, the concentrations of methylmercury in the muscle and liver of the iliophagous fish presented a significant difference (p=7 x 10^{-6}) and the average ratio liver MeHg: muscle MeHg was 26.9. Nevertheless, iliophagous fish showed lower % MeHg in the liver (9%) than carnivorous (26%) and omnivorous (22%) fishes, as shown in table 1.

In all cases, the gonad presented the lowest T-Hg, and the percentage of MeHg to T-Hg was around 100.

Micropogonias furnieri, a carnivorous benthic fish, which foods on bottom fauna mainly macro and microcrustacea (copepod), polychaeta, mollusc and fish (Vazzoller [25]), showed a higher capacity to accumulate total mercury and methylmercury than the other studied fish species with different feeding habits and lower trophic levels, not considering total length and weight differences.

Fish feeding on copepods ingested more methylmercury than inorganic mercury owing to the larger fraction of methylmercury found in the soft tissues of the copepods Lawson and Mason [26].

Carnivorous species, which are on top end of the aquatic food chain, are a good indicator of mercury in fish (Malm et al [27]). Of these, all analysed specimens presented the highest total mercury concentration in their muscle tissue, always showing the values below the maximum limit of 1.0 µg Hg.g^{-1} wet wt. established for human intake of predatory fish by Brazilian legislation (Brasil [28]).

Table 1: Average total mercury concentration (T-Hg) and the percentage of methylmercury as T-Hg (% MeHg) in the tissues of the fish (muscle, liver, and gonad) collected at Guanabara Bay.

Fish species	Total length (mm)	Weight (g)	Tissues	T-Hg (µg.g^{-1} wet wt.)	% MeHg
Micropogonias furnieri (carnivorous fish) – N=65	401	720	Muscle	0.14 ± 0.10	99
			Liver	0.21 ± 0.13	26
			Gonad	0.03 ± 0.02	100
Bagre spp. (omnivorous fish) – N=14	390	660	Muscle	0.07 ± 0.03	97
			Liver	0.29 ± 0.10	22
			Gonad	*	*
Mugil liza (iliophagous fish) – N=31	315	584	Muscle	0.01 ± 0.008	52
			Liver	0.12 ± 0.08	9
			Gonad	0.01 ± 0.007	100

* not analysed.

Total mercury concentrations in carnivorous fish muscle ranged from 0.043 to 0.27 µg.g^{-1} wet wt (N=30) in 1990 (Kehrig et al [10]), from 0.063 to 0.56 µg.g^{-1} wet wt. (N=20) in 1999 (Kehrig et al [5]) and from 0.040 to 0.28 µg.g^{-1} wet wt. (N= 15) in 2003 respectively. The average of the methylmercury to total mercury ratios was 99%, indicating that MeHg was the predominant form of Hg in carnivorous fish muscles (table 1). The methylmercury concentration at higher trophic levels reflects uptake at low trophic levels and other factors, such as diet and growth (Watras et al [29]).

Bagre spp., an omnivorous benthic fish, which feeds mainly on organic detritus, small fishes and invertebrates (Blaber [30]), showed lower T-Hg and MeHg in its muscle than carnivorous fish. The concentrations in *Bagre* spp. muscle (N=14) ranged from 0.046 to 0.18 µg T-Hg.g^{-1} and from 0.049 to 0.16 µg MeHg.g^{-1} and the average of MeHg was 97%.

In this study, *Mugil liza*, an iliophagous fish, which feeds mainly on benthic diatoms (Blaber [30]), showed the lowest concentrations of T-Hg and MeHg and also the lowest % MeHg in the muscle. Total mercury concentrations in iliophagous fish muscle ranged from 0.002 to 0.027 µg.g^{-1} wet wt (N=16) in

1999 (Kehrig *et al* [5]) and from 0.014 to 0.12 µg.g^{-1} wet wt. (N= 15) in 2003 respectively. The average of the percentage of MeHg was 52.

Highly significant differences were observed between the average of MeHg concentrations in the muscle of the carnivorous and omnivorous fishes (p=1 x 10^{-8}) and also between the ones in the same tissue of the carnivorous and iliophagous fishes (p=1 x 10^{-8}).

In the year 1999, the samples of carnivorous fish presented the highest T-Hg concentrations in the muscle, 0.17 ± 0.09 µg.g^{-1} wet wt. (fig. 1a). In this year, all carnivorous specimens showed sexual maturity and also the highest total length (350 mm - 577 mm). The sexual maturity in *Micropogonias furnieri* occurred at around 450 mm or 4 years old (Vazzoler *et al* [31]). However, the specimens of carnivorous fish colleted in the years 1990 and 2003 presented similar T-Hg in their muscle, 0.11 ± 0.046 µg.g^{-1} wet wt and 0.096 ± 0.082 µg.g^{-1} wet wt respectively (fig 1a). In theses years, the fish specimens were smaller and younger than the ones sampled in 1999, and also did not present sexual maturity. The total length of the specimens sampled in 1990 and in 2003 ranged from 190 mm to 525 mm and 340 mm to 430 mm respectively. A significant difference (p << 0.05) was observed between T-Hg in the muscle of the organisms sampled in the years 1999 and 2003.

Iliophagous fish presented higher T-Hg concentrations in the muscle tissue, 0.019 ± 0.004 µg THg.g^{-1} wet wt., in the samples collected in the year 1999 than in the ones from 2003, 0.004 ± 0.002 µg THg.g^{-1} wet wt (fig 1b). In the year 2003, all the specimens of iliophagous fish presented lower total length (290 mm – 355 mm) than the ones from the year 1999 (370 mm – 500 mm). The sexual maturity of *Mugil liza* occurred at around 400 mm.

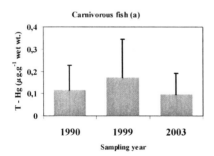
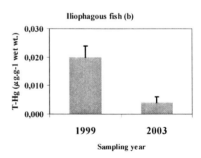

Figure 1: Average total mercury in the muscle tissue of the carnivorous fish (a) and iliophagous fish (b) collected at Guanabara Bay in different years.

In all sampling years, carnivorous fish always presented higher T-Hg in the muscle than iliophagous fish in the same tissue, due to they occupy different trophic levels in the food chain. Carnivorous fish are at the top level and the iliophagous fish at the bottom level of the Guanabara Bay food chain. So, it is important to note that regarding mercury dietary intake, which can influence mercury uptake in marine organisms; there is a marked difference between the concentrations of total mercury and methylmercury in fishes with different feeding habits, carnivorous, omnivorous and iliophagous, and also occupying different trophic levels in the food chain.

In this study, the amplification of the total mercury and methylmercury concentrations and also the percentage of methylmercury in the muscle tissue of the fishes probably are related to the increase of the trophic level position in the food chain. This could be indicating that biomagnification might be occurring throughout the Guanabara Bay food chain.

However, mercury in aquatic environments is not simply transferred from prey to predator tissues; it is accumulated by complex mechanisms. Methylmercury is largely responsible for the bioaccumulation of mercury in organisms and the transfer of mercury from one trophic level to another, causing biomagnification.

A variety of biotic and abiotic factors have been identified which can affect the efficiency that marine animals accumulate metals and metalloids in their tissues (Reinfelder *et al* [32]). Many parameters may affect the mercury accumulation: such as specimen size, sexual maturity, sensitivity to seasonal, feeding habit, trophic position, water quality and environmental contamination (Huchabee *et al* [9], Kehrig *et al* [10]).

A significant positive relationship was observed between T-Hg in the muscle and liver of carnivorous fish ($R=0.91$; $p \ll 0.05$). However, mercury content in the tissues, muscle and liver, of the omnivorous and iliophagous fishes did not present any relationship ($p \gg 0.15$).

A significant positive relationship was found between T-Hg concentrations in the muscle of carnivorous ($R = 0.56$; $p \ll 0.05$) and iliophagous ($R = 0.82$; $p \ll 0.05$) fishes and their total length (L). The concentration of mercury in the muscle of fish showed a linear increase with the total length, presenting the following relations as eqn (1) for carnivorous and eqn (2) for iliophagous.

$$[T - Hg] = 0.042L - 0.98 \tag{1}$$
$$[T - Hg] = 0.0064L - 0.17 \tag{2}$$

However, the T-Hg concentrations in the muscle of omnivorous fish did not show any relationship with its total length ($p \gg 0.05$).

No significant difference ($p \gg 0.05$) was observed between the average of T-Hg concentration in the female fish specimens and in males of all analysed fish species.

References

[1]   IPCS, Environmental Health Criteria 86, Mercury-Environmental Aspects, WHO: Geneva, p. 116, 1989.

[2]   Lindqvist, O., Johnasson, K., Aastrup, M., Andersson, A., Bringmark, L., Hovsenius, G., Håkanson, Å., Meili, M. & Timm, B., Mercury in the

Swedish environment-recent research on causes, consequences and corrective methods. Water, Air and Soil Pollution, **55,** pp. 1-251, 1991.

[3]   Fisher, N.S. & Reinfelder, J.R., The trophic transfer of metals in marine system. Metal speciation and bioavailability in aquatic systems, ed. A. Tessier & D.R. Turner, John Wiley & Sons: Chichester, pp. 363-406, 1995.

[4]   Joiris, C.R., Das, H.K. & Holsbeek, L., Mercury accumulation and speciation in marine fish from Bangladesh. Marine Pollution Bulletin, **40(5),** pp. 454-457, 2000.

[5]   Kehrig, H.A., Costa, M., Moreira, I. & Malm, O., Total mercury and methylmercury in a Brazilian estuary, Rio de Janeiro. Marine Pollution Bulletin, **44,** pp. 1018-1023, 2002.

[6]   Kehrig, H.A., Lailson-Brito Jr., J., Malm, O. & Moreira, I., Methyl and total mercury in the food chain of a tropical estuary-Brazil. RMZ-Materials and Geoenvironment, **51(1),** pp. 1099-1102, 2004.

[7]   Holsbeek, L., Siebert, U. & Joiris, C.R., Heavy metals in dolphins stranded on the French Atlantic coast. The Science of the Total Environment, **217,** pp. 241-249, 1998.

[8]   Endo, T., Haraguchi, K. & Sakata, M., Mercury and selenium concentrations in the internal organs of toothed whales and dolphins marketed for human consumption in Japan. The Science of the Total Environment, **300,** pp. 15-22, 2002.

[9]   Huchabee, J.W., Elwood, J.W. & Hildebrand, S.C., Accumulation of mercury in freshwater biota. The biogeochemistry of mercury in the environment, ed. J.O. Nriagu, Elsevier: Amsterdam, pp. 277-302, 1979.

[10] Kehrig, H.A., Malm, O. & Moreira, I., Mercury in a widely consumed fish Micropogonias furnieri (Demarest, 1823) from four main Brazilian estuaries. The Science of the Total Environment, **13,** pp. 263-271, 1998.

[11] FEEMA, Projeto de recuperação gradual do ecossistema da Baía de Guanabara: Rio de Janeiro, parts 1 and 2 pp.203, 1990.

[12] Carreira, R.S., Wagener, A.L.R., Readman, J.W., Fileman, T.W., Macko, S.A. & Veiga, A., Changes in the sedimentary organic carbon pool of a fertilized tropical estuary, Guanabara Bay, Brazil: an elemental, isotopic and molecular marker approach. Marine Chemistry, **79,** pp. 202-227, 2002.

[13] Rebello, A.L., Ponciano, C. & Melges, L.H.F., Primary production and availability of nutrients in Guanabara Bay. Anais da Academia Brasileira de Ciências, **60,** pp. 419-430, 1988.

[14] Costa, M., Paiva, E. & Moreira, I., Total mercury in Perna perna mussels from Guanabara Bay-ten years later. The Science of the Total Environment, **261,** pp. 69-73, 2000.

[15] Bastos, W.R., Malm, O., Pfeiffer, W.C. & Cleary, D., Establishment and analytical quality control of laboratories for Hg determination in biological and geological samples in the Amazon, Brazil. Ciência e Cultura, **50 (4),** pp. 255-260, 1998.

[16]   Akagi, H. & Nishimura, H., Speciation of mercury in the environment. Advances in mercury toxicology, ed. T. Suzuki, I. Nobumassa & T.W. Clarkson, Plenum Press: New York, pp. 53-76, 1991.

[17]   Kehrig, H.A. & Malm, O., Methylmercury in fish as a tool for understanding the Amazon mercury contamination. Applied Organometallic Chemistry, **13**, pp. 687-696, 1999.

[18]   Marcovecchio, J.E., The use of Micropogonias furnieri and Mugil liza as bioindicators of heavy metals pollution in La Plata river estuary, Argentina. The Science of the Total Environment, **323**, pp. 219-226, 2004.

[19]   Niencheski, L.F., Windom, L., Baraj, B., Wells, D. & Smith, R., Mercury in fish from Patos and Mirim Lagoons, Southern Brazil. Marine Pollution Bulletin, **42 (12)**, pp. 1403-1406, 2001.

[20]   Sant'Anna Jr, N.; Costa, M. & Akagi, H. Total and methylmercury levels of a coastal human population and of fish from the Brazilian Northeast. Environmental Science and Pollution Research, **8(4)**, pp. 280-284, 2001.

[21]   Pinho, A.P., Guimarães, J.R.D., Martins, A.S., Costa, P.A.S., Olavo, G. & Valentin, J., Total mercury in muscle tissue of five shark species from Brazilian offshore waters: effects of feeding habit, sex, and length. Environmental Research, **89**, pp. 250-258, 2002.

[22]   Mol, J.H., Ramlal, J.S., Lietar, C. & Verloo, M., Mercury contamination in freshwater, estuarine, and marine fishes in relation to small-scale gold mining in Suriname, South America. Environmental Research, **86**, pp. 183-197, 2001.

[23]   Viana, F., Huertas, R. & Danulat, E., Heavy metal levels in fish from coastal waters of Uruguay. Archives of Environmental Contamination and Toxicology, **48**, pp. 530-537, 2005.

[24]   Frodello, J.P., Roméo, M. & Viale, D., Distribution of mercury in the organs and tissues of five toothed-whale species of the Mediterranean. Environmental Pollution, **108**, pp. 447-452, 2000.

[25]   Vazzoler, G., Distribuição da fauna de peixes demersais e ecologia dos Sciaenidae da plataforma continental brasileira, entre as latitudes 29°21'S (Tôrres) e 33°41'S (Chuí). Boletim do Instituto de Oceanografia, S. Paulo, **24**, pp. 85-169, 1975.

[26]   Lawson, N.M., Mason, R.P., Accumulation of mercury in estuarine food chains. Biogeochemistry, **40**, pp. 235-247, 1998.

[27]   Malm, O., Branches, F.J.P., Akagi, H., Castro, M.B., Pfeiffer, W.C., Harada, M., Bastos, W.R. & Kato, H., Mercury and methylmercury in fish and human hair from the Tapajós river basin, Brazil. The Science of the Total Environment, **175(2)**, pp. 141-150, 1995.

[28]   Brasil, Agência Nacional de Vigilância Sanitária-Portaria n° 685 de 27/08/1998.

[29]   Watras, C.J., Back, R.C., Halvorsen, S., Hudson, R.J.M., Morrison, K.A. & Wente, S.P., Bioaccumulation of mercury in pelagic freshwater food webs. The Science of the Total Environment, **219**, 183-208, 1998.

[30]   Blaber, S.J.M., Fish and Fisheries in Tropical Estuaries. Fish and Fisheries Series, **22**, Chapman & Hall: London, 1997.

[31]   Vazzoler, A.E., Zaneti, E.M. & Kamakami, E., Estudo preliminar sobre o
       ciclo de vida dos Scianidae, Programa Rio Grande do Sul II, Brasil, pp.
       240-291, 1973.
[32]   Reindfelder, J.R., Fisher, N.S., Luoma, S.N., Nichols, J.W. & Wang,
       W.X., Trace element trophic transfer in aquatic organisms: a critique of
       the kinetic model approach. The Science of the Total Environment, **219**,
       pp. 117- 135, 1998.

Use of urinary porphyrin profiles as an early warning biomarker for Monomethylarsonous acid (MMA^{III}) exposure

M. Krishnamohan[1], S.-H. Huang[1], R. Maddalena[1], J.-P. Wang[1],
P. Lam[2], J. C. Ng[1] & M. R. Moore[1]
[1]*National Research Centre for Environmental Toxicology,
The University of Queensland, Brisbane, Australia*
[2]*City University of Hong Kong, Kowloon,
Hong Kong, People's Republic of China*

Abstract

Although it is well known that arsenic is toxic and that Arsenic is carcinogenic, the mechanism underlying this carcinogenesis is unknown. Our laboratories have established a model that produces multi-organ tumours in mice following extended exposure to arsenic in drinking water. Until recently the metabolism of arsenic was thought to be a detoxification process. Recent studies have shown that Monomethylarsonous acid (MMA^{III}) is the toxic intermediate of arsenic metabolism. It is a more potent cytotoxin and genotoxin than As^{III} and As^{V}, and is believed to be the proximal carcinogen. Exposure to arsenic is known to affect the activity of the enzymes of haem biosynthesis. We evaluated the use of urinary porphyrin profiles as an early warning biomarker for arsenic carcinogenicity. Young female mice were given drinking water containing arsenic as MMA^{III} *ad libitum*. 24h urine samples were collected at various time intervals for up to 48 weeks for urinary arsenic accumulation by HPLC-ICPMS and urinary porphyrin measurement by HPLC. Dimethylarsinic acid (DMA^{V}) was the major metabolite excreted and it showed significant dose-dependent increase at each time point and exposure dependent increase over 48 weeks. Porphyrin levels appeared to be age dependent. The results indicate that the urinary porphyrin concentration has the potential for use as an early warning indicator of chronic arsenic exposure prior to the onset of arsenic carcinogenesis.
Keywords: arsenic, porphyrins, haem synthesis, Monomethylarsonous acid, biomarkers, cancer.

WIT Transactions on Biomedicine and Health, Vol 10, © 2006 WIT Press
www.witpress.com, ISSN 1743-3525 (on-line)
doi:10.2495/ETOX060191

1 Introduction

Over one hundred million people globally are risk of developing cancer or other arsenic-related disease as a result of exposure to arsenic in drinking water or in food. The commonest form of drinking water exposure occurs when groundwaters become acidic, from acid sulphate soils, and dissolve environmental arsenic. Such exposures commonly occur in countries such as Bangladesh China and Taiwan. Studies show arsenic exposure causes non carcinogenic effects including black foot disease (NRC [1]), diabetes (Rahman et al. [2]) and ischemic heart disease (Tseng et al. [3]). Epidemiological evidence shows that arsenic exposure produces tumors of the skin, lung, and urinary bladder (ATSDR [4]; IPCS [5]; IARC [6]). Although arsenic is toxic to both humans and animals, and arsenic is carcinogenic, the mechanism underlying this is unknown.

Arsenic undergoes stepwise reduction of pentavalent arsenic to trivalent arsenic followed by oxidative addition of a methyl group to the trivalent arsenic (Cullen et al. [7]). Until few years back the methylation of inorganic arsenic was considered to be a detoxification mechanism (Vahter and Marafante [8]). Methylated arsenicals were thought to be less genotoxic than the inorganic arsenic (Moore et al. [9]). From recent studies by Petrick et al. [10] showing that MMA^{III} is more toxic than As^{III} and As^{V}, the hypothesis of methylation as a detoxification pathway has been questioned by Cullen and Dodd [11] who first demonstrated the toxicity of MMA^{III} in the yeast *Candida humicola*. MMA^{III} was more cytotoxic than inorganic arsenite in human hepatocytes (Petrick [10]). In a study by Mass et al [12] both MMA^{III} and DMA^{III} were 77 and 386 times more potent genotoxins than iAS^{III}, respectively.

Arsenic is also known to affect the activity of about 200 enzymes (Li and Rossman [13]), and some of these may have potential for use as biomarkers of effects. One of the important effect is that arsenic interferes with the function of the group of enzymes responsible for haem biosynthesis (Moore et al. [14]) including inhibition of coproporphyrinogen oxidase and haem synthase (Woods and Fowler [15]). Several studies have reported that exposure to arsenic affects the haem biosynthesis pathway in humans and animals (Woods and Fowler [15]; Ng et al. [16]). There were increases in protoporphyrin IX, coproporphyrin III, coproporphyrin I in the blood, liver, kidney and in the urine of rats 24-48hr after a single dose of arsenic (Wang et al. [17]). Urinary porphyrins were higher in the people of coal-borne arsenicosis endemic area in southwest of PR China. Greater increases in the urinary arsenic and porphyrins were found in women, children, and older people. The reason could be that they tend to spend more time indoors than the males (Wang et al. [18]). The synthesis of harderoporphyrin and the alteration of the porphyrin profile in the harderian glands of rodents is a highly sensitive biomarker for both single sub-lethal and chronic arsenic exposure (Ng et al. [19]). The aim of this study was to investigate arsenic metabolism and the alteration of urinary porphyrin profiles in mice chronically exposed to low doses of MMA^{III} and see whether these profiles could be used as biomarkers prior to the onset of carcinogenesis.

2   Materials and methods

2.1 Reagents

MMA[III] was purchased from Bill Collin, British Columbia University, Canada. Mixed arsenic standards for HPLC-ICP-MS analysis including arsenobetaine (AsB), sodium arsenite (As[III]), dimethylarsinic acid (DMA[V]), disodium methyl arsenate (MMA[V]) and sodium arsenate (As[V]) were obtained from TRI chemicals Lab Inc. (USA), BDH chemicals (UK), Sigma (USA), Chem Service (USA) and Ajax Chem. Ltd. (Australia), respectively.

Porphyrin acid chromatographic marker kit (a mixture of standards) consisting of uroporphyrin III, 7-carboxyporphyrin, 6-carboxyporphyrin, 5-carboxyporphyrin and 4- carboxyporphyrin (coproporphyrin I), protoporphyrin IX, coproporphyrin III dihydrochloride, uroporphyrin I dihydrochloride and the internal standard meso-porphyrin IX dihydrochloride were purchased from Porphyrin Product (Logan, Utah, USA). All.

2.2 Animals and treatment

Animal experimental protocols were approved by the Institutional Animal Ethics Committee (AEC No. NRC 3/02/19). Female C57Bl/6J mice, aged 4-weeks-old, were divided into groups of 70, 5 mice per cage, and were given drinking water containing 100μg, 250μg or 500μgAs[III]/L as monomethylarsonous acid (MMA[III]) ad libitum for 12 months. A group of 105 control mice was given demineralised water containing <0.1μg As/L. The 50ppm MMA[III] stock solution was prepared and kept at 4°C for a month and the working solution of 100, 250 and 500 ppb was prepared every 2 days. The animal care facility was operated at controlled temperature set at 21–23°C, 13 filtered air changes per hour, 12/12 h light/dark cycle and year round relative humidity of about 60%. All animals were kept in standard polypropylene cages with stainless steel wire-mesh tops equipped with polycarbonate plastic drinking bottles with stainless steel sip-tubes and given a commercial rodent diet ad libitum (Norco P/L, Brisbane, Australia).

2.3 Urinary arsenic speciation by HPLC-ICP-MS

The arsenic concentrations of the drinking water and feed were monitored by HPLC–ICP–MS. The volumes of the drinking water consumed by the mice in each cage and body weight of each mouse were measured weekly. Six cages of mice out of 14 cages from every group were selected and each cage of five animals was kept in a metabolic cage for 24 h pooled urine collection on time zero, week 1, 2, 4, 8 and every 8 weeks after that till 48 weeks. During urine collection, mice were kept in metabolic cages and given powdered rodent diet and water containing MMA[III] at the original exposure level. Urine was collected on dry ice to preserve the urinary arsenicals and porphyrins in their excreted forms. Urine samples were stored away from light and at -80°C until analysis.

2.4 HPLC analysis of porphyrins in urine

Porphyrins were analyzed by HPLC with fluorescence detection (Wang et al. [17]). Briefly, an aliquot of urine sample (750 µL) with the addition of 10M hydrochloric acid (50µL) in a glass test tube was vortexed thoroughly and allowed to stand in the dark for 1/2 h. After centrifugation at 13,200 rpm in a bench top microfuge for 10 min, the supernatant (200µL) was diluted with 1M hydrochloric acid (800 µL) containing mesoporphyrin (73.32 nM) as the internal standard.

This solution was filtered using a 0.45 µm pore size syringe filter (Nalgene, USA), and the porphyrins were analyzed by injecting an aliquot (40 µL) onto the C18 reverse phase HPLC column (Radialpak, Novapak C18, 8mm × 100 mm, 4µm, and Novapak guard column, Waters Associates, USA) coupled to a Shimadzu HPLC system (LC-10A, Shimadzu, Japan); a fluorescence detector (LC-240, Perkin–Elmer, USA) set at 395 nm excitation wavelength and 615 nm emission wavelength for detection. The concentration of each porphyrin was calculated from a standard curve using the peak height mode.

2.5 Statistics

Statistical analysis was performed using SPSS 11.0 software. Differences in the level of urinary arsenic species and porphyrins were analyzed by ANOVA and post analysis LSD.

3 Results

No tumors were observed in any of the animals in the control or treatment groups as expected. Previous study from our lab shows the earliest solid tumors observed in the C57BL6/J mice exposed to 500µg As/L in drinking water were after about 18 months of exposure. No abnormal appearance or behaviour was observed.

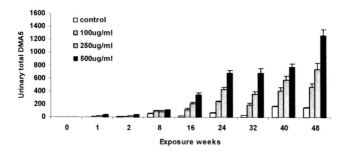

Figure 1:   Average 24h pooled total urinary arsenic excretion (ng/24h) for the control and treatment groups. Vertical lines represent standard errors (N=5).

3.1 Urinary arsenic methylation profile

Twenty four hour urinary excretion pattern of DMA^V is shown in Fig.1. Limits of detection (LOD) for AsB, As^{III}, DMA^V, MMA^V and As^V were 0.5, 0.6, 0.6, 0.8 and 0.8 µg/L, respectively. AsB and MMA^V were below the LOD in all urine samples. There is a positive correlation between the urinary total arsenic and MMA^{III} concentrations in the drinking water. In addition, levels of total arsenic were significantly dose related within each age group.

3.2 Urinary Porphyrin profiles

Uroporphyrin showed dose-dependent increase after 4 and 16 weeks of exposure to MMA^{III}. There is a significant difference in all the test groups compared to the control after 8 weeks of exposure. The level of uroporphyrin in control group is significantly different from 250 and 500µg/L after 32 weeks and control group and 100µg/L group are significantly different from 500µg/L after 40 weeks of exposure.

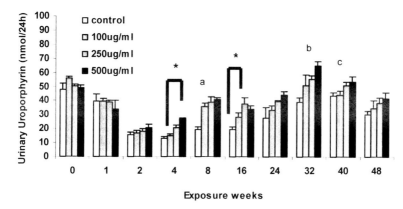

Figure 2:   Average 24h pooled total urinary uroporphyrin excretion (nmol/24h) for the control and treatment groups. Vertical lines represent standard errors. * indicates a significant dose-response relationship ($p < 0.01$). "a" - indicates a significant difference in the uroporphyrin level between control and test groups ($p < 0.001$) "b" – indicates the level of uroporphyrin of control group is significantly different from 250 and 500 µg/L ($p < 0.02$). "c" – indicates level of uroporphyrin of control and 100 µg/L groups are significantly different from 500 µg/L group ($p < 0.05$) (n=5).

Urinary coproporphyrin-III concentrations also increased in most of the treatment groups compared to the control groups except time zero and week 1. The level of coproporphyrin III in the control group is significantly different from all the test groups after 8, 24, 32 and 40 weeks of exposure. Control, 100, and 500µg/L groups showed dose-response relationship after 24 months of exposure.

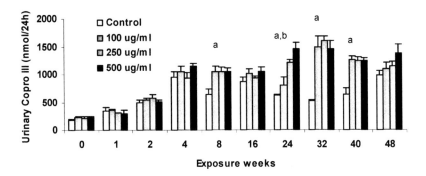

Figure 3: Average 24h pooled total urinary coproporphyrin III excretion (nmol/24h) for the control and treatment groups. Vertical lines represent standard errors. "a" - indicates the level of copro III in control group is significantly different from all the test groups (p<0.01). "b" - Control, 100 and 500 μg/L shows significant dose response relationship (p<0.05).

4 Discussion

Urinary levels of arsenic species can be used as reliable biomarkers of exposure to inorganic arsenic rather than total urinary arsenic, which often contains high levels of organo-arsenic compounds derived from dietary intake of seafood. Average arsenic from three representative samples of mouse food was found to be 385μg/kg, of which the inorganic As^{III} and As^{V} contents were 18.5% and 81.5%, respectively.

In animals a major interspecies difference in the arsenic methylation rate has been observed. Liver cystosol studies from marmoset and tamarin monkeys (Zakharyan et al. [20]) and guinea pigs (Healy et al. [21]) shows that these species are deficient in methyltransferase activity compared to other species like rabbit. Studies by Healy et al [22] using cystosol prepared from liver, lung, kidney and testes of male B6C3F$_1$ mice, shows all the tissues had the capacity to methylate arsenite. These results suggest that although the major part of arsenite methylation is happening in the liver, extrahepatic methylation may also be significant.

C57BL6/J mice exposed to 100, 250 and 500 μg arsenic as sodium arsenate showed DMA^{V} as a major urinary metabolite (Wu et al. [23]). In our experiment though the animals are exposed to MMA^{III} the major metabolite excreted in the urine was DMA^{V}. Studies have found monomethylarsonous acid (MMA^{III}) and dimethylarsenous acid (DMA^{III}) in human urine (Aposhian et al. [24]). Our result shows a dose-response relationship between the MMA^{III} concentrations in the drinking water and urinary DMA^{V} concentration. It also shows exposure dependent increase in the urinary excretion of DMA^{V}.

Since age is a confounding factor for the urinary porphyrin profile, it is important to use animals of similar age when conducting biomarker study. In our study urinary uroporphyrin showed dose-response relationship after 4 and 16 weeks of exposure period and a significant difference between control and test groups after 8, 32 and 40 weeks. This result agrees with the human study in Xing Ren of Guizhou Province of pr China where individuals chronically exposed to arsenic from burning arsenic contaminated coal for heating, cooking and drying food purpose had a significant increase in urinary uroporphyrin compared to the non-exposed population (Wang et al. [18]). In contrast, the same strain of mice (C57BL6/J) exposed to sodium arsenate at concentrations of 20mg/L (Woods and Fowler [15]) and 100, 250 and 500 ppb (Wu et al. [25]) showed no significant increase in uroporphyrin. However, C57Bl/6J male mice sub-chronically exposed to drinking water containing sodium arsenite or sodium arsenate showed significant increase in uroporphyrin (Garcia-Vargas et al. [26]). Significant increase in the protoporphyrin IX, uroporphyrin and Copro III were recorded in Wistar rats dosed by oral gavage with either a solution of sodium arsenite or sodium arsenate at a dose rate of 5 mg As/kg body weight but no significant increase in Copro I excretion (Wang et al. [17]).

Even though the concentration of MMA^{III} used was very low in this study, significant differences were observed between control and test groups for Coproporphyrin III. The findings here support our previous observations in mice which were chronically exposed to inorganic sodium arsenate. In conclusion, as shown in the human study of Ng and Qi [27] coproporphyrin and uroporphyrin could be used as biomarkers for arsenicosis in humans, particularly in the young age group (age < 20 years), our study also demonstrates that the Coproporphyrin III and uroporphyrin can be used as early warning biomarkers before the onset of cancer. Since MMA^{III} is a more potent cytotoxin, genotoxin and potential carcinogen than other arsenic species, it would be interesting to evaluate the overall porphyrin profile over the whole life span of these animals and the effect of MMA^{III} on enzymic control of haem biosynthesis.

References

[1]   NRC (1999). National Research Council: Arsenic in drinking water. National Academy Press, Washington, pp. 1 - 310.

[2]   Rahman, M. et al. (1999). "Relations between exposure to arsenic, skin lesions and glucosuria." Occupational and Environmental Medicine 56: 277-281.

[3]   Tseng, C.-H. et al. (2002). "Epidemiologic evidence of diabetogenic effect of arsenic." Toxicology Letters 133(1): 69-76.

[4]   ATSDR (2000). "Toxicological profile of arsenic. US Department of Health and Human Services, Public Health Service, Agency for Toxic Substances and Disease Registry."

[5]   IPCS (2001). Environmental Health Criteria 224, Arsenic and Arsenic Compounds. Geneva, WHO.

[6]   IARC (2004). Some drinking-water disinfectants and contaminants, including arsenic. Lyon, France, IARC.

[7]   Cullen, W. R. et al. (1984). "The reduction of trimethylarsine oxide to trimethylarsine by thiols: a mechanistic model for the biological reduction of arsenicals." J. Inorg. Biochem. 21: 45-60.

[8]   Vahter, M. and E. Marafante (1983). "Intracellular distribution and metabolic fate of arsenite and arsenate in mice and rabbits." Chem-Biol Interact 47: 29-44.

[9]   Moore, M. M. et al. (1997). "Relative genotoxic potency of arsenic and its methylated metabolites." Mutat. Res. 386: 279-290.

[10]   Petrick, J. S. et al. (2000). "Monomethylarsonous acid (MMA(III)) is more toxic than arsenite in Chang human hepatocytes." Toxicology and Applied Pharmacology 163: 203-207.

[11]   Cullen, W. R. and M. Dodd (1989). "Arsenic Speciation in Clams of British Columbia." Applied Organometallic Chemistry 3: 79-88.

[12]   Mass, M. J. et al. (2001). "Methylated trivalent arsenic species are genotoxic." Chemical Research in Toxicology 14: 355-361.

[13]   Li, J. H. and T. G. Rossman (1989). "Mechanism of comutagenesis of sodium arsenite with n-Methyl-n-nitrosourea." Biological Trace Element Research 21: 373-380.

[14]   Moore, M., K. McColl, A. Goldberg, C. Rimington. (1987). "Disorders of porphyrin metabolism." Plenum Medical Book Company, New York & London: pp l-374.

[15]   Woods, J. S. and B. A. Fowler (1978). "Altered regulation of mammalian hepatic heme biosynthesis and urinary porphyrin excretion during prolonged exposure to sodium arsenate." Toxicology and applied pharmacology 43: 361-371.

[16]   Ng, J. C. (2002). "Porphyrin Profiles in Blood and Urine as a Biomarkers for Exposure to Various Arsenic Species." Cellular and Molecular Biology 48: 111-123.

[17]   Wang, J. P. et al. (2002). "Porphyrins as early biomarkers for arsenic exposure in animals and humans." Cellular and Molecular Biology 48: 835-843.

[18]   Wang, J. P. et al. (2002). "A review of animal models for the study of arsenic car 21. Healy, S. M., R. A. Zakharyan, et al. (1997). "Enzymatic methylation of arsenic compounds: IV. in vitro and in vivo deficiency of the methylation of arsenite and monomethylarsonic acid in the guinea pig." Mutat. Res. 386: 229-239.

[19]   Ng, J. C. et al. (2002). "HPLC measurement of harderoporphyrin in the harderian glands of rodents as a biomarker for sub-lethal or chronic arsenic exposure." Toxicology Letters 133: 93-101.

[20]   Zakharyan, R. A. et al. (1996). "Enzymatic methylation of arsenic compounds: III. the marmoset and tamarin, but not the rhesus, monkeys are deficient in methyltransferases that methylate inorganic arsenic." Toxicology and applied pharmacology 140: 77-84.

[21] Healy, S. M. et al. (1997). "Enzymatic methylation of arsenic compounds: IV. in vitro and in vivo deficiency of the methylation of arsenite and monomethylarsonic acid in the guinea pig." Mutat. Res. 386: 229-239.
[22] Healy, S. M. et al. (1998). "Enzymatic methylation of arsenic compounds, V. arsenite methyltransferase activity in tissues of mice." Toxicology and applied pharmacology 148: 65-70.
[23] Wu, H., Krishnamohan, M., Lam, P., Huang, S., Wang, J. P. & Ng, J. C. 2004, 'Urinary arsenic speciation and porphyrins in C57B1/6J mice chronically exposed to low doses of sodium arsenate', *Toxicology Letters*, 154, pp. 149 - 157.
[24] Aposhian, H. V. et al. (2000). "Occurrence of monomethylarsonous acid in urine of humans exposed to inorganic arsenic." Chemical Research in Toxicology 13(8): 693-697.
[25] Wu, H., Krishnamohan, M., Lam, P. & Ng, J. C. 2004, 'Urinary biomarkers for chronic arsenic exposure to C57BL/6J mice', *Toxicology and Applied Pharmacology*, 197, pp. 242 - 243.
[26] Garcia-Vargas, G. et al. (1995). "Time-dependent porphyric response in mice subchronically exposed to arsenic." Human & Experimental Toxicology 14: 475-483.
[27] Ng, J. C. and L. Qi (2001). "Mutations in C57Bl/6J and Metallothionin Knock-out Mice ingested Sodium Arsenate in Drinking Water for Over Two years." Book chapter in Arsenic: Exposure and Health Effects. Eds. WR Chappell, CO Abernathy & RL Calderon. Elsevier Science, Oxford,. pp 231-242.

Gastropod molluscs as indicators of the cadmium natural inputs in the Canarian Archipelago (Eastern Atlantic Ocean)

R. Ramírez, C. Collado, O. Bergasa, J. J. Hernández
& M. D. Gelado
Department of Chemistry, BIOGES,
University of Las Palmas de Gran Canaria, Spain

Abstract

Nowadays, gastropod molluscs are being utilized more and more as bioindicator organisms. Similarly, harmful metals on human health such as cadmium have been widely studied. The Canarian Archipelago (specifically the eastern islands) is constantly bathed by the African coastal upwelling, provoking oceanographic and biological differences between the islands. This process could assume an increase in the Cd concentration in their coastal waters and in the biota. Thus, in order to assess this fact, we measured cadmium concentrations in the soft parts of two species of limpets (*Patella rustica* and *Patella candei crenata*) and in a topshell snail (*Osilinus atrata*). Metal determination was performed using atomic absorption spectrometry (AAS). We found significant differences for metal concentrations between the eastern islands and the western islands for all species. *P. rustica*, *P. c. crenata* and *O. atrata* presented values ranging from 7.71, 2.11 and 7.56 μg g^{-1} dry wt. (eastern islands) to 1.38, 0.4 and 1.08 μg g^{-1} dry wt. (western islands) respectively. Therefore, we concluded that limpets and topshell snails seem to be suitable indicators of the cadmium concentrations in the coastal waters of the Canary Islands.
Keywords: Canary Islands, heavy metals, cadmium, gastropod molluscs, Patella rustica, Patella candei crenata, Osilinus atrata.

WIT Transactions on Biomedicine and Health, Vol 10, © 2006 WIT Press
www.witpress.com, ISSN 1743-3525 (on-line)
doi:10.2495/ETOX060201

1 Introduction

Cadmium (i.e. Cd) is a non-essential metal for life and it can be toxic even at low concentrations. Its harmful effects on humans have made it one of the most studied metals. In the oceans, its concentration is related with the nutrients phosphate and nitrate concentrations [1]. Besides, all of them present higher concentrations in deep waters than in the ocean surface. In contrast to this, Cd can be found at higher concentrations in the surface waters when a coastal up-welling process has happened [1]. Furthermore, any increase in the Cd load in the nearshore waters can have an immediate effect on the levels of this metal in the biota [2]. In this sense, organisms can be reflecting into their bodies an increase in the local bioavailability of the metal due to natural and human inputs. Thus, molluscs represent one of the best groups of the animal kingdom to be utilized as bioindicators [3].

The Canary Islands lie in a transition zone between the oligotrophic open ocean and the northwest African upwelling region (so-called Northwest African Coastal Transition Zone [NACTZ]) [4]. Quasi-permanent filaments of these waters reach the eastern islands (Chinijo Archipelago, Lanzarote, Fuerteventura and Gran Canaria) of the Canarian Archipelago. As a result, eastern and western islands present differences in their oceanographic characteristics [5, 6]. In addition, in the rocky coasts of the Canary Islands exist several species of gastropods molluscs, like limpets and topshell snails. These could be utilized as bioindicators for metal entrances in the coastal environments, such as what has been done in other geographical areas (e.g., [7–11]). We must take into account that food is one of the major entrance of Cd into humans [2]. In our case, the limpets and topshell snails are widely harvested and consumed by local people throughout the Canarian Archipelago [12, 13]. This could suppose a harmful metal transfer that, until now, has not been valued in the realized medical studies [14].

Therefore, our objective was to determine if the limpets (*Patella rustica* and *Patella candei crenata*) and the topshell-snail (*Osilinus atrata*) can be used as indicators for the natural inputs of Cd in the Canarian Archipelago.

2 Material and methods

2.1 Study area and sampling

Our study was carried out in the Canary Islands (13-19° W, 27-30° N), in March 2003 (Figure 1). Organisms were randomly handpicked from each location (three per islands) according to their availability. Specimens of *Patella candei crenata* were not found in Fuerteventura Islands and *Patella rustica* was not found in El Hierro Islands. Samples (*P. rustica*, n=104, *P. candei crenata*, n=121, *Osilinus atrata*, n=112) were placed in polypropylene bags and transported to the laboratory where each individual was measured (total length), rinsed with deionized water (Mili-Q, Millipore, 18 MΩcm), and frozen in new bags until the digestion processed [15, 16].

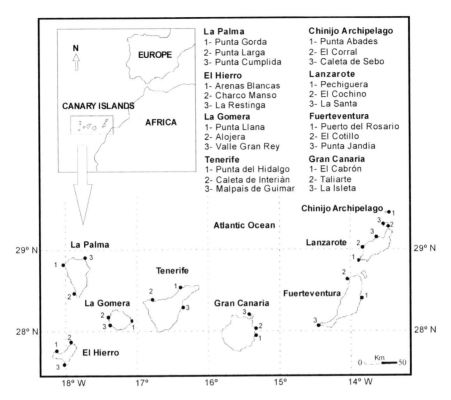

La Palma
1- Punta Gorda
2- Punta Larga
3- Punta Cumplida

El Hierro
1- Arenas Blancas
2- Charco Manso
3- La Restinga

La Gomera
1- Punta Llana
2- Alojera
3- Valle Gran Rey

Tenerife
1- Punta del Hidalgo
2- Caleta de Interián
3- Malpaís de Guimar

Chinijo Archipelago
1- Punta Abades
2- El Corral
3- Caleta de Sebo

Lanzarote
1- Pechiguera
2- El Cochino
3- La Santa

Fuerteventura
1- Puerto del Rosario
2- El Cotillo
3- Punta Jandía

Gran Canaria
1- El Cabrón
2- Taliarte
3- La Isleta

Figure 1: Map of the study area which shows the locations where the mollusc samples were recollected.

2.2 Analysis

Metal determination requires careful sample handling, storage and analysis. Therefore, all plastic materials were previously treated with an acid cleaning. It consisted in successive rinsing with diluted nitric (0.3 M) and hydrochloric (0.1 M) acid solutions for 72 hours, and washing with water of analytic quality (Millipore Milli-Q system, 18 MΩ.cm) [15, 16].

We proceed to extract the soft tissues from their shells and to clean them to remove the salts and impurities. To obtain the dried weight, the samples were dried with a IR lamp until constant weight (48 h). Then, the samples digestion was carried out on a hot plate for 3-4 h at 120°C, using a mixture of nitric and perchloric acids with a 2:1 ratio (4ml:2ml). Finally, the residual acids solutions were diluted up to 100 ml with Mili-Q deionized water.

The metal concentrations analysis was calculated using atomic absorption spectrometry (AAS). In particular, the Cd concentration was performed by graphite furnace atomic absorption spectrometry (GFAAS). Besides, for its determination a matrix modifier was used ($NH_4H_2PO_4$). Data quality control was provided by a separate comparative study of a standard reference material (BCR

CRM 278-mussel tissue). This was satisfactory showing a recovery of 86.50%. Each sample, blanks and the certified reference material were carried out in triplicate.

2.3 Statistical analysis

Mean metal concentrations together with standard errors were calculated for overall registered data. To check up on the differences in the metal levels the Kruskal-Wallis test was previously used for each species. Non-parametric Mann-Whitney U-tests were then conducted to test the significance of the differences in the metal concentrations among islands. Associations between size, weight and Cd concentration were studied by means of non-parametric Spearman correlation coefficient. All analysis were performed using the package SPSS version 12 © software for Windows XP.

3 Results

The mean total concentrations of Cd for *Patella rustica*, *Patella candei crenata* and *Osilinus atrata* were 3.61 ± 0.45, 0.71 ± 0.10 and 2.55 ± 0.45 $\mu g \ g^{-1}$ dry wt. (mean $\pm$ S.E) respectively. Similarly, the mean total size and weight from overall individuals analysed were 27.37 ± 0.40, 38.08 ± 0.58 and 16.48 ± 0.35 (mm $\pm$ S.E) for the size, whereas the weight were as follows 0.32 ± 0.02, 0.33 ± 0.02 and 0.11 ± 0.01 (g $\pm$ S.E) for *P. rustica*, *P. candei crenata* and *O. atrata* respectively. With respect to the islands, the maximum concentrations were found in the Chinijo Archipelago (7.71 ± 1.39 $\mu g \ g^{-1}$ dry wt., *P. rustica*; 7.56 ± 0.83 $\mu g \ g^{-1}$ dry wt., *O. atrata*) and Lanzarote (2.11 ± 0.73 $\mu g \ g^{-1}$ dry wt., *P. c. crenata*) (Figure 2). We also observed a high value for *O. atrata* (4.34 ± 0.83 $\mu g \ g^{-1}$ dry wt.) in La Palma. In contrast to this, the minimum values were registered in La Palma for the limpet *P. rustica* (1.38 ± 0.48 $\mu g \ g^{-1}$ dry wt.) and in Gran Canaria for *P. c. crenata* and *O. atrata* (0.25 ± 0.16 and 0.58 ± 0.07 $\mu g \ g^{-1}$ dry wt. respectively) (Figure 2).

For each species, we have detected significance differences ($p < 0.01$) between the studied islands (Kruskal-Wallis test). The results obtained with the Mann-Whitney U-tests are reported in the table 1. The Cd showed a tendency from eastern (high concentrations) to western islands (low concentrations) with great significant differences (Table 1).

Only positive correlations were found between the size and the Cd concentrations for *Patella candei crenata* ($r = 0.219$, $p = 0.02$) and *Osilinus atrata* ($r = 0.292$, $p < 0.01$). However, for *Patella rustica* we did not registered any relationship between this biometric parameter and the Cd concentrations. Strong correlations were detected between the size and weight of the three species (0.689, *P. rustica*; 0.746, *P. c. crenata* and 0.971, *O. atrata*; $p < 0.0001$). On the other hand, for all species the weight did not show any relationship with the measured Cd concentrations.

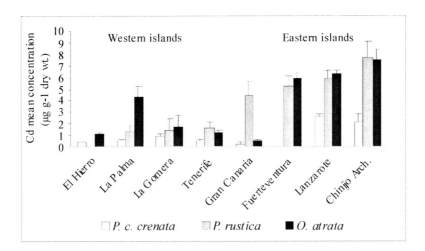

Figure 2:   Mean concentration of Cd ($\mu g\ g^{-1}$ dry wt. ± S.E) in the three studied gastropods throughout the Canarian Archipelago.

Table 1:   Differences on metal concentration obtained for the different islands using the U-Mann Whitney tests.

	Level of significance		
	$0.01 < p < 0.05$	$0.001 < p < 0.01$	$p \leq 0.001$
P. rustica	CA-F, CA-GC, F-G, GC-P	CA-G, GC-T	CA-T, CA-P, L-T, L-G, L-P, F-T, F-P
P. c. crenata	CA-GC, CA-G, T-G	CA-T, G-P	CA-P, CA-H, L-T, L-G, L-P, L-H, G-H
O. atrata			CA-GC, CA-T, CA-G, CA-H, L-GC, L-T, L-G, L-H, F-GC, F-T, F-G, F-H, GC-T, GC-G, GC-P, T-P, G-P, P-H

CA: Chinijo Archipelago, L. Lanzarote, F: Fuerteventura, GC: Gran Canaria, T: Tenerife, G: La Gomera, P: La Palma; H: El Hierro

4 Discussion

Our results have shown a clear tendency of the Cd concentrations from the eastern to western islands. Besides, this accumulation pattern at regional scale is constant for the three studied species of gastropod molluscs. We have not found a clear reason to explain the relatively high peak of Cd that was found for *Osilinus atrata* in La Palma. Differences between the oceanographic characteristics (e.g., Tª, salinity, nutrients concentration, etc) in the Canarian Archipelago have been shown in multitude of works (e.g., [5, 6, 17]). This variation can imply changes at biological level of the populations (e.g., abundance) (e.g., [18, 19]). Moreover, our work show that the upwelling process which takes place in the northwest coast of Africa, provoke physiological alterations under a chemical point of view. In this sense, there are evidences that associate Cd concentrations with these kinds of events, such as what have occurred in Baja California and the Moroccan coast for two different species of mussels [20, 21].

On the other hand, Cd concentrations obtained in the eastern islands were similar for *P. rustica* and *O. atrata*, whereas *P. c. crenata* showed lower concentrations. It is known that the gastropods related species may exhibit distinct accumulation strategies for heavy metals [22–25]. The studied gastropods present some differences regarding to their vertical distributions (tidal height of habitat) in the intertidal zone of the Canary Islands [12]. It could affect to their food supply (type of microalgae) and consequently provoke differences in the metals uptake [11, 26, 27]. Thus, an enhance in local bioavailability of Cd, whether dissolved or in the diet, can cause an increase in the uptake rate of that metal into the body [24, 28]. In consequences, Cd variations into their bodies seem to be due to external facts more than to the individuals' characteristics.

Finally, biometrics parameters showed light or none relationships with the Cd levels found in the studied molluscs. Despite to this, the correlations were always positive as it was observed in the Mediterranean area with similar gastropod species [11]. This means that the largest individuals presented higher lightly Cd concentrations than those found in the smaller individuals. Even though, we need more information to clarify the actual accumulation patterns of these species, taking into account other parameters such as season variations, sex, reproductive stage, etc and to include new possible species as bioindicators.

5 Conclusion

We concluded that limpets (*Patella rustica and Patella candei crenata*) and topshell snails (*Osilinus atrata*) could be suitable organisms to be used as indicators in the Canary Islands. They clearly reflect the Cd natural inputs in the Canarian Archipelago coming of the African up-welling zone.

References

[1]   Bruland, K. W. Trace elements in sea-water, in Riley, J. P. & Chester, R. (Eds.), Chemical Oceanography. Academic Press, London. Vol. 8. 157 220 pp. 1983.

[2]   Nriagu, J. O. Cadmium in the environment. Part I. Ecological cycling. John Wiley & Sons, Inc, Canada. 682 pp. 1980.

[3]   Oehlmann, J. & Schulte-Oehlmann, U. Molluscs as bioindicators, in Markert, B. A., Breure, A. M., Zechmeister, H. G. Bioindicators and biomonitors. Principles, concepts and applications. Elsevier, Oxford. 941 pp. 2003.

[4]   Pacheco, M. & Hernández-Guerra, A. Seasonal variability of recurrent phytoplankton pigment patterns in the Canary Islands area. International journal of remote sensing, 20, 1405-1418. 1999.

[5]   Llinás, O., Rueda, M.J., Pérez-Martell, E. Características termohalinas y nutrientes en aguas de las plataformas insulares canarias a finales de primavera. Boletín Instituto Español de Oceanografía, 10 (2), 177-189. 1994.

[6]   Davenport, R., Never, S., Helmke, P., Perez-Moreno, J. & LLinás, O. Primary productivity in the northern Canary Islands region as inferred from seawifs imagery. Deep Sea Research II, 49, 3481-3496. 2002.

[7]   Ramelow, G. J. A study of heavy metals in limpets (*Patella sp*) collected along a section of the southeastern Turkish Mediterranean coast. Marine Environmental Research, 16, 243-253. 1985.

[8]   Miramand, P. and Bentley, D. Heavy metal concentrations in two biological indicators (*Patella vulgate* and *Fucus serratus*) collected near the French nuclear fuel reprocessing plant of La Hague. The science of the total environment, 111, 135-149. 1992.

[9]   Miguel, C. S., Machado, L. M., & Bebianno, M. J. Concentraçones de Cd, Cu e Zn em Mexilhoes *Mytilus galloprovincialis* e Lapas *Patella aspera*, ao Longo da Costa Algarvia (Sul de Portugal). Ecotoxicology and Environmental Restoration, 2 (1), 1-6. 1999.

[10]   Campanella, L., Conti, M. E., Cubadda, F. & Sucapane, C. Trace metals in seagrass, algae and molluscs from an uncontaminated area in the Mediterranean. Environmental Pollution, 111, 117-126. 2001.

[11]   Cubbada, F., Conti, M. E. & Campanella, L. Size-dependent concentrations of trace metals in four Mediterranean gastropods. Chemosphere, 45, 561-569. 2001.

[12]   Navarro, P. G., Ramírez, R., Tuya, F., Fernández-Gil, C., Sánchez-Jerez, P. & Haroun, R. J. Hierarchical analysis of spatial distribution patterns of patellid limpets in the Canary Islands. Journal of Molluscan Studies, 71, 67-73. 2005.

[13]   Ramírez, R., Tuya, F., Sánchez-Jerez, P., Bergasa, O., Haroun, R. J. & Hernández-Brito, J. J. Estructura poblacional y distribución espacial de los moluscos gasterópodos *Osilinus atrata* (Wood, 1828) y *Osilinus sauciatus*

(Koch, 1845) en el intermareal rocoso de las Islas Canarias (Atlántico centro–oriental). Ciencias Marinas, 31 (4), 697-706. 2005.

[14] Rubio, C., Hardisson, A., Reguera, J. I., Revert, C., Lafuente, M. A. & González-Iglesias, T. Cadmium dietary intake in the Canary Islands, Spain. Environmental Research, 100, 123-129. 2006.

[15] Moody, J. R. & Lindstrom, R. M. Selection and cleaning of plastic containers for storage of trace element samples. Analytical Chemistry, 49, 2264-2267. 1977.

[16] Capodaglio, G., Barbante, C., Turetta, C., Scarponi, G. & Cescon, P. Analytical quality control: sampling procedures to detect trace metals in environmental matrices. Mikrochimica Acta, 123, 129-136. 1996.

[17] Mittelstaedt, E. The ocean boundary along the northwest African coast: circulation and oceanographic properties at sea surface. Progress in Oceanography, 26, 307-355. 1991.

[18] Tuya, F., Ramírez, R., Sánchez-Jerez, P., Haroun, R. J., González-Ramos, A. J & Coca, J. Coastal resources exploitation can mask bottom-up mesoscale regulation of intertidal populations. Hydrobiologia, 553, 337-344. 2006.

[19] Tuya, F. & Haroun, R. J. Spatial patterns and response to wave exposure of shallow water algal assemblages across the Canarian Archipelago: a multi-scaled approach. Marine Ecology Progress Series, 311, 15-28. 2006.

[20] Segovia-Zavala, J. A., Delgadillo-Hinojosa, F., Muñoz-Barbosa, A., Gutiérrez-Galindo, E. A. & Vidal-Talamantes, R. Cadmium and silver in *Mytilus californianus* transplanted to an anthropogenic influenced and coastal upwelling areas in the Mexican northeastern pacific. Marine Pollution Bulletin, 48, 458-464. 2004.

[21] Banoui, A., Chiffoleau, J-F, Moukrim, A., Burgeot, T., Daaya, A., Auger, D. & Rozuel, E. Trace metal distribution in the mussel *Perna perna* along the Moroccan coast. Marine Pollution Bulletin, 48, 378-402. 2004.

[22] Rainbow, P. S. Philips, D. & Depledge, M. H. The significance of trace metal concentrations in marine invertebrates: a need for laboratory investigation of accumulation strategies. Marine Pollution Bulletin, 21, 321-324. 1990.

[23] Rainbow, P. S. & Philips, D. Cosmopolitan biomonitors of trace metals. Marine Pollution Bulletin, 26, 593-601. 1993.

[24] Rainbow, P. S. Trace metal concentrations in aquatic invertebrates: why and so what?. Environmental Pollution, 120, 497-507. 2002.

[25] Cravo, A., Bebianno, M. J. & Foster, P. Partitioning of trace metals between soft tissues and shells of *Patella aspera*. Environmental International, 30, 87-98. 2004.

[26] Van Roon, M. Availability, toxicity and uptake of heavy metals by marine invertebrates. A review with reference to the Manukau Harbour, New Zealand. Department of Planning. Working Paper Series, 99-2, pp. 17. 1999.

[27] Cubbada, F., Enrique Conti, M. & Campanella, L. Size-dependent concentrations of trace metals in four Mediterranean gastropods. Chemosphere, 45, 561-569. 2001.
[28] Boyden, C. R. Effect of size upon metal content of shellfish. Journal of Marine Biological Associations United Kingdom, 57, 675-714. 1977.

Prolactin levels are positively correlated with polychlorinated biphenyls (PCB) in cord serum

L. Takser[1], J. Lafond[2] & D. Mergler[3]
[1]*Departement Obstétrique et Gynecologie, Université de Sherbrooke, Sherbrooke, Quebec, Canada*
[2]*Laboratoire de Physiologie Materno-Fœtale, Université du Québec à Montréal, Montréal, Quebec, Canada*
[3]*CINBIOSE, Université du Québec à Montréal, Montréal, Quebec, Canada*

Abstract

Polychlorinated biphenyl compounds (PCB) are global environmental contaminants that cause the disruption of endocrine and nervous systems in animals and humans. However, little evidence exists showing their potential interference with nervous system development following *in utero* exposure to low PCB levels. The aim of this study is to examine cord serum prolactin (PRL) levels in relation to cord blood concentrations of 14 PCB congeners in healthy women recruited during pregnancy. Our results showed a significant increase of cord serum PRL with an increasing level of PCB-153, the most prevalent PCB congener. Thus, knowing that PRL release is under inhibitory dopamine regulation, our results suggest that low level exposure to a mixture of persistent environmental contaminants could interfere with dopaminergic transmission in the fetal brain.
Keywords: PCB, prolactin, cord blood, environment.

1 Introduction

PCB are global pollutants characterized by their resistance to metabolic degradation, their persistence contributes to their accumulation at the top of food chain. Potential mechanisms underlying PCB-induced developmental neurotoxicity are still unclear, but significant effects on dopamine metabolism and thyroid function have been reported [1, 2]. *In vitro* studies suggest that signal transduction pathways can be disrupted by PCB [3]. Recently, an interaction of

WIT Transactions on Biomedicine and Health, Vol 10, © 2006 WIT Press
www.witpress.com, ISSN 1743-3525 (on-line)
doi:10.2495/ETOX060211

PCB with dopamine transporters and vesicular monoamine transporters, as well as oxidative mechanisms of dopaminergic toxicity, have been reported [4, 5]. Unlike to the non coplanar congeners, the coplanar and *mono-ortho*-substituted PCB have been shown to alter the signaling pathway *via* an aryl hydrocarbon receptor (AhR) mechanism [6].

Although it exists an increasing evidence of PCB effects on neurodevelopment, there is no sufficient data about congener-specific PCB neurotoxic effects in human [7]. The objective of our exploratory study was to examine the relationship between low level of environmental exposure to PCB and cord serum PRL level, which is used as an indirect marker of dopaminergic transmission. In humans, fetal PRL secretion begins early (7.5-10 weeks) during fetal development, is independent of maternal PRL levels and is dopamine dependent [8]. Considering the structure-specific mechanisms of PCB toxicity, we analysed and evaluated the relationship between PRL and PCB structure, distinguishing levels of coplanar (*mono-ortho*-congeners) or non-coplanar PCB congeners separately.

2 Materials and methods

2.1 Studied population

The women participating in this study were recruited during their first prenatal visit at the Centre for Local Community Services (CLSC, part of the National Public Health System) in the Southwestern region of Québec. The study was approved by Ethical Committee of Université du Québec à Montréal. Inclusion criteria for recruitment were no history of workplace exposure to toxic chemicals and no history of previous chronic illness, such as diabetes and/or other endocrine pathologies, arterial hypertension, or cancer. After signing a consent form, an interview administered questionnaire, which contained general socio-demographic data and information on residency, medical history, drinking and smoking habits and diet, was filled out and blood samples were obtained.

From 310 pregnant women invited to participate, 246 (79%) have accepted to participate, 42 were excluded according exclusion criteria and 47 were also excluded because their advanced gestational age. Finally, from 157 pregnant women recruited in two CLCS, 6 had spontaneous abortions and 101 gave birth in the hospital selected for the final step of the study. Two weeks following birth, a 2nd questionnaire was interview administered to all women who delivered at the selected hospital or not. This second questionnaire included information on medical and obstetrical history, birth data, as well as smoking and drinking habits during pregnancy. The characteristics of women who had not delivered in the selected hospital were not significantly different.

2.2 Biological sampling

Maternal blood and cord blood samples were collected at the hospital at delivery. Serum samples were kept at $-20°C$ until contaminant or hormone assays (within 4 months maximum).

2.3 PCB determination

Laboratory analyses were performed by the Centre de Toxicologie du Québec (CTQ) by a GC-MS method, using Gas Chromatograph 6890 and mass spectroscopy detector 5973 from Agilent (Mississauga, Ontario). Two ml of blood (maternal or cord) were extracted using an ammonium sulfate/ethanol/hexane mixture, cleaned-up on Florisil columns and taken to a final volume of 100 µL. Routine checks of accuracy and precision were accomplished using reference materials from National Institute of Standards and Technology. Also, periodic evaluations were accomplished by the participation of CTQ to two external proficiency testing programs. The detection limit for all PCB congeners, according the chosen method, was 0.02 µg/L.

2.4 Lipid determination

Total cholesterol (TC), free cholesterol (FC), triglyceride (TG) and phospholipid (PL) levels were determined in serum by enzymatic methods (Randox CH 202, Crumlin, UK) using the Technicon RA-500 analyzer (Bayer Corporation Inc, Tarrytown, NY). Serum total lipids were calculated using the summation method: total lipids = 1.677 (TC-FC) + FC + TG + PL [9].

2.5 PRL determination

Cord serum PRL levels were determined by an automatic enzymo-immunologic assay using Elecsys automate (Roche Diagnostics). The maximum limit level for PRL was 500 ng/mL.

2.6 Statistical analyses

All statistical analyses were performed using SAS version 8.12 [10]. Only 86 subjects have complete data (PCB, PRL and questionnaire), and all presented results were obtained on these 86 subjects. In order to include subjects with non-detected values in statistical analysis, the half of the detection limit was attributed to these subjects. Relationships between PCB and PRL were tested using ANCOVA (GLM procedure) and correlation analysis (CORR procedure) taking into account maternal tobacco smoking during pregnancy and month of delivery (because known PRL seasonal variation [11]). Total lipids concentrations were introduced to all models as fixed model parameters. The normality of distribution of residuals was checked for all models. The relationships between PCB and PRL were presented for adjusted for lipids models and those without adjustment.

Previously, we reported for the same population, a relationship between cord blood manganese and PRL [12]. Thus, manganese exposure variable was introduced in all models, but it did not modify our results (data not shown).

The following PCB exposure variables were used in analysis: - total PCB (sum of all congeners), - *mono-ortho*-coplanars (sum of CB-105, CB-118, and

CB-156) and – non-coplanar PCB congeners (CB-138, CB-153, CB-180) individually. CB-101 and CB-28 were excluded from statistical analysis because 100% were undetected values.

3 Results

3.1 Population characteristics and PCB levels

The maternal average age was 26 years old (range 17 – 39); 23 women (27 %) smoked during pregnancy and nine (10%) consumed alcohol moderately (0.5-2 drink/week, 4-30g alcohol/week). During pregnancy, ten women (10%) had gestational diabetes, among them 2 had pregnancy induced hypertension; 10 women had pregnancy induced hypertension without gestational diabetes, and among these ten, two had proteinuria. Eight (9 %) women delivered by cesarean, peridural anesthesia was performed in 65 cases and sedative therapy was used in 36 deliveries. Four deliveries (4.6 %) occurred before 37 weeks of pregnancy, the average birth weight was 3.3 kg (range 2.1 – 5.0 kg) and 51 % of newborns were boys.

The PCB data are given in Table 1. Total PCB in cord blood is highly correlated with maternal serum PCB (Spearman r=0.74 and 0.73, p<0.0001 for PCB values uncorrected and corrected with total lipids, respectively). Maternal PCB values uncorrected for total lipids were significantly higher than those from cord serum (paired t=11.4, p<0.0001). However, this relation was inverse for PCB values corrected with total lipids (paired t=-10.6, p<0.0001).

Table 1:   Cord serum PCB concentrations, µg/L.

Congeners	Median	5th-95th percentiles	Maximum value
Σ PCB	0.16	nd-0.35	0.38
Σ *mono-ortho-coplanars* (CB-105, CB-118, CB-156)	0.04	nd-0.07	0.13
CB-138	0.02	nd*-0.06	0.07
CB-153	0.02	nd-0.08	0.10
CB-180	nd	nd-0.05	0.06

*undetected value.

The median value of cord serum PRL was 372.3 ng/mL, ranging from 146 ng/mL to the maximal possible level (500 ng/mL) according the assay conditions; 25 samples were equal or above this level. Cord serum PRL levels were not significantly related to mode of delivery, pregnancy pathologies or medication.

3.2 Relation of cord blood PCB to PRL

The relationships between cord blood PCB exposure and PRL levels are presented in Table 2. Sum of PCB is positively correlated to cord serum PRL levels (partial Pearson r=0.22, p<0.05). *Mono-ortho*-chlorinated PCB congeners are not related to PRL levels. Among non-coplanar congeners, only CB-153 reaches the significance level (partial Pearson r=0.23, p<0.05).

Table 2:   PRL levels in relation to PCB concentrations in cord blood (General Linear Model estimates).

Cord blood PCB	PRL, ng/mL	
	Unadjusted for lipids models	Adjusted for lipids models*
ΣPCB		
β	492.9	379.5
p	0.04	0.07
Σ*mono-ortho*-coplanars		
β	1251.4	1155.9
p	0.10	0.12
CB-138		
β	1139.3	1014.6
p	0.12	0.17
CB-153		
β	1284.0	1191.9
p	0.03	0.04
CB-180		
β	1270.2	1102.4
p	0.25	0.32

* The lipid concentrations were introduced in linear models as fixed variable. All models were adjusted for tobacco smoking during pregnancy and month of birth (degree of freedom was 3;80 for unadjusted for lipids models and 4;78 for adjusted for lipids models).

4 Discussion

Our study is the first to show a relationship between PRL levels in cord serum and PCB exposure *in utero*. The finding that cord serum PRL levels increase in relation to exposure to environmental PCB suggests that these contaminants can interfere with fetal dopamine metabolism. Dopamine is the primary neuroendocrine factor that regulates the PRL secretion. Thus, removal of endogenous dopaminergic tone by catecholamine depletion or blockade of the dopamine (D_2) receptors on lactotrophs results in a significant elevation of circulating PRL levels [13]. Although many other peptides, such as thyrotropin-releasing hormone (TRH), vasoactive intestinal peptide (VIP), oxytocin or endothelin [14], are known to regulate PRL secretion, their actions are not primary responsible for the increase of PRL secretion. Thus, TRH stimulation mechanism could be hypothesized, but in our population, no significant relation was observed between PCB and thyroid-stimulating hormone (TSH) levels (regulated by TRH) (data not shown). Considering the available literature, the only explanation for our finding remains the dopaminergic mechanism.

According Richardson and Miller [5], the observed relationship between PRL and PCB was more significant in relation to CB-153, a non coplanar congener, than to *mono-ortho*-chlorinated congeners, which do not interact with dopaminergic transmission. Moreover, *in vitro* [1] and *in vivo* studies on dopaminergic toxicity induced by PCB, especially with non coplanar congeners, were reported in adult [15, 16], in developing animal and in low doses prenatal exposure (without systemic effects) [17]. However, the sum of PCB was more significantly related to PRL, than individual congeners, suggesting the additive effect of individual congeners in PCB mixture. In addition, high PCB concentrations were found in caudate nuclei from *post mortem* Parkinson disease patients, suggesting their causal role in dopamine loss in humans [18].

Few experimental data are available about chronic *in utero* exposure to PCB and their effects on dopamine or PRL secretion. Lyche et al. (2004) observed no significant effect on PRL levels of female goat prenatally exposed to low doses of CB-126 and CB-153 (similar to environmental doses) [19]. However, acute exposure to non persistent non coplanar congeners leads to transitory rise of blood PRL levels and to decrease of hypothalamic dopamine levels in developing rats [20]. In adult man, high dietary exposure to PCB and pesticides, from fish consumption, was not linked to circulating levels of PRL [21]. All this information does not exclude our suggestion that PCB exposure *in utero* can be related to the increase of PRL level in human fetuses. In fact, inter- and intra-species differences should be considered regarding the neurotoxic susceptibility in relation to PCB exposures, as well as to the differences in prenatal development (singleton pregnancy in primates, length of gestation, i.e. exposure duration, type of placentation).

Our study presents some limitations. Firstly, for 25 samples, the PRL concentration was above the maximum limit of detection, of our PRL assay and a value of 500 ng/mL was assigned to these samples. The literature data report many different PRL values in cord blood, but Parker et al. (1986) reported an average cord serum PRL concentration in full term newborns that is comparable to our results. But it is important to note that some authors also reported lower levels [2]. Secondly, retrospectively calculated statistical power of study sample was not sufficient: 56% calculated for alpha error 0.05 and observed correlation of 0.23.

In addition, exposure levels to PCB congeners in our population were 3-45 times lower than in previous reported studies reviewed by Longnecker et al. [22].

And, only CB-153, the most prevalent congener, as well as the sum of PCB congeners, were significantly related to PRL. The lack of statistical association with other PCB congeners can be related to very low levels of exposure or to lack of statistical power. Thus, there is a need for further investigations using newborns from more exposed populations.

The present study, despite limitations, is of major interest given limited scientific literature available about biomarkers of neurotoxicity in human fetuses. Knowing that developmental exposures to dopaminergic disruptors can result in long term behavioral modifications, it is important to develop strategies to determine *in vivo* biomarkers of dopaminergic toxicity induced by pollutants present in general population environment.

References

[1]   Shain, W., Bush, B. & Seegal, R., Neurotoxicity of polychlorinated biphenyls: structure-activity relationship of individual congeners. *Toxicol Appl Pharmacol,* **111**, pp. 33-42, 1991.

[2]   Seegal, R.F., The neurotoxicological consequences of developmental exposure to PCBs. *Toxicol Sci,* **57**, pp. 1-3, 2000.

[3]   Inglefield, J.R., Mundy, W.R. & Shafer, T.J., Inositol 1,4,5-triphosphate receptor-sensitive Ca(2+) release, store-operated Ca(2+) entry, and cAMP responsive element binding protein phosphorylation in developing cortical cells following exposure to polychlorinated biphenyls. *J Pharmacol Exp Ther,* **297**, pp. 762-773, 2001.

[4]   Lee, D.W. & Opanashuk, L.A., Polychlorinated biphenyl mixture aroclor 1254-induced oxidative stress plays a role in dopaminergic cell injury. *Neurotoxicology,* **25**, pp. 925-939, 2004.

[5]   Richardson, J.R. & Miller, G.W., Acute exposure to aroclor 1016 or 1260 differentially affects dopamine transporter and vesicular monoamine transporter 2 levels. *Toxicol Lett,* **148**, pp. 29-40, 2004.

[6]   Van den Berg, M., Birnbaum, L., Bosveld, A.T., Brunstrom, B., Cook, P., Feeley, M., Giesy, J.P., Hanberg, A., Hasegawa, R., Kennedy, S.W., Kubiak, T., Larsen, J.C., van Leeuwen, F.X., Liem, A.K., Nolt, C., Peterson, R.E., Poellinger, L., Safe, S., Schrenk, D., Tillitt, D., Tysklind, M., Younes, M., Waern, F. & Zacharewski, T., Toxic equivalency factors (TEFs) for PCBs, PCDDs, PCDFs for humans and wildlife. *Environ Health Perspect,* **106**, pp. 775-792, 1998.

[7]   Schantz, S.L., Widholm, J.J. & Rice, D.C., Effects of PCB exposure on neuropsychological function in children. *Environ Health Perspect,* **111**, pp. 357-576, 2003.

[8]   Freemark, M., Ontogenesis of prolactin receptors in the human fetus: roles in fetal development. *Biochem Soc Trans,* **29**, pp. 38-41, 2001.

[9]   Patterson, D.G., Jr., Isaacs, S.G., Alexander, L.R., Turner, W.E., Hampton, L., Bernert, J.T. & Needham, L.L., Determination of specific polychlorinated dibenzo-p-dioxins and dibenzofurans in blood and

adipose tissue by isotope dilution-high-resolution mass spectrometry. *IARC Sci Publ*, 299-342, 1991.

[10] SAS Institute Inc., *SAS/STAT User's Guide, version 8*. 1999: Cary, NC. 3884.

[11] Houghton, D.C., Young, I.R. & McMillen, I.C., Photoperiodic history and hypothalamic control of prolactin secretion before birth. *Endocrinology,* **138**, pp. 1506-1511, 1997.

[12] Takser, L., Mergler, D., de Grosbois, S., Smargiassi, A. & Lafond, J., Blood manganese content at birth and cord serum prolactin levels. *Neurotoxicol Teratol*, **26**, pp. 811-815, 2004.

[13] Lamberts, S.W. & Macleod, R.M., Regulation of prolactin secretion at the level of the lactotroph. *Physiol Rev*, **70**, pp. 279-318, 1990.

[14] Samson, W.K., Taylor, M.M. & Baker, J.R., Prolactin-releasing peptides. *Regul Pept*, **114**, pp. 1-5, 2003.

[15] Seegal, R.F., Bush, B. & Brosch, K.O., Comparison of effects of Aroclors 1016 and 1260 on non-human primate catecholamine function. *Toxicology*, **66**, pp. 145-163, 1991.

[16] Seegal, R.F., Bush, B. & Brosch, K.O., Sub-chronic exposure of the adult rat to Aroclor 1254 yields regionally-specific changes in central dopaminergic function. *Neurotoxicology*, **12**, pp. 55-65, 1991.

[17] Seegal, R.F., Brosch, K.O. & Okoniewski, R.J., Effects of in utero and lactational exposure of the laboratory rat to 2,4,2',4'- and 3,4,3',4'-tetrachlorobiphenyl on dopamine function. *Toxicol Appl Pharmacol*, **146**, pp. 95-103, 1997.

[18] Corrigan, F.M., Murray, L., Wyatt, C.L. & Shore, R.F., Diorthosubstituted polychlorinated biphenyls in caudate nucleus in Parkinson's disease. *Exp Neurol*, **150**, pp. 339-342, 1998.

[19] Lyche, J.L., Skaare, J.U., Larsen, H.J. & Ropstad, E., Levels of PCB 126 and PCB 153 in plasma and tissues in goats exposed during gestation and lactation. *Chemosphere*, **55**, pp. 621-629, 2004.

[20] Khan, M.A., Lichtensteiger, C.A., Faroon, O., Mumtaz, M., Schaeffer, D.J. & Hansen, L.G., The hypothalamo-pituitary-thyroid (HPT) axis: a target of nonpersistent ortho-substituted PCB congeners. *Toxicol Sci*, **65**, pp. 52-61, 2002.

[21] Hagmar, L., Bjork, J., Sjodin, A., Bergman, A. & Erfurth, E.M., Plasma levels of persistent organohalogens and hormone levels in adult male humans. *Arch Environ Health*, **56**, pp. 138-143, 2001.

[22] Longnecker, M.P., Wolff, M.S., Gladen, B.C., Brock, J.W., Grandjean, P. & Jacobson, J.L.e.a., Comparison of polychlorinated biphenyl levels across studies of human neurodevelopment. *Environ Health Perspect*, **111**, pp. 65-70, 2003.

Section 7
Interactive effects
of chemicals

Toxicity assessment of fosthiazate, metalaxyl-M and imidacloprid and their interaction with copper on *Daphnia magna*

A. Kungolos[1], V. Tsiridis[1], H. Nassopoulos[2], P. Samaras[3] & N. Tsiropoulos[4]
[1]*Department of Planning and Regional Development, University of Thessaly, Volos, Greece*
[2]*Department of Civil Engineering, University of Thessaly, Volos, Greece*
[3]*Department of Pollution Control Technologies, Technological Educational Institute of West Macedonia, Kozani, Greece*
[4]*Department of Agriculture, Crop Production and Rural Environment, University of Thessaly, Nea Ionia-Volos, Greece*

Abstract

The toxic effects of three agrochemicals (fosthiazate, metalaxyl-M and imidacloprid) and copper were investigated in this study on the crustacean *Daphnia magna*. Copper was the most toxic tested compound with the lowest EC_{50} value, equal to 0.11 mg/L. Among the three agrochemicals, fosthiazate was the most toxic with an EC_{50} value equal to 0.32 mg/L, about two orders of magnitude lower than the EC_{50} of metalaxyl-M and imidachloprid. The interactive effects of binary mixtures of fosthiazate, metalaxyl-M or imidacloprid and copper were also investigated. The interactive effects between agrochemicals and copper were antagonistic in most cases, while additive effects were observed for binary mixtures of metalaxyl-M and copper for low concentration combinations.
Keywords: pesticides, agrochemicals, bioassays, Daphnia magna, interactive effects.

WIT Transactions on Biomedicine and Health, Vol 10, © 2006 WIT Press
www.witpress.com, ISSN 1743-3525 (on-line)
doi:10.2495/ETOX060221

1 Introduction

The contamination of the aquatic and terrestrial environment with various agrochemicals poses a direct risk of toxic effects on the ecosystem. An integrated evaluation of the environmental impact of agrochemicals requires performance of both physicochemical and ecotoxicological analyses. Ecotoxicological analysis is an important issue for the assessment of chemical contamination and its potential impact on water quality and biota of receiving aquatic ecosystems [1]. Bioassays provide a more direct measure of environmentally relevant toxicity of contaminated systems than chemical analyses, since aquatic and terrestrial ecosystems are impacted by numerous chemicals. In addition, single chemicals almost never occur alone in nature; even with products or wastes containing only a few chemicals, the ultimate toxicity is not the sum of the individual toxicities. Phenomena such as e.g. bioavailability, synergistic or antagonistic effects indeed always dictate the ultimate hazard of the mixtures [1-3].

Each year during the last five decades an enormous amount of pesticides has been used in agricultural activities. The environmental effect of this use is that the pesticides can be found in important concentrations in surface waters and ground waters because of their physicochemical properties such as Log octanol/ water partition coefficient, hydrolysis, photolysis, adsorption/desorption, mobility, volatility [4, 5]. These concentrations can have an adverse effect on aquatic organisms. Furthermore, pesticides are never observed in aquatic environment alone. Various pesticides that are used in agriculture can be found together in various concentrations in water close to agricultural fields [5]. The existence in the same time of different pesticides may have an additive, antagonistic or synergistic effect on the aquatic life [2].

The objective of this study is the evaluation of the toxic effects of three pesticides (fosthiazate, metalaxyl-M and imidacloprid) alone or in combination with copper on the crustacean *D. magna*.

2 Materials and methods

The toxicity tests on *D. magna* were carried out using a commercial toxicity test Daphtoxkit F magna [6] provided by Microbiotest Inc., Belgium. Experiments with Daphtoxkit F are based on the immobilization of the crustacean *D. magna* due to the action of toxicants. *D. magna* were hatched from dormant eggs (ephippia) in 3 days under continuous illumination (6000 lux) at $20^{\circ}C$. Neonates (younger than 24 h) were exposed to the samples for 24 h at a temperature of $20^{\circ}C$ in darkness. Twenty neonates were used for each concentration or combination of concentrations examined in a series of 4 well; each well containing 10 mL toxicant solution and 5 neonates. The agrochemicals examined for their toxic properties were: fosthiazate, metalaxyl-M and imidacloprid using their commercial formulations of Nemathorin, Ridomil and Confidor, respectively. The toxicity of copper was also evaluated using copper chloride dehydrate, $(CuCl_2 \cdot 2H_2O)$ provided by J.T. Baker, Holland. The toxicities of the tested compounds were evaluated by preparing certain dilution of each compound for the estimation of the EC_{50} values. The effective concentrations (EC_{50} values) causing 50% immobilization to *D.* magna was calculated according to Probit model [7].

The interactive toxic effects between agrochemicals and copper on *D. magna* were also investigated and the evaluation of the results was performed by statistical analysis. The concentrations of the tested compounds used, were obtained from the EC_{50} estimation experiments of each compound.

A mathematical model was used for the prediction of the expected effect of the combined mixtures [2, 3]. According to this model, if P_1 is the inhibition rate caused by a certain concentration of chemical A_1 and P_2 the inhibition rate caused by a certain concentration of chemical A_2, then, the theoretically expected additive inhibition rate $P(E)$, when the same concentrations of the two chemicals are applied together, is given by the Equation:

$$P(E) = P_1 + P_2 - P_1P_2/100 \tag{1}$$

With regard to the significance of the differences between theoretically predicted and observed experimental values, one sample hypothesis for testing percentages was used for the *D. magna* test [8]. The null hypotheses were that the observed values were higher or lower than the theoretically predicted ones, for synergistic and antagonistic effects respectively. The result was considered to be antagonistic or synergistic, only if the observed effect was significantly lower or higher respectively than the theoretically predicted one at the 0.05 level of significance.

3 Results and discussion

The toxicities of the single agrochemicals and copper were first examined by applying several dilutions for each one of them and the EC_{50} value was evaluated as an endpoint. The EC_{50} values and the corresponding confidence ranges calculated using Probit model for the tested compounds are presented in Table 1, while the dose/response curves are given in Figure 1.

Table 1: EC_{50} values and the corresponding confidence ranges for the tested compounds on *D. magna*.

Compound	EC_{50} (mg/L)	Confidence range (mg/L)
Fosthiazate (Nemathorin)	0.32	0.25 – 0.41
Metalaxyl-M (Ridomil)	27.6	21.1 – 38.0
Imidacloprid (Confidor)	64.6	43.3 – 122.5
Copper ($CuCl_2\,H_2O$)	0.11	0.09 – 0.13

As it is shown in Table 1, copper was the most toxic tested compound with the lowest EC_{50} value, equal to 0.11 mg/L. Among the three agrochemicals, fosthiazate was the most toxic with an EC_{50} value equal to 0.32 mg/L, about two orders of magnitude lower than the EC_{50} of metalaxyl-M and imidachloprid. The toxicity of the three agrochemicals to *D. magna* was in the order: fosthiazate> metalaxyl-M> imidacloprid.

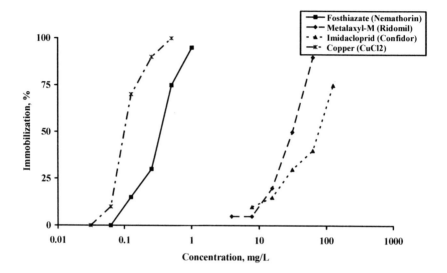

Figure 1:   *D. magna* immobilization caused by the tested concentration of agrochemicals and copper (dose/response curves).

After having tested the effects of the compounds alone on *Daphnia magna*, the interactive effect between the agrochemicals and copper was evaluated. The theoretical expected interactive effects, as calculated by Equation 1 and the observed effects for the combine action of fosthiazate and copper are illustrated in Figure 2. It can be seen that in most cases the observed effect was lower than the theoretical expected, indicating their antagonistic action. The theoretically expected and observed effects for the combined action of metalaxyl-M and copper on *D. magna* are shown in Figure 3. The mode of interaction was generally antagonistic with few cases of additive action for the low concentration combinations A, B and C (Figure 3). The theoretical expected and observed effects for the combined action of imidacloprid and copper are presented in Figure 4. The interactive effects for these binary mixtures were antagonistic for all concentration combinations.

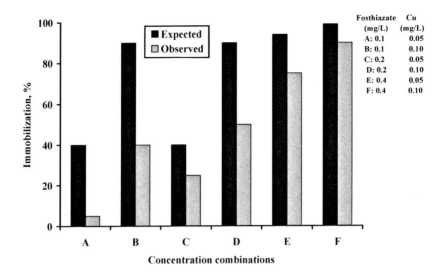

Figure 2: Comparison between theoretically expected and observed immobilization for the combined effect of fosthiazate and copper on *D. magna*.

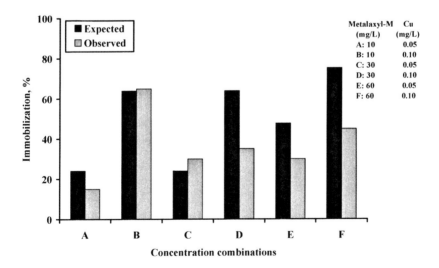

Figure 3: Comparison between theoretically expected and observed immobilization for the combined effect of metalaxyl-M and copper on *D. magna*.

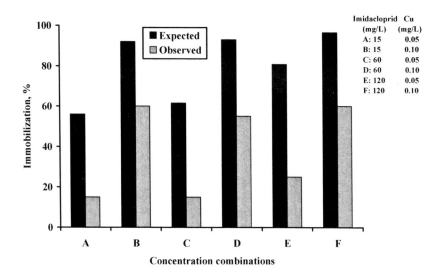

Figure 4:   Comparison between theoretically expected and observed immobilization for the combined effect of Imidacloprid and copper on *D. magna*.

4   Conclusions

The crustacean *D. magna* was used in this study in order to examine the effect of three agrochemicals, single and in combination with copper. Fosthiazate was the most toxic agrochemical. The interactive effect between fosthiazate or imidacloprid and copper was antagonistic for all tested concentration combinations. The results of the interactive effect between metalaxyl-M and copper were generally either synergistic or additive, depending on the concentration combinations. Little is known about the mechanisms that make the interactive effects of toxicants additive, synergistic or antagonistic. The results of this study showed that the combination of the tested agrochemicals with copper may result to a toxicity reduction, due to antagonistic actions.

References

[1] Persoone G., Goyvaerts M., Janssen C., De Coen W. and Vangheluwe M., Cost-effective acute hazard monitoring of polluted waters and waste dumps with the aid of Toxkits. Final Report, Commission of the European Communities, Contract ACE 89/BE 2/D3, 1993.

[2] Kungolos A., Samaras P., Kipopoulou A.M., Zoumboulis A. and Sakellaropoulos G.P., Interactive toxic effects of agrochemicals on aquatic organisms' Water Science and Technology, 40 (1), 357-364, 1999.

[3] Hadjispyrou S., Kungolos A. and Anagnostopoulos A., Toxicity, bioaccumulation and interactive effects of organotin, cadmium and

chromium on Artemia franciscana. Ecotoxicology and Environmental Safety, 49, 179-186, 2001.

[4] Tsiropoulos N.G., Bakeas E.B., Raptis V. and Batista S.S., Evaluation of solid sorbents for the determination of fenhexamid, metalaxyl-M, pyrimethanil, malathion and myclobutanil residues in air samples Application to monitoring malathion and fenhexamid dissipation in greenhouse air using C-18 or Supelpak-2 for sampling. Analytica Chimica Acta , in press, 2006.

[5] Konstantinou I.K., Hela D.G., and Albanis T.A., The status of pesticide pollution in surface waters (rivers and lakes) of Greece. Part I. Review on occurrence and levels. Environmental Pollution, 141, 555-570, 2006.

[6] Daphtoxkit F magna, Crustacean toxicity screening test for freshwater, Standard Operational Procedure, Creasel, Deinze, Belgium, 16 pages, 1996.

[7] EPA, Methods for measuring the acute toxicity of effluents and receiving waters to freshwater and marine organisms. US Environmental Protection Agency, EPA/600/4-90/027F, Fourth Edition, 1993.

[8] Sanders D.H., Eng R.J. and Murph A.F., Statistics, a fresh approach. McGraw-Hill Company, New York, 1985.

Set-up of a dynamic *in vitro* exposure device for the study of indoor air pollutant effects on human derived cells

F. Pariselli, M. G. Sacco & D. Rembges
European Commission, Joint Research Centre,
Institute for Health and Consumer Protection,
Physical and Chemical Exposure Unit, Ispra, Italy

Abstract

Exposure to air pollutants such as volatile organic chemicals (VOCs) is recognized as a potential cause of allergies, asthma, mucous irritation, headaches and tiredness and may substantially contribute to the increase of cancer incidence in the population. Their presence is not limited to urban and workplace environment but in homes, offices, schools, and hospitals at concentrations often much higher than outdoors. As the number of chemical products created is still increasing, the study of the toxic effects of VOC mixtures requires the development of an effective and reproducible technique for *in vitro* exposure of cell culture to air pollutants as an alternative tool to *in vivo* tests. The aim of this work was to develop an *in vitro* exposure set-up for the assessment of the toxicity of single volatile chemicals and their mixtures on representative VOCs target tissues: skin and lung. We used a dynamic exposure module, named CULTEX®, where human derived cell lines from lung epithelium (A549) and keratinocytes (HaCaT) were exposed in an air/liquid interface to a low flow of air pollutants mixture as a close simulation of *in vivo* exposure. Exposure to toluene or benzene concentrations showed reproducible and dose-related direct toxic effects on both cell lines. Moreover, benzene or toluene induced an inflammatory response. The results obtained so far show the sensitivity and specificity of the overall CULTEX® exposure module that could allow one to conduct further investigations on the effects of mixture of air pollutants on these representative target tissues.
Keywords: dynamic exposure device, inflammatory response, in vitro cytotoxicity, volatile organic compounds.

WIT Transactions on Biomedicine and Health, Vol 10, © 2006 WIT Press
www.witpress.com, ISSN 1743-3525 (on-line)
doi:10.2495/ETOX060231

1 Introduction

Volatile organic compounds (VOCs) are important indoor air contaminants that are evaporated into the atmosphere at room temperature. Recent studies highlight the fact that exposure to these volatile organic compounds are more important indoors that outdoors. Emission sources of VOCs are stored household products (white spirit, paint, glue…) and common household activities (e.g., painting, staining, varnishing) that can produce air concentrations of VOCs as high as 1000 mg/m^3. Benzene is a well-known carcinogen that has been causally linked to leukaemia. The major source of exposure to benzene arises from cigarette smoke that accumulates inside homes, offices and vehicles [1, 2]. Although toluene and xylene are not currently classified as carcinogens, it was reported an increase of cancer incidence in paint industry workers with long-term exposure to these organic solvents. Furthermore, increased risk for esophageal cancers was suspected to be related to occupational exposure to toluene and xylene. Hence, there is a pressing need to find "stress indicators" to rigorously evaluate the impact of these xenobiotics on biological processes [3].

The aim of this work was to set-up an *in vitro* exposure model in order to evaluate toxicological properties of single volatile chemicals and their mixtures characteristic of indoor-air environment. Since the target tissues for VOCs toxicity are lungs and the skin, the cell lines chosen for this purpose were human tumor bronchial epithelial cells (A549) and human keratinocytes (HaCaT). This study started with the application of priority air pollutants, which were selected in the frame of the European INDEX project ("Critical appraisal of the setting and implementation of European INDoor EXposure limits") [4]. Toluene and benzene were chosen as first air pollutants to set-up the exposure technique as they are known to induce lung and dermal toxicity.

2 Materials and methods

2.1 Cell culture conditions

Experiments were carried out with two human derived cells lines: lung epithelial carcinoma cell line (A549) and keratynocyte derived cell line (HaCaT). They were both cultivated in RPMI 1640 medium, supplemented with fetal calf serum (FCS) 10%, 100 unit/ml penicillin and 100 µg/ml streptomycin (Invitrogen; USA) at 37°C under humidified atmosphere containing 5% CO_2. For exposure experiments, cells were harvested with trypsin, counted and seeded onto cell culture inserts with track-etched polyethylene teraphtalate (PET) membranes permeable to culture medium (Becton Dickinson; USA). At confluence, the culture medium was removed and cells were seeded with RPMI 1640 medium without serum for 24h. Before exposure, insert were washed two times and transferred into the exposure device. Here inserts are only immersed in medium to feed the cells from the basal side with medium without FCS and exposed to atmosphere at their apical side until the end of the experiment (1 to 2 hours). After exposure experiments, inserts were transferred into conventional plates, fed on both side by medium supplemented with 1% FCS and antibiotics and incubated 24h before further measurements.

2.2 Chemical compounds

As test atmospheres we used synthetic air (80% nitrogen, 20% oxygen), benzene (1 ppm, balanced with nitrogen) or toluene (1 ppm, balanced with nitrogen) purchased from Air Liquide (France). Chemicals and reagents for biochemical analysis were obtained from Sigma (USA).

2.3 Cell exposure device: CULTEX® system

The CULTEX® system (Vitrocell, Germany) [5] is based on the use of cell culture inserts which offer the unique possibility of culturing cells on membranes permeable to culture media and to expose them to a dynamic flow of atmosphere at their air/liquid interface as it occurs in the *in vivo* environment. This system is made out of glass and stainless steel and consists of two parts. The lower part is a culture chamber maintained at 37°C that can hold separately three inserts and can provide medium supply. Here inserts are only immersed in medium; the medium on the apical side was removed and cells were exposed to the test atmospheres at their air/liquid interface until the end of the experiment. The upper part of the device houses the tube system for the transport of test atmosphere and seals the lower part. Figure 1 shows how insert cell culture is exposed to test atmosphere when the device is sealed; the gases are conducted by a trumpet-shaped tube on the surface of cell monolayer and leave each insert from corresponding hole by negative pressure. This system guarantees a continuous flow and an equal distribution of gases in the three inserts and on the entire surface of the culture.

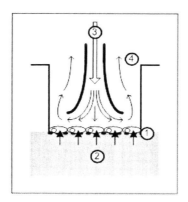

Figure 1:   Cell exposure scheme. 1: Cells grown into inserts on microporous membrane, 2: culture medium supply, 3: exposure gas flow conducted very close to the cells, 4: gas outlet drove by negative pressure.

For exposure experiments, the CULTEX® device was connected to a system of test atmosphere generation and conduction (Figure 2). The test atmosphere was generated by mixing synthetic air with different doses of air pollutants and conducted trough the CULTEX® (1 to 2 ml/min/inserts) by a negative pressure set by vacuum pump and a mass-flow controller (Bronkhorst, Netherlands).

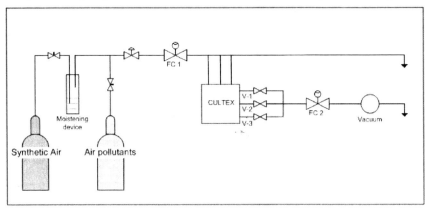

Figure 2:   Scheme of generation and conduction of test atmospheres. (FC: flow controller, V: valve).

2.4 Endpoints

Cell viability was evaluated with two different methods. The dosage of the lactate deshydrogenase (LDH) released from cells only upon damage of the cytoplasmique membrane is an index of cell membrane integrity and for extension of cytotoxicity. LDH was measured by a spectrophotometric method, reading the kinetics of NADH oxidation as a decrease of absorbance at 340 nm. The percentage of toxicity was calculated as percentage of LDH released in the medium on the total LDH (i.e. medium + cells). The cell proliferation assay WST-1 (Roche Applied Science, USA) contains tetrazolium salt cleaved to a soluble formazan dye by mitochondrial enzyme active only in viable cells. Total activity of this mitochondrial enzyme in a sample rises with the increase of viable cells. As the increase of enzyme activity leads to an increase of the production of formazan dye, the absorbance of formazan dye in the medium is proportional to the number of viable cells. The formation of formazan is measured after 3hours of incubation in the medium at 37°C by a spectrophotometer at 405nm. In order to evaluate inflammatory response after exposure with priority pollutants, Interleukin-8 (IL-8) has been chosen as a pro-inflammatory biomarker. It was measured in medium 24h after exposure by an ELISA test (Pierce, USA).

2.5 Controls

As cell integrity control we used inserts with cells in normal culture conditions (with media on both sides) left during the exposure time in the incubator (37°C,

5% CO2, 100% humidity). Synthetic air exposed inserts were used as control for the air pollutant exposures. Inserts left in the incubator during 1 hour without media on the apical side gave the same results than those with medium on both sides (data not shown).

2.6 Atmosphere test analysis

Before and after each exposure experiment, sampling of test atmosphere were made before entering the CULTEX® and analysed by gas chromatography/mass spectrometry (GC-MS HP 5890, Finnigan).

2.7 Statistical analysis

Exposures were always carried out in triplicate and experiments were repeated at least three times. Data are presented as the mean ± S.E.M. and statistically analysis was carried out with the Student t test.

3 Results

3.1 Set-up of the exposure conditions

A particularity attention was made on healthy conditions of the cell culture during exposure experiments. In order to keep humidity level and pH of medium constant, we used a moistening device connected between the synthetic air delivery and the exposure device (see figure 2) and an HEPES buffered medium. Moreover, the CULTEX® device housed a warmed water jacket in order to keep cells and culture medium at 37°C.

Flow rate of test atmosphere and exposure time were also determined in order to minimize mechanical stress or dehydration of the cell monolayer and to keep cell viability as higher as possible after exposure. Several experiments with different flow rates (1 to 8 ml/min) and different exposure times (from 60 to 120 minutes) were performed with the A549 cell line. We obtained the best results evaluated with LDH and WST-1 assay (viability >90%) with a synthetic airflow rate from 1 to 2 ml/min in each insert for an exposure time from 60 to 120 minutes (data not shown).

Table 1: Cell viability and cytotoxicity of human derived cells 24h after exposure to synthetic air with WST-1 and LDH assays.

| Cell line | Treatments | Cell viability (WST-1 assay) | | | Cytotoxicity (LDH assay) | | |
		% ± S.E.M.	n	t test	% ± S.E.M.	n	t test
A549	Control	100 0.0	14		5.7 0.8	9	
	Synthetic air exposure	98 1.0	27	NS	8.0 1.0	27	NS
HaCaT	Control	100 0.0	17		7.1 0.2	4	
	Synthetic air exposure	85 5.0	35	S	14.3 1.5	4	S

NS: not significant; S: significant p<0.005

In order to validate culture and exposure conditions with the CULTEX® device, we performed repeated synthetic air exposure with a 2ml/min flow during 1 hour on both cell lines and evaluated cytotoxicity endpoints after 24h. Results are presented in table 1. Exposure of A549 cell line to synthetic air shows no statically significant differences ($p>0.05$) with respect to the control with both tests. With the HaCaT cell line, we reached about 85% of survival (WST-1 assay) for cells exposed to synthetic air; moreover, the results obtained with LDH assay show statistically significant differences between control and exposed cells.

3.2 Effects of toluene exposure

Cells were exposed under airlifted conditions with a mixture of synthetic air and toluene with a 2 ml/min flow for 1 hour in the CULTEX® device. At the end of the exposure, cells were left on the inserts to recovery for 24 hours in the CO_2 incubator with fresh medium (with 1% FCS) on both sides before LDH assay and IL-8 released determination. The range of toluene concentration determined with Tenax® sampling and gas chromatography was ranging between 0.1 and 0.6 ppm. The average cytotoxicity of the A549 and HaCaT cell lines exposed to toluene are summarized in figure 3. For both cell lines, exposure to toluene concentration ranging from 0.1 to 0.6 ppm reduced significantly the viability after 24 hours when compared with the control. Toluene exposure stimulated significantly the IL-8 release from both cell lines. These results are summarized in figure 4.

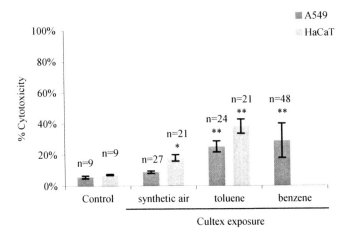

Figure 3: LDH cytotoxicity assay on A549 and HaCaT cell lines exposed to synthetic air, toluene (0.1 to 0.6 ppm) or benzene (0.1 to 0.3 ppm) in the CULTEX® device. *Statistically significant from control ($p<0.005$). **Statistically significant from control and synthetic air exposure ($p<0.0005$).

3.3 Effects of benzene exposure

Cells were exposed to benzene with the same experimental conditions used for toluene. Here only A549 cell line was tested. The range of benzene concentration was between 0.1 to 0.3 ppm. Cytotoxic effect and IL-8 released induced by benzene exposure were summarized in figure 3 and 4.

Exposure to benzene reduced significantly the viability of A549 cell line after 24 hours and increased significantly the release of IL-8 in the culture medium.

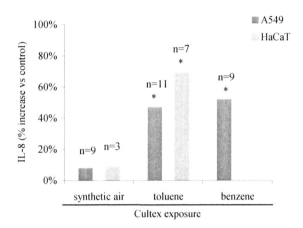

Figure 4:   IL-8 release determination with A549 and HaCaT cell lines exposed to synthetic air, toluene (0.1 to 0.6 ppm) and benzene (0.1 to 0.3 ppm) with the CULTEX® device. *Statistically significant from control (p<0.005).

4 Conclusion

The aim of this work was to set-up a dynamic *in vitro* exposure model and to evaluate toxicological properties of priority air pollutants (VOCs) as single components in order to extend further studies on the effects of the indoor air pollutants mixtures on human derived cells. CULTEX® device was used to expose to chemicals in air/lifted interface representative of VOCs' target tissues: human lung and dermal cell lines (A549 and HaCaT).

In order to use healthy cells in optimal conditions and to obtain reproducible results several requirements were met for the exposure experiments [6]: The system is designed in order to keep cells in optimal conditions during their maintenance in an air/liquid interface; i.e. temperature and pH of the medium should be constant and the atmosphere must be humid, so that extended exposure is possible. The system has a precise control of atmosphere generation and measurements with mind on the purity of the air supply. The gases are in contact with cells as closely as possible to mimic *in vivo* conditions, avoiding the presence of medium, since gases could react with its components and toxic effects may be masked. Sterile conditions are maintained during exposure, since bacterial or fungal contaminations could easily affect cellular metabolism and lead to erroneous conclusions. For these reasons several culture conditions parameters were defined before and during the exposure (airflow, exposure time, pH buffer, moistening) allowing us to keep cells in better conditions and to decrease some mechanical effects which can damage them. This was achieved for A549 cell line, while further improvements are necessary for HaCaT cell line. Exposure experiments performed with toluene or benzene showed reproducible direct toxic effects on cell cultures for both cell lines. The increasing LDH cytotoxicity from 0.1 ppm toluene exposure to 0.6 ppm gave first indications of a dose related response. Similar results were obtained for A549 cells exposed to benzene (0.1 to 0.3 ppm). Further exposure experiments with different concentrations of toluene and benzene have to be worked out for validating this hypothesis. In addition, we demonstrated the ability of toluene and benzene to induce an inflammatory response (IL-8 stimulation), according to data reported in literature [7, 8].

In vitro exposure systems have several advantages: toxic agents that cannot be used in animal for ethical and financial reasons can be tested, individual cell type response to pollutants permits better determination of the independent contribution of this cell type to a particular response, a number of human cell lines, tissues or primary cell lines can be used and as the exposure conditions were rigorously controlled they can easily be reproduced [9]. Moreover, this dynamic exposure system has the advantage to test volatile chemicals alone or in a complex mixture, soluble or insoluble and particularly in experimental conditions that closely simulate the exposure of target tissues *in vivo*.

Summarizing, all these results show the sensitivity and specificity of the over all CULTEX® exposure set-up and this will be the basis to perform exposure treatments to air pollutants alone or in mixtures.

Acknowledgements

The authors would like to acknowledge the technical assistance of Mr. T. Krebs from the Vitrocell Company, Mr. A. Collotta for his skillful technical assistance and Mr. P. Leva, A. Katsogiannis and O. Geiss from the PCE unit (JRC, IHCP) for all air pollutants determination.

References

[1] Wallace, L., Environmental exposure to benzene: an update. *Environ Health Perspect*, **104(6)**, pp. 1129-1136, 1996.
[2] Wallace, L., Major sources of benzene exposure. *Environ Health Perspect*, **82**, pp. 165-169, 1989.

[3] Sikkema, J., de Bont J.A. and Connell, D.W., Interactions of cyclic hydrocarbons with biological membranes. *J Biol Chem*, **269**, pp. 8022-8028, 1994.

[4] Kotzias, D., Koistinen, K., Kephalopoulos, S., Schlitt, C., Carrer, P., Maroni, M., Jantunen, M. Cochet, C., Kirchner, S., Lindvall, T., McLaughlin, J., Mølhave, L., de Oliveira Fernandes, E., Seifert, B. Final Report of the INDEX project, Critical Appraisal of the Setting and Implementation of Indoor Exposure Limits in the EU. EUR 21590 EN 2005, 2006. www.jrc.cec.eu.int/pce/documentation_reports.htm

[5] Aufderheide, M. and Mohr, U., CULTEX – a new system and technique for the cultivation and exposure of cells at the air/liquid interface. *Exp Toxicol Pathol*, **51**, pp. 489-490, 1999.

[6] Rassmussen, R.E., In vitro systems for exposure of lung cells to NO_2 and O_3. *J Toxicol Environ Health*, **13(2-3)**, pp. 397-411, 1984.

[7] Chou, C.C., Riviere J.E. & N.A., The cytotoxicity of jet fuel aromatic hydrocarbons and dose-related interleukin-8 release from human epidermal keratinocytes. *Arch Toxicol*, **77**, pp. 384-391, 2003.

[8] Kawasaki, S. et al., Benzene-extracted components are important for the major activity of diesel exhaust particles: effect on interleukin-8 gene expression in human bronchial epithelial cells. *Am J Respir Cell Mol Biol*, **24(4)**, pp. 419-426, 2001.

[9] Wallaert, B., Fahy, O., Tsicopoulos, P., Gosset, P. and Tonnel, A.B., Experimental systems for mechanistic studies of toxicant induced lung inflammation. *Toxicol Lett*, **112-113**, pp. 157-163, 2000.

Section 8
Hazardous waste environmental effects. Monitoring and remediation

Bioindication of microcystins toxicity by germinating seeds

Z. Romanowska-Duda B.[1], M. Grzesik[2], J. T. Mankiewicz[3]
& M. Zalewski[3]
[1]*Department of Ecophysiology and Plant Growth Development,
University of Lodz, Banacha 12/16, 90-237 Lodz, Poland*
[2]*Department of Ornamental Nursery and Seed Science,
Research Institute of Pomology and Floriculture, Skierniewice, Poland*
[3]*International Centre for Ecology, Polish Academy of Science,
Lodz, Poland*

Abstract

The effects of microcystins, produced by *Microcystis aeruginosa* PCC 7820, on seed germination, seedling growth, activity of selected enzymes and membrane permeability were evaluated, in order to elaborate the biotests for natural solution toxicity indication in water reservoirs. Commercial seeds of four selected species were allowed to germinate at 20°C on filter paper moistened with: (i) water (control), (ii) three concentrations of microcystins cultures produced by *Microcystis aeruginosa* PCC 7820 or (iii) medium BG11 in which these cultures were conducted. To evaluate the sensitivity of these seeds to microcystins the number of the germinated seeds were counted and the length of roots and hypocotyls were measured 1, 2 and 3 days after sowing, while activity of RN-ase, phosphatase (pH 6.0 and pH 7.5), dehydrogenases and electrolyte leakage were measured after 8 and 24 hours of seed exposition in these compounds. The results showed that microcystins did not decrease significantly germination percentage and mean germination time of the tested seeds, except of *L. sativa* where the number of the germinated seeds was slightly reduced. A toxic effect of these solutions was exhibited by decreased root length in all germinated seeds, while hypocotyls length was not significantly affected by them. Microcystins also decreased the activity of the investigated enzymes and increased electrolyte leakage measured 8 and 24 hours after sowing. The research showed that growth of roots and activity of RN-ase, phosphatase (pH 6.0 and pH 7.5) or dehydrogenases, as well as electrolyte leakage from germinating seeds of selected species could be used for fast monitoring of the microcystin toxicity in water.
Keywords: germination, microcystins, monitoring, toxicity, seeds, Sinapis alba, Cardamine sp., Lactuca sativa, Sorghum saccharatum.

WIT Transactions on Biomedicine and Health, Vol 10, © 2006 WIT Press
www.witpress.com, ISSN 1743-3525 (on-line)
doi:10.2495/ETOX060241

1   Introduction

Blue-green algal toxins present in water during algae blooming may cause serious diseases and fatal intoxication of humans as well as farm and wild animals [4]. Microcystins belong to these toxins and they are the most common hepatotoxins produced mainly by *Cyanophyta* from the genus *Microcystis,* as well as *Anabaena*. Their toxicity is much higher than that of strychnine and sodium cyanide [4]. It was showed that regular taking by human body of small doses of hepatotoxins from water reservoir, where algae are blooming, causes dyspepsia and may initiate liver cancer, bleeding, necrosis, as well as apoptosis in hepatocyte cells [5].

Increasing pollution of water environment with blue-green algal toxins has brought about the need to monitor their harmfulness, which would enable to undertake the actions focused on protection of conditions of human life. Most of the present methods of water pollution evaluation are based on chemical analyses of hazardous compounds. They are expensive, difficult to conduct and often indicate only some specific pollutants without information about their negative effect on organisms. Thus, in the preservation of human health and water quality, it is important to evaluate ecotoxicological effects of contaminated water to complement the techniques of analytical chemistry [12]. Resent reports show that bioindication has become the leading method in the world for monitoring pollution in water environment. This method utilises living organisms and their reactions as an indication of environmental pollution. In contrast to chemical tests bioindication assesses total toxicity of the environment and its summary effect. As a result of pollutants influence on the marker organisms variations from control organisms (untreated) are observed including morphological changes, retardation or induction of growth and mobility, occurrence of diseases and mortality rate.

The research has suggested that for these purposes animals, plants and germinated seeds that enable for rapid evaluation of pollution level of water can be employed [8, 9, 10, 13, 14]. Seed germination and root elongation technique seems to be an easy and inexpensive screening test and it can be easily monitored. Since the metabolism of seeds can be influenced by environmental conditions, phytotoxicity tests based on germination and root elongation to assess the potential contamination by metal ions were proposed recently [1, 8]. Suggested plant species for phytotoxicity indication include cucumber, lettuce, radish, red clover, wheat [3, 6, 18], oats, corn, cabbage, carrots, soybean, tomato, ryegrass, onions, beans [16], as well as rice, sorghum, mustard, rapeseed, turnip, vetch, fenugreek [2, 10, 17]. The used seeds can be exposed by floating in test solution, placed on filter paper soaked with the solution or sowed on sediments and tested soil, as well as deployed on a solid substrate and hydrated with reference water or test solution. They can be also exposed to aqueous extracts from soil or sediments [2].

Although the mentioned literature data indicate possibility to use seeds to monitor the toxicity of different compounds in water or soil, information concerning the exact sensitivity of seeds of particular species to stress caused by chosen pollutants is very scant. This applies also to the toxicity of blue-green algal toxins dissolved in water. Therefore, the aim of the presented research is to evaluate the effects of microcystins produced by *Microcystis aeruginosa* PCC 7820 on seed germination, selected physiological properties and growth of roots and hypocotyls of white mustard, cuckoo, lettuce and sorghum, in order to elaborate a fast and inexpensive biological test for the microcystin toxicity monitoring in water.

2 Material and methods

2.1 Materials

Commercial seeds of *Sinapis alba* (white mustard), *Cardamine sp.* (cuckoo), *Lactuca sativa* (lettuce), produced in the Central Poland and *Sorghum saccharatum* (sorghum), cultivated in South Italy, were used in experiment following the method described by Grzesik and Romanowska-Duda [8].

2.2 Seed treatments

For each experiment, 6 ml of distilled water (control) or medium BG11 or three concentrations of microcystins cultures (100 000, 200 000, 300 000 cells/ml, majority of MC-LR and MC-RR, identified with HPLC), produced by *Microcystis aeruginosa* PCC 7820 and cultivated in medium BG11, were added to a polystyrene Petri dishes (60 x 10 mm) containing two Whatman No 1 filter paper discs. Seeds of the investigated species were placed in dishes containing the tested solutions. Seeds number in each dish or number of replicates for each tested sample depended on method of their evaluation. The plates with the moistened seeds were incubated at 20°C in the dark for 3–5 days.

2.3 Estimation of microcystins toxicity

To assess the sensitivity of the germinating seeds to the tested medium and microcystins, the percentage of the germinated seeds, mean germination time (time to reach 50% germination), length of roots and hypocotyls, were evaluated. Activity of RN-ase, phosphatase (pH 6.0 and pH 7.5), dehydrogenases and electrolyte leakage from seeds was observed additionally in *Sinapis alba, Cardamine sp.* and *Lactuca sativa,* which have been chosen to these measurements due to their fast germination, as compared to *Sorghum saccharatum.*

2.3.1 Germination percentage and mean germination time
To evaluate the germination percentage and mean germination time, the number of germinated seeds, exposed to microcystins, was counted every day on the base of three replicates of 50 seeds each. A seed was considered germinated, when ridicule emerged from the pericarp and was at least 1 mm long. Percentage of germinated seeds and mean germination time were calculated with Seed

Calculator 3.0 programme developed by CPRO-DLO, Wageningen in the Netherlands.

2.3.2 Length of roots and hypocotyls

The length of roots and hypocotyls were measured 1, 2 and 3 days after sowing of 3 x 50 seeds to dishes containing microcystins cultures, by means of a ruler to the closest millimeter.

2.3.3 RN-ase activity

For RN-ase activity evaluation, 3 x 200 mg of seeds, 8 and 24 hours after sowing on filter paper moistened with distilled water (control), medium BG11 and microcystins cultures, were ground in mortar and homogenised in 6 ml of 0.1 M Tris-HCl buffer (pH 7.45) containing 0.4 g of PVP (polynylpyrrolidone). The homogenate was centrifuged at 4500 g for 20 minutes. Supernatant was used for enzyme assays. All stages of the extraction procedure were carried out at 1-4°C. Enzyme activity were defined as the amount of RN-ase, which at 260 nm released one absorbance unit of soluble nucleotide per gram of fresh matter, during 1 minute at 30°C [11].

2.3.4 Phosphatase activity

Acid and alkaline was evaluated using 3 x 200 mg of seeds. Seeds, 8 and 24 hours after sowing on filter paper in dishes containing the tested solutions/cultures, were homogenized at 1–4°C in 6 ml of 0.05 M Tris-HCl buffer (pH 7.45) containing 0.4 g of PVP (polynylpyrrolidone). The homogenate was centrifuged at 4500 g for 20 minutes. Supernatant was used for enzyme assays. Constitutive phosphatase (pH 6.0 and pH 7.5) activity was measured according to Knypl and Kabzińska [11]. One unit of activity was defined as the amount of enzyme, which hydrolyzed 1 µmol of substrate p-NPP during 1 minute at 30°C.

2.3.5 Dehydrogenase activity

For total dehydrogenase activity evaluation, fifty seeds, in four replicates, were imbibed for 8 or 24 hours in Petri dishes on cotton layer, moistened with distilled water (control) or with the tested solutions/cultured. Then, samples of 0.2 g seeds were placed in Eppendorf tubes, grinded and incubated at 25°C for 24 hours in 1 ml 0.1 M sodium phosphate buffer, pH 7.2 containing 0.7% (w/v) of 2,3,5-triphenyl tetrazolium chloride. After that time samples were centrifuged (5 min, 5000 x g) and the pellet were extracted six times with 1 ml acetone. The solution absorption was measured at 488 mm. A standard curve was prepared from known concentration of formazan [7].

2.3.6 Electrolyte leakage

Electrolyte leakage was measured on the base of four replicates of 50 seeds each. Seeds, after 8 and 24 hours of exposition on filter paper to the tested compounds/cultures, were thoroughly dried between filter paper. Than, they were soaked in 3 ml of water and incubated at 20°C for 2 and 4 hours and boiled for 10 minutes. Electrical conductivity (in $\mu S \cdot 50$ seeds$^{-1} \cdot 3$ ml H$_2$O^{-1}) was measured after 2 and 4 hour of soaking and after boiling. A microcomputer conductivity meter CC-551 (ELMETRON, Poland) was used to measure the electrical conductivities of the leachate from seeds. Results were expressed as the percentage of the total leakage from seeds.

For the obtained results, the standard errors (±SE) for all parameters were calculated from values of the replicates.

3 Results

The obtained results showed that microcystins (mainly MC-LR and MC-RR), produced by *Microcystis aeruginosa* PCC 7820 and cultivated in medium BG11 were toxic to the germinating seeds and decreased some metabolic processes, dependently on plant species. The used concentrations of these compounds affected the germinating seeds similarly.

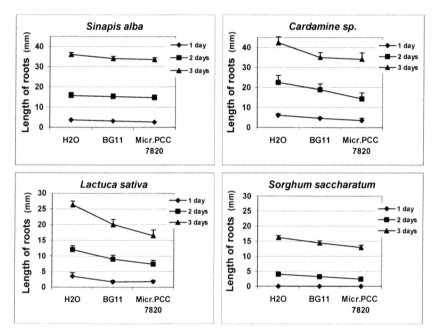

Figure 1:   Length of 3-day-old roots grown at 20°C, on filter paper moistened with distilled water (control), medium BG11 or microcystins produced by *Microcystis aeruginosa* PCC 7820 and cultivated in medium BG11. Error bars represent ± SE.

The tested microcystins did not decrease significantly germination percentage and mean germination time (time to reach 50% of germination) of the tested seeds, which in normal conditions germinated in high degree and fast (all over 90% and during one day, *Sinapis alba*, *Cardamine sp.*) or in lower percentage and slow (all 72% and during 3–4 days, *Sorghum saccharatum*). In *Lactuca*

sativa seeds, which germinated also over 90% and during one day, germination percentage was slightly reduced under exposition to medium BG11 and to higher extend due to the microcystins treatments.

Speed of germination, exhibited by mean time of germination of seeds of four tested species was not affected by microcystins and by medium BG11, in which *Microcystis aeruginosa* (producing these toxic compounds) were cultivated.

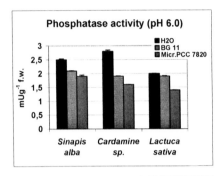
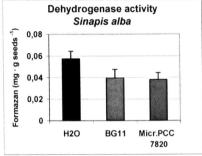
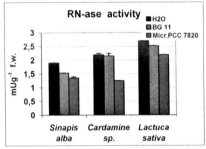
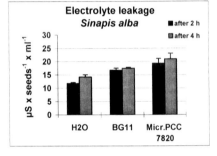

Figure 2: Activity of RN-ase, acid (pH 6.0) phosphatase or dehydrogenases and electrolyte leakage from embryos soaked for 2 and 4 hours in water, as affected by seed imbibition for 8 hours in distilled water (control), medium BG11 or cultures of microcystins produced by *Microcystis aeruginosa* PCC 7820 and cultivated in medium BG11. Error bars represent ± SE.

Toxic effect of the investigated microcystins to seeds of all four tested species was exhibited by decreased root length. The reductions in the root length, affected by microcystins, were mostly visible in *Lactuca sativa* and *Cardamine sp.* After 3 days of exposition to microcystins the length of roots in these both species was decreased by about 40% (from 27 to 17 mm) and 20% (from 43 to 34 mm), respectively, in comparison to control. Inhibiting influence of these compounds on root growth was also observed 2 and even 1 day after seed sowing. However, these reductions in the 2nd and especially in the 1st day were much less evident than after 3 days. Length of roots of *Sinapis alba* was reduced from 36 mm (in control) to 32 mm, 3 days after seed sowing to microcystins cultures, while in *Sorghum saccharatum* this reduction was from 16 to 13 mm. Reduced root length of the last both species, 2 and 1 day after sowing was no or barely noticeable (Fig. 1). The hypocotyl length was not affected significantly by the used solutions.

Cystins, produced by *Microcystis aeruginosa* PCC 7820, decreased activity of selected enzymes: RN-ase, acid (pH 6.0) or alkaline (pH 7.5) phosphatase, and dehydrogenases, in seeds exposed to these compounds/cultures for 8 and 24 hours. It was observed in *Sinapis alba, Cardamine sp.* and *Lactuca sativa* which have been chosen to these measurements due to their fast germination, as compared to *Sorghum saccharatum*. Activity of these enzymes was also reduced by medium BG11, in which *Microcystis aeruginosa* PCC 7820 (producing the evaluated microcystins) was cultivated. However, these reductions were always lower than in the case of microcystins and in some treatments they were very little (Fig. 2).

The evaluated microcystins evidently increased electrolyte leakage from seeds. The conductivity of water, in which seeds were soaked after previous exposition to microcystins for 8 hours, was much higher than in control.

Conductivity of water increased when time of soaking increased from 2 to 4 hours (Fig. 2). Similar dependencies were found when seeds were treated with microcystins for 24 hours.

4   Discussion

The presented research showed that toxicity of microcystins (mainly MC-LR and MC-RR), produced by *Microcystis aeruginosa* PCC 7820 cultivated in medium BG11 can be elevated by some physiological processes which premise or are subsequent to germination of *Sinapis alba, Cardamine sp. Lactuca sativa* and *Sorghum saccharatum* seeds.

The obtained results indicated that germination percentage and speed of germination of seeds of the tested species, expressed by mean germination time, can not be used for measuring of the toxic effects of microcystins in spite of the slightly decreased germination of *Lactuca sativa* seeds. This corresponds well with other our findings, when germination of seeds of the mentioned species was also not affected by salts of cadmium, copper, zinc and lead at concentration of 5 – 680 mg L^{-1} [8] and with the ascertainment of Kapustka and Reporter [10] that seed germination is rather insensitive to many toxic substances. Data concerning microcystins toxicity measurement with germinating seed it is hard to find in spite of several literature information in the toxicology field [9, 10, 13–15].

Toxic effects of the investigated microcystins to seeds of the four tested species can be elevated by decreased root length, which is very easy to measure and it is recommended for measurement of toxicity of some heavy metals [1, 8]. The research shows that for the microcistins toxicity measurements the germinating seeds of *Lactuca sativa* and *Cardamine sp.* are more useful, due to their fast germination in normal conditions and higher sensitivity to microcystins, than *Sinapis alba* and *Sorghum saccharatum*. Because of uniform and rapid germination of *Lactuca sativa* and *Cardamine sp.*, in just one day, root elongation values show excellent repeatability and for majority of samples significant differences between repetitions did not exceed a few millimeters. The length of roots exposed to microcystins was significantly decreased 3 days after seed sowing, although the first symptoms of inhibitions of the root growth can be observed 1 and 2 days after sowing. The sensitivity of the elongated roots of *Lactuca sativa* to some toxic compounds was also reported by Kapustka and Reporter [10].

Toxic effects of microcystis can also be monitored by measurements of activity of RN-ase, acid (pH 6.0) or alkaline (pH 7.5) phosphatase, dehydrogenases and by increased electrolyte leakage from seeds of these both species tested. These measurements show the toxic effects after 8 hours of seed exposition to microcystins. This indicates that using seeds of *Lactuca sativa* and *Cardamine sp.* it is possible to elevate the toxicity of microcystins during 8 hours and these observations can be confirmed by the root length measurements after 1, 2 and 3 days of exposition to toxins. The possibility to use the activity of some enzymes for phytotoxicity assessment, such as activity of peroxidase and some isoenzymes was indicated also by some authors cited by Kapustka and Reporter [10]. Seeds of *Sinapis alba* seem to be a little less useful in the microcystins toxicity evaluation because the growing roots are less sensitive to toxic effects of these cultures. However, these seeds germinate in height percentage during one day. *Sorghum saccharatum* seeds germinate much slower than the mentioned three species, up to 3-4 days, and thus the results of measurements are delayed, when compared to *Lactuca sativa* and *Cardamine sp.* Similarly, hypocotyls length measurement seems to be less useful for testing of the microcystins toxicity than root length evaluation, because of their late growth. Toxic effects of microcystins produced by *Microcystis aeruginosa* PCC 7820 can be also indicated by *Spirodela oligorrhiza*, as it was found by Romanowska-Duda et al. [14] and Romanowska-Duda [15].

The presented research indicates that fast biomonitoring of toxic pollution in water by microcystins produced by *Microcystis aeruginosa* PCC 7820 can be conducted first of all by the following measurements in germinating seeds of *Cardamine sp.* and *Lactuca sativa*:

-length of roots, measured 3 days after seed sowing on filter paper moistened with the tested microcystins cultures
-electrolyte leakage in seeds treated with microcystins for 8 or 24 hours
-activity of enzymes: RN-ase, phosphatase (pH 6.0 and pH 7.5) or dehydrogenases in seeds exposed to microcystins for 8 or 24 hours.

Lactuca sativa and *Cardamine sp.* seeds are cheap and easy to use for rapid evaluation of toxicity and thus they can be useful to monitor water pollution with the mentioned microcystins.

Acknowledgement

This study was supported by grant from the State Committee for Scientific Research, 3 PO4G 059 25 and 2 PO4F 044 27.

References

[1]   Barbero, P., Beltrami M., Baudo R. & Rossi D., Assessment of Lake Orta sediments phytotoxity after the liming treatment. *J. Limnol,* **60(2),** pp. 269-276, 2001.

[2]   Beltrami, M., Rossi, D. & Raudo, R., Phytotoxity assessment of Lake Orta sediments. *Aquatic Ecosystem Health and Management.* **2** pp. 391-401, 1999.

[3]   Brusick, D.J., & Young, R.R., IERL-RTP procedure manual: level I, environment assessment biological tests. *US Environmental protection Agency, Washington, D.C.: EPA-600/8,* pp. 81-02, 1981.

[4]   Carmichael, W.W., Toxic microcystins and the environment. *Toxic microcystins,* ed M.F. Watanabe: CRS Pres pp. 1-10, 1996.

[5]   Falconer, J.R., An overview of problems caused by toxic blue-green algae (Cyanobacteria) in drinking and recreational water. *Environ. Toxicol.* **14,** pp. 5-12, 1999.

[6]   Fletcher, J.S., Muhitch, M.J., Vann, D.R., McFarlane, J.C. & Benenati, F.E., PHYTOTOX database evaluation of surrogate plant species recommended by the U.S. Environmental Protection Agency and the Organization for Economic Cooperation and Development. *Environ Toxicol Chem.* **4,** pp. 523-532, 1985.

[7]   Górnik, K. & Grzesik, M., Effect of Asahi SL on China aster 'Aleksandra' seed yield, germination and some metabolic events. *Acta Physiologiae Plantarum* **24(4)**, pp. 379-383, 2002.

[8]   Grzesik, M. & Romanowska–Duda, Z., Effects of heavy metal salts on seed germination and seedling growth of white mustard and cuckoo flower (pol). *X Ogólnopolski Zjazd Naukowy Hodowców Roślin Ogrodniczych.* Skierniewice, 14-15.02.2005, pp. 256-262, 2005.

[9]   Hardy, D., *10th International Symposium on Toxicity Assessment (ISTA 10).* Quebec City, August 26 – 31: 2001, 2001.

[10]   Kapustka, L.A. & Reporter, M., Terrestrial Primary Producers. *Handbook of Ecotoxicology.* ed. P. Calow Ed. Oxford Blackwell Scientific Publications London Edinburgh Boston. **I**, pp. 278-299, 1994.

[11]   Knypl, J.S. & Kabzińska, E., Growth, phosphatase and ribonuclase activity in phosphate deficient *Spirodela oligorrhiza* cultures. *Biochem Physiol Pflanzen* **171**, 279-287, 1977.

[12]   Mendonca, E. & Picado, A., Ecotoxicological monitoring of remediation in a coke oven soil. *Environ Toxicol.,***17 (1),** pp. 74-9, 2002.

[13]   Romanowska-Duda, Z. & Tarczyńska, M., The influence of Microcystin-LR and hepatotoxic cyanobacterial extract on waterplant (*Spirodela oligorrhiza*). *Environm. & Toxicology.* John Wiley and Sons, Inc., **17 (3),** pp. 383-390, 2002.

[14]   Romanowska-Duda, Z., Jurczak, T., Grzesik, M. & Tarczyńska, M., Allelopatic relations between metabolities from Cyanobacteria against water plant (*Spirodela oligorrrhiza*). *VI International Conference on*

Toxic Cyanobacteria. Bergen, Norvay 21-27.06.2004. Abstracts, pp. 56, 2004.

[15] Romanowska-Duda, Z., *Spirodela oligorrhiza* (Kurz.) Hegelm. as bioindycator of toxicity of land water by hepatotoxins of cyanobacteria and heavy metals (pol). *Habilitation thesis.* University of Lodz, /in print/, 2005.

[16] US EPA., Ecological Effects Test Guidelines. OPPTS 850.4200. *Seed germination/Root Elongation Toxicity Test.* EPA 712-C-96-154, 1996.

[17] OECD, Terrestrial Plants: Growth test. *OECD Guidelines for testing of chemicals.* Paris, No. 208, 1984.

[18] Wang, W. & Wiliams, J.M., Screening and biomonitoring of industrial effluents using phytotoxity tests. *Environ. Toxicol Chem* **7**, pp. 645-652, 1988.

Method for selective determination of polybrominated diphenylethers (BDE-47 and BDE-99) in landfill leachates by capillary gas chromatography and electron-capture detection

D. Odusanya[1], J. Okonkwo[1] & B. Botha[2]
[1]*Department of Environmental Science,
Tshwane University of Technology, Pretoria, South Africa*
[2]*Department of Chemistry and Physics,
Tshwane University of Technology, Pretoria, South Africa*

Abstract

The last few decades have seen dramatic growth in the scale of production and the use of polybrominated diphenylethers (PBDEs) as flame retardants. Consequently, PBDEs such as BDE-47, BDE-99, BDE-153 and BDE-209 have been detected in various environmental matrices. These compounds have also been linked with the disruption of the endocrine systems of man and wildlife. The present research work is aimed at determining the concentration of (PBDEs) in landfill leachates in Tshwane Municipality. A simulated landfill leachate was used to optimise various chromatograph parameters such as oven temperature programme, injector and detector temperature, carrier gas flow rate, and the limit of detection. A Varian GC coupled with ECD detector was used for the analysis. An effective method of recovery was developed using petroleum ether instead of hexane/acetone as reported by the literature. Recoveries of BDE-47 -99 and 209, each at a fortification level of 5ng and 6ng respectively, were in the range of 97.5% to 123% and relative standard deviation of 6 to 12 (n=3). The method developed was applied to leachate samples collected from two landfill sites producing leachates (Temba and Soshanguve) in the Municipality. Results obtained are 0.90 and 0.13mg/l for BDE-47 and 0.48 and 0.21 mg/l for BDE-99 respectively. BDE-209 was not detected. Compared to other studies these concentrations are significantly higher than those reported for Japanese landfill sites. Therefore, there is cause for concern if these leachates were to infiltrate into groundwater.
Keywords: flame-retardants, GC-ECD, landfill, leachates, PBDEs, waste.

WIT Transactions on Biomedicine and Health, Vol 10, © 2006 WIT Press
www.witpress.com, ISSN 1743-3525 (on-line)
doi:10.2495/ETOX060251

1 Introduction

The presence of persistent man-made chemicals in our environment has been a common problem since a large number of these chemicals have been identified in environmental samples. Apart from heavy metals, the group of chlorinated hydrocarbon pollutants that are regarded as major environmental problems include: polychlorinated biphenyls (PCBs), polychlorinated dibenzodioxins (PCDDs) and polychlorinated dibenzofurans (PCDFs). These chemicals are known to be harmful to man and the environment [1–7]. Their toxicity and presence in certain food items, mostly of animal origin, have resulted in the introduction of dietary restrictions and recommendations by food administrations in different countries [2, 3].

There are relatively new generation chemicals used in the industries as flame retardants called brominated flame retardants (BFR). Brominated flame retardants are chemical compounds that are added to polymers used in plastics, textiles, electronic circuitry and other materials to prevent fires [3]. Polybrominated diphenyl ethers (PBDEs) are important group of brominated flame retardants. PBDEs have been produced since 1970s and were first reported as environmental contaminants in River Viskan, in the early 1980s [8]. Since then, they have been found in most environmental matrices, including aquatic and terrestrial ecosystem [9–12,]. Recently, relatively high levels of PBDEs have been found in whales and dolphins stranded alive around the Dutch coast [13]. Studies have shown increased levels of PBDEs in environmental samples and human tissues, although the levels are still lower than those for PCBs and DDT [14, 15]. There is little data available on the toxicology of the PBDEs flame retardants, the main findings published so far are changes in liver weight accompanied by historical alterations in animals given relatively large dose [15]. In humans, PBDEs like many halogenated organics, imbalance the level of the hormone thyroxin on exposure, consequently they have been classified as endocrine disrupting chemicals [12, 16, 17].

Wastes from plastic, textile and other products with BFRs are either incinerated before being disposed into landfill sites or are disposed directly as they are with other municipal wastes. Since some BFRs such as PBDEs are additives that are not chemically bound to the plastic, it may be released more readily into the landfill environment. Bergman, [2] Meironyte-Guvenius and Noren [18] stressed that leaching is expected to be greater from additive flame-retardants as compared to reactive flame-retardants. Studies on the leaching of PBDEs from landfill sites are scanty.

However, de Witt [19] and Oberg et al. [20] reported the distribution and levels of BFRs in sewage sludge of several treatment facilities in Sweden. In Minnesota, Oliaei et al. [21] surveyed the occurrence and concentrations of PBDEs (BDE-47, -99, -100, -153, -154, -207 and -208 in the leachate from five landfill sites with concentrations ranging from not detectable to thousands of pico-gram per litre. Masahiro et al. [16] reported the concentrations of BDE-47, -99 and -100 from seven different leachate samples to range from not detectable to 4000pg/l for raw and treated leachate respectively. So far, there is still paucity of data on the presence of PBDEs in landfill leachates in developing countries including South Africa, where the environmental impact of PBDEs is yet to be determined.

The introduction of gas-liquid chromatography (GC) in 1952 [22] stimulated a rapid development of this technique. Over the years gas chromatography has firmly been established as a powerful and versatile method of analysis. Although, nowadays, gas chromatography-mass spectrometry (GC-MS) and gas chromatography-gas chromatography-mass spectrometry (GC-GC-MS) are the detection techniques generally used for the analysis of most halogenated compounds because they enable qualitative and qualitative analysis simultaneously, gas chromatography-electron capture detector (GC-ECD), is still used extensively for the same purpose because of its high specificity, sensitivity and resolution at very low concentration range [23–27]. Furthermore, the operation and maintenance of GC-ECD is a lot easier and cheaper than GC-MS and GC-GC-MS. Consequently, GC-ECD is very suitable for performing routine analysis, especially when the appropriate capillary column is used.

The present study was carried out using gas chromatography coupled with electron capture detector (GC/ECD), which is equally sensitive and efficient for environmental monitoring of organochlorine pesticides [4, 28]. Although most researchers have used hexane and acetone as extracting solvents this work presents the evaluation of efficiency of solvent extraction methods for common PBDEs (BDE-47, BDE-99 and BDE-209). The aim of this work was to validate an analytical method for the determination of PBDEs in landfill leachates from municipal landfills in Tshwane, South Africa.

2   Materials/method of analysis

2.1 Reagents

All the solvents used: hexane, dichloromethane (DCM), petroleum ether (PE), and acetone were of analytical grade. BDE-47, BDE-99 and BDE-209 standards were purchased from Industrial Analytical Pty. Ltd. South Africa and the rest of the PBDEs standards were donated by Dr. Jacob de Boer of Animal Sciences Group, Institute for Fisheries Research, Netherlands.

2.2 Prepared simulated leachates

2-liters of low and high concentration of leachates were prepared using de-ionised water according to USEPA standards [29].

2.3 Determination of retention time, limit of detection, response factors and standard mixture

From the stock solution, standards were prepared by serial dilution for individual standard as well as the mixture. 1.0μl of each standard at different concentrations was injected into the GC for necessary optimum output. The response factor (RF) of the PBDEs standards relative to the internal standard (I.S.), pentachloronitrobenzen (PCNB) were carried out by injecting 1.0µl of the appropriate mixture into the GC-ECD system. The response factor was calculated based on the equation below:

Response factor = Peak area of the PBDEs standard/Peak area of the internal standard

For the determination of standard mixture, 1.0µl of standard mixture containing 250ng/ml of BDE-47, 66.7ng/ml of BDE-99 and 33.3ng/ml of internal standard was injected into the GC. These concentrations were selected based on the individual standard response to the GC conditions used.

2.4 Liquid –liquid extraction (spiking)

The validation of the LLE method was carried out by spiking simulated leachate with mixture of BDE-47 and BDE-99 at the fortification value of 100 and 120ng/ml respectively, and then extracting with 3 x 15ml of each of the extracting solvents (hexane, DCM, PE, and acetone). Petroleum ether/ acetone / hexane / (1:1:2) combination was also used. The extracts were dried with anhydrous sodium sulphate and concentrated to about 2ml using liquid nitrogen. Blank extraction of unspiked simulated leachate was carried out using the same solvents. The recoveries of the PBDEs congeners were first investigated in the low and high concentration of simulated leachates to check the effect of matrix on the extraction efficiencies. These recoveries were thereafter calculated.

Recoveries of the PBDEs standards were also investigated in raw leachate samples collected from Temba and Soshanguve landfill sites using the same PBDEs standards at the same fortification levels as for simulated leachates, to check the extraction efficiencies of the solvents on environmental samples.

2.5 Silica gel column chromatography

The extracts obtained were subjected to column chromatography by packing about 6.0g slurry of silica gel to a column (30cm x 8mm I.D.). Before use the silica gel was activated overnight. Approximately $0.5cm^3$ of anhydrous sodium sulphate was placed at the top of the column. The column was pre-eluted with another 10ml of the extracting solvent and just before the exposure of the anhydrous sodium sulphate layer to air, the reduced extract from the liquid-liquid extraction step above was placed in the column and allow to sink below the drying agent. The column was then eluted with 2 x 15ml portions of the solvents under investigation. The combined elute of each of the solvent was collected into a 250ml RBF glass and anhydrous $NaSO_4$ was added to remove water. After which nitrogen gas was bubble into the combined elute to concentrate it to about 2ml and anhydrous $NaSO_4$ was added again before spiking with 0.5ml of 10ng/ml of the internal standard (PCNB). 1.0µl of the extract was injected into the GC-ECD under the optimised instrumental condition.

2.6 Capillary gas chromatographic analysis GC-ECD

Separation and determination of the PBDEs congeners were carried out with: Varian CP-3800 Gas Chromatograph fitted with a CP-8400 Auto-sampler and equipped with Ni^{63} electron capture detector, with ZB-5 Capillary column, 5% phenyl and 95% dimethylpolysiloxane (30m x 0.25mm I.D. x 0.25µm) purchased from Phenomenex® (Separations) Randburg, South Africa. For data capturing star chromatography workstation version 6 was used.

The injector and detector temperatures were maintained at 250°C and 300°C respectively. The oven temperature was initially maintained at 90°C for 1min and then programmed at 30°C/min to 210°C and finally to 290°C at 10°C/min. The BOC gases, 99.99% ultra pure helium and nitrogen, purchased from Afrox (South Africa) were used as the carrier and make-up gases respectively. The carrier gas flow-rate was 3ml/min. while the make-up gas flow was set at 30ml/min for optimum performance. 1.0µl of the standards and processed samples were injected into the GC in the split-less vent ratio of 40:1 after 2mins of injection for all analyses.

2.7 Real leachate analysis

To determine the profile of landfill leachates, samples were collected at two of the selected project sites namely: Temba, and Soshanguve landfill. An outline of the sites describing years of operation, site area and monthly classification of waste is shown below.

Table 1: Detailed description and general information on selected landfill sites.

Name	Age Yrs.	Site Area (Ha)	% Monthly waste Classification in Tonnage				
			Total	Building	Garden	Household	Industrial
Temba	10	+3.7	9363	10	10	70	10
Soshanguve	10	19.5	12200	5	10	80	5

Source: Adapted from daily/monthly reports of The Department of Solid Waste Management City of Tshwane 2005 [30].

3 Results and discussion

The gas chromatogram of the mixture of the PBDEs standard congeners plus the I.S. (pentachloronitrobenzen) is shown in Fig.1; the congeners were well resolved as shown below.

Table 2 shows the retention times, limit of detection and the response factors for the congeners. It was observed that the response of BDE-47 is higher than that of BDE-99 and no response for BDE-209.

Fig. 2 shows the chromatogram of a typical spiked simulated leachate. The efficiency of extraction of the PBDEs congeners from simulated leachates by LLE with hexane, DCM, PE and acetone in a 30m ZB-5 column is presented in Table 3 below. From the result of the recovery analysis, the mean percentage recovery of PBDEs congeners with different extracting solvent ranged from 13.4 to 124.6%. The high standard deviations obtained from the recovery analysis in Table 3 below may be due to matrix effect (high value of simulated leachate).

Most of these solvents have been widely used in LLE of PBDEs congeners in environmental samples. The mean percentage recovery of BDE-47, -99 and -209 congeners with petroleum ether as extracting solvent gave the best result (97.5and 94.1%) in low value simulated leachate with acceptable repeatability, followed by hexane/acetone (91.4 and 69.0%). Hence, petroleum ether was used for LLE of PBDEs congeners in environmental samples. The optimum condition for LLE recoveries of PBDEs congeners from environmental samples could be obtained by introducing a cleanup procedure.

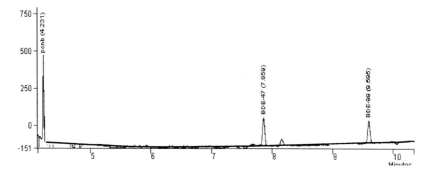

Figure 1: Chromatogram of 1μl of PBDE Mixture BDE-47 (7.85min), BDE-99 (9.59min) and PCNB (4.23min) as internal standard.

Table 2: Retention time ± SD, limit of detection and response factor of PBDEs standards.

Standards	Retention time (min)	LOD (μg/l)	Response factor
BDE-47	7.867 ± 0.03	0.1	0.44 ± 0.29
BDE-99	9.596 ± 0.44	1	0.19 ± 0.11
BDE-209	None	None	-
PCNB	4.234	0.01	

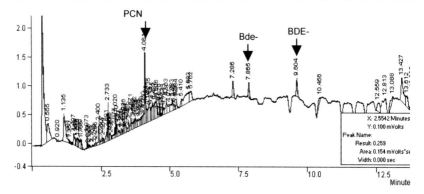

Figure 2: A typical chromatogram for spiked simulated leachate with BDE-47 and BDE-99.

Table 3: Mean percentage recoveries ± S.D of PBDEs standards added to simulated leachates by LLE with different solvents.

Extracting solvents	BDE-47		BDE-99	
	LOW	HIGH	LOW	HIGH
Hexane	63.7± 12.9	98.9± 8.8	91.7± 27.6	103.1± 97.6
Acetone	23.8± 13.9	47.2± 6.2	87.9± 88.0	108.7± 30.1
Petroleum ether	97.5± 6.0	123.6±12.1	94.1± 9.5	117± 7.4
Dichloromethane	63.7± 14.3	13.4±11.5	124.6±21.8	33.3± 26.3
Pet. Ether/DCM	41.9± 29.0	96.6±26.1	22.6± 39.1	66.2± 64.3
Pet. Ether/Acetone	51.8± 13.9	22.5± 3.4	63.8± 27.7	37.8± 7.3
DCM/Acetone	51.8± 13.9	22.5± 3.4	72.6± 7.5	37.2± 7.4
Hexane/Acetone	91.4± 17.9	96.5± 31.9	69.0± 60.0	60.7± 57.4

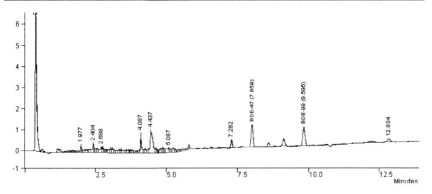

Figure 3: A typical chromatogram for raw leachate sample (Temba) with BDE-47 and BDE-99.

Table 4: The concentration (mg/l) of PBDEs (BDE-47,-99) in real leachates.

PBDEs Congeners	Temba raw leachate		Soshanguve raw leachate	
	BDE-47	BDE-99	BDE-47	BDE-99
Concentration mg/l	0.90	0.13	0.48	0.21

Fig. 3 is a typical chromatogram of raw leachate sample (Temba). The results of PBDEs levels in real leachates are presented in Table 4. The levels of BDE-47 and BDE-99 found in the leachates of Temba (0.90 and 0.13) mg/l and Soshanguve (0.48 and 0.21) mg/l are relatively higher than what has been reported in landfill leachates [16].

4 Conclusion

The present study shows LLE with petroleum ether as a good and reliable methodology of extracting BDE-47 and BDE-99 congeners from aqueous samples. An optimised methodology was developed, based on the injection of

1.0μl of the samples into 30m ZB-5 column installed to a GC-ECD, in the splitless vent ratio of 40:1 after 2mins of injection for all analyses. BDE-209 congener was not detected at all during this study. This may be due to its photo / biodegradation problems [22].

The validation parameter shows a satisfactory recovery of BDE-47 and BDE-99 for simulated leachate. The high levels of the PBDEs congeners detected in real leachate samples from these two sites, shows that landfill leachate may be contributing a significant amount of BFRs into the environment.

References

[1]   Allchin, C. R., Law, R. J. and Morris, S. 1999. Polybrominated diphenylethers in sediments and biota downstream of potential sources in the UK, *Environmental Pollution* 105, p197-207

[2]   Andersson O, Blomkvist G. 1981. Polybrominated aromatic pollutants found in fish in Sweden. Chemosphere 10(9):1051-1060.

[3]   Bergman, A., 1989, 'Brominated fire-retardants in a global environmental perspective, in: proceedings Workshop on Aromatic Fire- retardants'. Skokloster, Sweden, pp.13-23

[4]   Bras, I, Santos, L., Alves, A, 2000, Monitoring organochlorine pesticides from landfill leachates by gas chromatography- electron-capture detection after solid-phase microextraction, journal of chromatography A, 891 pp. 305-311

[5]   Burreau S, Zeb HR, Ishaq R, et al. 2000. Comparison of biomagnification of PBDEs in food chains from the Baltic Sea and the North Atlantic Sea, Organohalogen Compounds 47:253-255.

[6]   Darnerud, P. O. and Thuvander, A. 1998. Studies on immunological effects of polybrominated diphenyl ether (PBDE) and polychlorinated biphenyl (PCB) exposure in rats and mice, *Organohalogen Compounds* 35, p415-418

[7]   Darnerud, P. O., Atuma, S., Aune, M., Cnattingius, S. and Wernroth, M.-L. 1998. Polybrominated diphenyl ethers (PBDEs) in breast milk from primiparous women in Uppsala County, Sweden. *Organohalogen Compounds* 35, p411-414

[8]   Darnerud, P., Gunnar, S., Torkell, J., Poul, B. & Martti, V. 2001. Polybrominated Diphenyl Ethers: Occurrence, Dietary Exposure, and Toxicology, Environmental Health Perspectives Vol. 109, pp 49-67.

[9]   de Boer J. 1989. Organochlorine compounds and bromodiphenylethers in livers of Atlantic cod (*Gadus morhua*) from the North Sea, 1977-1987. Chemosphere 18(11/12):2131-2140.

[10]   de Boer, J., Wester, P. G., Pastor i Rodriguez, D., Lewis, W. E. and Boon, J. P. 1998. Polybrominated biphenyls and diphenylethers in sperm whales and other marine mammals- a new threat to ocean life, *Organohalogen Compounds* 35, p383-386

[11] de Boer, J., Allchin, C.R., Law, R., Zegers, B.N., Boon, J.P., 2001. Method for the analysis of polybrominated diphenylethers in sediments and biota. Trends Anal. Chem pp 591-599.

[12] Eriksson, P., Jakobsson, E. and Fredriksson, A. 1998. Developmental neurotoxicity of brominated flame retardants, polybrominated diphenyl ethers and tetrabromo-bis-phenol A. *Organohalogen Compounds* 35, p375-377

[13] Fernlof G, Gadhasson I, Podra K, et al. 1997. Lack of effects of some individual polybrominated diphenyl ether (PBDE) and polychlorinated biphenyl congeners on human lymphocyte functions in vitro, Toxicol Lett 90(2-3):189-197.

[14] Hallgren, S. and Darnerud, P. O. 1998. Effects of polybrominated diphenyl ethers (PBDEs), polychlorinated biphenyls (PCBs) and chlorinated paraffins (CPs) on thyroid hormone levels and enzyme activities in rats, *Organohalogen Compounds* 35, p391-394

[15] Lindström, G., Wingfors, H., Dam, M. and Bavel, B. v. 1999. Identification of 19 polybrominated diphenyl ethers (PBDEs) in long-finned Pilot Whale (*Globicephala melas*) from the Atlantic. *Archives of Environmental Contamination and Toxicology* 36, p355-363

[16] Masahiro, O., Yong-Jin, K., Shin-ichi, S., 2004. Leaching of brominated flame retardants in leachates from landfills in Japan, Chemosphere 57, 1571-1579

[17] Meironyte-Guvenius DM, Noren K, Bergman A. 1999. Analysis of polybrominated diphenyl ethers in Swedish human milk. A time related trend study, 1972-1997. J Toxicol Environ Health A 58(6):329- 341.

[18] Meironyte-Guvenius DM, Noren K. 1999. Polybrominated diphenyl ethers in human liver and adipose tissues. A pilot study. Organohalogen Compounds 40:372-382.

[19] de Witt CA. 2002. An overview of brominated flame retardants in the environment. Chemosphere 46(5):583-624.

[20] Oberg K, Warman K, Oberg T. 2002. Distribution and levels of brominated flame retardants in sewage sludge. Chemosphere 48(8):805-809.

[21] Oliaei F, King P, Phillips L. 2002. Occurrence and concentrations of polybrominated diphenyl (PBDEs) in Minnesota environment. Organohalogen Compounds 58:185-188.

[22] Sellström U, Soderstrom G, de Wit C, et al. 1998. Photolytic debromination of decabromodiphennyl ether (DeBDE). Organohalogen Compounds 35:447-450.

[23] Sjödin, A., Hagmar, L., Klasson-Wehler, E., Kronholm-Diab, K., Jakobsson, E. and Bergman, A. 1999. Flame retardant exposure: Polybrominated diphenylethers in blood from Swedish workers. *Environmental Health Perspectives* 107, p643-648.

[24] Troitzsch, J., H., 1999. Flammability and fire behaviour of TV sets Presented at the Sixth International Symposium of Fire Safety Science, 5 – 9 July, 1999 in Poitiers, France

[25] Watanabe I, Tatsukawa R. 1987. Formation of brominated dibenzofurans from the photolysis of flame retardant decabromobiphenyl ether in hexane by UV and sunlight. Bull Environ Contam Toxicol 39:953-959.

[26] Watanabe I, Tatsukawa R. 1990. Anthropogenic aromatics in the Japanese environment. Workshop on brominated aromatic flame retardants, Skokloster, Sweden. KEMI, National Council Inspectorate, Solna, Sweden, 1990, 63-71.

[27] Watanabe I, Kashimoto T, Tatsukawa R. 1986. Confirmation of the presence of the flame retardant decabromiphenyl ether in river sediment from Osaka, Japan. Bull Environ Contam Toxicol 36(6):839-842.

[28] Yamamoto H, Okumura T, Nishikawa Y, et al. 1997. Determination of decabromodiphenyl ether in water and sediment samples by gas chromatograpy with electron capture detection. J AOAC Int 80(1):102-106.

[29] U.S. Environmental Protection Agency (1996), *Test Methods for Evaluating Solid Waste - Physical/Chemical Methods (SW-846)*, USEPA, Washington, DC.

[30] Daily/monthly reports of The Department of Solid Waste Management City of Tshwane 2005

Study of metals in leached soils of a municipal dumpsite in Tampico, Tamaulipas, Mexico: preliminary results

P. F. Rodríguez-Espinosa[1], D. Chazaro Mendoza[1],
J. A. Montes de Oca[1], G. Sánchez Torres E.[2] & A. Fierro Cabo A.[3]
[1]CICATA-UA-Instituto Politécnico Nacional, México
[2]II-FI-Universidad Autónoma de Tamaulipas, México
[3]DEPI-FI- Universidad Autónoma de Tamaulipas, México

Abstract

The Zapote dumpsite measures 420000 m^2 and is 28 years old; an estimated 2.5 millions tons of waste have accumulated on the site (household waste, clinical waste, commercial waste). The thickness of the waste is 3 to 9 meters. Since operations began, no control regulations have existed on the residues received. The Zapote dumpsite is located within a salt-marsh between a system of channels and river lagoons of brackish water, located in a tropical sedimentary environment in the urban zone of Tampico, Tamaulipas, Mexico. Recently, the Zapote has been closed and work is presently underway in its rehabilitation since a geo-environmental perspective. The present investigation integrates information of preliminary results of metals (Pb, Ni, Cd, Cu, Mg, Fe and Al) contained in sediments that underlie the Zapote dumpsite. In laboratory research the metals of the sediment were correlated with the metals contained in samples of leachate from the Zapote dumpsite. The concentration of metals Pb, Ni, Cd, Cu, Mg, Fe and Al were analyzed in samples of sediments that underlie the body of the dumpsite in layers of 10 cm, reaching a depth of 1.5 m under the interface waste-soil. The results denote high concentrations of metals in layers that are in contact with waste that decreased until reaching 60 to 80 cm of depth. The proportions of the concentrations of metals studied in the soil are comparable with that leached, until layers of 60 to 80 cm of depth are reached, and are then lost in the deepest layers. The high plastic characteristics of clay layers have stood in the way of metallic contaminants in sub layers of the Zapote dumpsite. The results were correlated with metal concentrations of natural and anthropogenic sediments of the region.
Keywords: metals in soils, sediments in a dumpsite, metal concentration.

WIT Transactions on Biomedicine and Health, Vol 10, © 2006 WIT Press
www.witpress.com, ISSN 1743-3525 (on-line)
doi:10.2495/ETOX060261

1 Introduction

The dumpsite of approximately 42 ha was in use from 1978 to 2003. An estimated 2.5 millions tons of waste accumulated on the site, mostly household waste, but clinical and commercial waste are also probable. At the time of the dumpsite opening no regulations existed in this matter and it was operated without a management plan. The object of this contribution is to prospectively assess the downward migration of waste originated metals through leachate.

2 Study area

The dumpsite is situated on a peninsula called "El Zapote", formed by the conjunction of Panuco and Tamesí river estuaries. The Zapote dumpsite is located in the Municipality of Tampico in Tamaulipas state, Northeastern Mexico, fig. 1, between 614481 N latitude, 2458816 W longitude to 615011 N latitude, 2458580 W longitude (UTM 12 Region, WGS84).

Figure 1:   Location of Zapote dumpsite in Tampico, Tamaulipas, Mexico.

The regime of rain and temperature corresponds to Aw Koeppen, which is tropical with rain in summer, like Tamayo [1]. Figure 2 shows a typical year.

The estuary has no dikes in this area, so it is influenced by floods and tides. The site is in a low, flat area with no geological deformations, dating from the mid or superior Eocene period. The wastes have been dumped over a layer of natural occurring clay, which is two to seven meters thick. This has been verified by three deep drillings of about 12 m. An additional 14 bore-holes of 1 m depth around the site indicate with fair confidence that the clay is ubiquitous beneath the dumpsite. The clay layer of high plasticity is believed to have its origins from sediment deposition of an ancient lake (low energy water dynamics allowing fine particles deposition). The permeability of this clay can generally be regarded as very low with a permeability coefficient ranging between 10^{-9} and 10^{-11}, UAT-DEP-FI-, DM-MT and CICATA-UA-IPN SES [2].

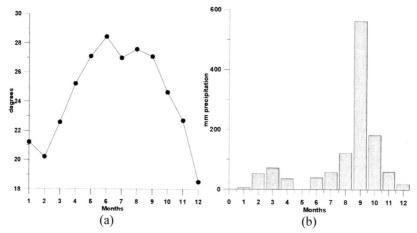

Figure 2:   Monthly mean temperature (a) and rain (b) for Tampico in 1998.
(Source Secretaria de Agricultura and Recursos Hidraulicos,
Mexico.)

3   Method

3.1 Leachate sampling

In 2004, samples of leachate were collected with a small electric pump in May
(dry season) from venting wells, and in October (rainy season) from small ponds
where leachate accumulates on the edges of the dumpsite. Samples were
collected after a clay cover was completed in April 2004. After collection,
samples were conserved, EPA 3051 [3], and cooled for their transportation to the
laboratory.

3.2 Soil sampling

Fifteen samples of soil beneath the dumpsite were collected in October 2004,
from a hole opened by an excavator machine after the removal of the waste
layer, in the southeast part of the dumpsite, considered the most active zone.
Samples were collected on the wall of the hole every 10 cm to a depth of 1,5 m.,
using small sections of PVC pipe (3.5cm diameter) drilled in horizontally.

3.3 Samples analysis

Metal content of leachate and soil was determined by Atomic Absorption
spectrometry, after acid digestion, EPA 3051 [3]. Results correspond to analyzed
metals and tracers with certified reference material in samples of leachate and to
results of the prospecting laboratory curve of for soil samples. Determinations in
soil and leachate were carried out with standards of primary type (CENAM) and
standards of international secondary type (High Purity).

4 Results

4.1 Leached

Leachate content of Pb, Ni, Cd and Cu are reported for dry and rainy seasons in 2004. Metal contents for the dry season are consistently of higher than those measured in the rainy season, fig. 3. The proportional concentration of metals is Ni> Pb > Cu >Cd.

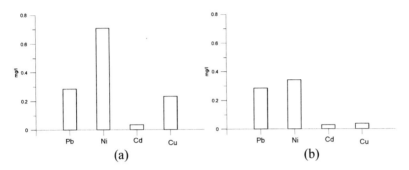

Figure 3:   a) Metal contents in leachate samples collected in dry (May) and b) rainy (October) seasons of year 2004.

4.2 Soil

Metal content (Pb, Ni, Cd, Cu, Cr, Fe, Mg and Mn) in soil beneath the dumpsite (layers 0-150 cm depth) generally decreases with depth, fig. 4, 5 and 6.

The observed decrease in metal concentrations with depth may be grouped in two general patterns. One for Pb, Cu, Cd with a pronounced decrease in the first layers to 60-80 cm, and no significant variation thereafter, fig. 4.

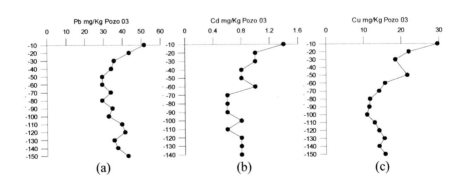

Figure 4:   (a) Pb, (b) Cd and (c) Cu contents in soil profile beneath the dumpsite.

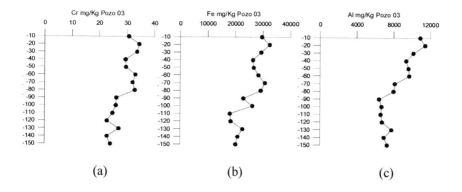

Figure 5:   (a) Cr, (b) Fe and (c) Al contents in soil profile beneath the dumpsite.

The second pattern is observed with Cr, Fe and Al where it is evident that a nearly linear decrease tendency of metal content with depth, fig. 5.

Variations of concentration with depth for Ni does not fit within these two patterns. Vanadium concentrations appears rather constant between 49.2-67 mg/kg., except in 60 and 90 cm depth layer. At this point a sudden increase was observed. The concentration of calcium contained in the sediments, its show an inversely proportional relation with Vanadium, which was observed high concentration in 60-90 cm depth, fig. 6(a) and 6(b).

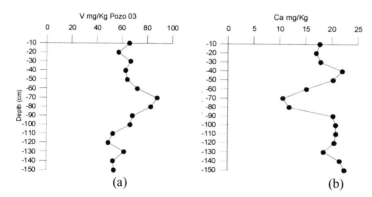

Figure 6:   (a) Vanadium and (b) Calcium contents in soil profile beneath the dumpsite.

5 Discussion

The comparison of leachate metal contents between dry and rainy seasons denotes a considerable dilution effect. The characteristic of the proportional concentration of leached metal concentration is compared with concentrations of these metals in Ehrig [4]. The author presents the metal concentration in typical leachate from a landfield. The average concentration presented for Pb, Ni, Cd and Cu, corresponds to 0.090, 0.200, 0.006, and 0.080 µg/l with the next expression: Ni>Pb>Cu>Cd. A similar proportion is reported to minimum and maximum conditions in metal concentrations in a German landfield. This condition is similar to our results of Zapote dumpsite, fig. 3a and 3b.

The comparison of the soil metal contents from Zapote dumpsite with "Screening Quick reference table for inorganic in solids", Buchman [5] gives an idea of the alteration level of our samples based on the reference data reported for freshwater and marine contaminated sediments. According to Buchman's criteria, the Zapote soil correspond to sediments with low to intermediate affectation in 0-50 cm layers for the metals analyzed here. It is important to mention that the proportional metal concentrations of background sediments correspond to Pb, 16000 ppb, Ni 13000 ppb, Cd $\cong$ 0.0 and Cu 17000 ppb or (Pb=Cu)>Ni>Cd, while contaminated sediments correspond to Pb 46700-218000 ppb, Ni 20900-51600 ppb, Cd 1200-9600 ppb and Cu 34000-270000 ppb or Pb>Cu>Ni>Cd.

The limits established for contaminated soils according to MENVIQ [6] correspond to class "A" (natural background concentrations) for the layers between 0-50 cm.

The results for layers between 0 and 60-90 cm depth, exceeds concentrations reported for contained metals in sediments of local lagoon systems. Villanueva et al. [7], Villanueva and Botello [8], Vázquez et al. [9], Alvarez-Rivera et al. [10] and Botello et al. [11]. The metal concentrations reported for layers 90 a 150 cm are close to reported concentrations of the metals, however, is higher than the criteria for sediments of freshwater and closer than natural marine sediments.

The reported data allows us to infer, according to the affectation of the sediments criteria of Buchman [5] correspondent zone 0-60 cm prevails affectation conditions from medium to low, here the condition is Pb>Cu>Ni>Cd. According to this criteria the soil in the 50-90 cm layers where the affectation conditions are less important, fig. 6, the condition Pb>Ni>Cu>Cd, can be considered as a transition zone. At last we deduct the proportion to finalize the metal concentration, which characterizes with Pb>(Cu=Ni)>Cd.

6 Conclusions

The conditions of dry and wet seasons impact the concentration of metals due to a dilution effect. We showed evidence of metal transfer from leachate downwards into soil layers beneath Zapote dumpsite. The higher concentrations observed in superficial layers (0-50 cm) may be evidence of clay both impermeability and adsorption capacity. The metal concentration in the 60-90 m and 90-150 m depth correspond to lower metal concentration zones closer to natural background.

This study is prospective in nature and will be complemented by a more intensive sampling protocol of both leachate and soil, either in time and space. However, our results shed the first light into to possible environmental impacts of this ample and long lived dumpsite unwisely opened and operated for 25 years in a marshland.

References

[1]   Tamayo J.I. Geografía Moderna de México, Editorial Trillas, Mexico, 398 p. 1975.

[2]   UAT-DEP-FI-, DM-MT and CICATA-UA-IPN SES Rehabilitation of the Isleta del Zapote Tampico, Mexico, Proposals for the rehabilitation of the closed municipal dumpsite, Tampico, Tamaulipas, Mexico 22 p, 2004.

[3]   Environmental Protection Agency Method 3051, 14 p., 1994.

[4]   J. Ehrig H.-J., Stau-und Sickerwasser-behandlungsverfahren Überblick. Had, 7. Erg.-Lfg.2.Aufl.,Mai pp 1-17 , 1997.

[5]   M. Buchman., Screening Quick Reference Table for Inorganics in Solids. NOAA, HAZMAT.1999

[6]   MENVIQ (Ministere de l'Environnement de Québec). Terrains contaminés, guide des methodes de conservation et d'analyses des echantillons déau et de sol. 93 p. 1990.

[7]   S. Villanueva-Fragoso, A. Botello A., F. Páez-Osuna., Evaluación de algunos Metales Pesados en Organismos del Río Coatzacoalcos y de la Laguna del Ostión, Ver., México. Contaminación Ambiental, 4:19-31, 1988.

[8]   S. Villanueva-Fragoso., A. Botello. Metales Pesados en la Zona Costera del Golfo de México: Una Revisión. Rev. Int. Contam Ambiental 8(1):47-61, 1992.

[9]   F. Vázquez., L. Aguilera., V. Sharma., Metals in Sediments of San Andres Lagoon,Tamaulipas, México. Bull. Environ. Contam. & Toxicol. 52:382-387, 1994.

[10] U. Alvarez-Rivera , L. Rosales-Hoz, A. Carranza-Edwards. Heavy Metals in Blanco River Sediments, Veracruz, México. An. Inst.Cienc.del Mar y Limnol.13:1-10, 1986.

[11] A. V. Botello, S. Villanueva-Fragoso y L. Rosales-Hoz, Distribución y contaminación de metales en el Golfo de México. en Caso M., I. Pisanty y E. Ezcurra (eds) Diagnostico ambiental del Golfo de México, SEMARNAT, INE, IEAC, Harte RIGOMS., Mexico pp 683-712. 2004.

The durability of stabilized sandy soil contaminated with Pb(NO$_3$)$_2$

S. R. Zeedan[1] A. M. Hassan[2] & M. M. Hassan[2]
[1]*Raw Materials and Processing Department,*
National Center for Housing and Building Research, Giza, Egypt
[2]*Sanitary and Environmental Engineering Department,*
National Center for Housing and Building Research, Giza, Egypt

Abstract

The purpose of this study was to investigate the effectiveness of cement kiln dust (CKD) and silica fume (SF) for the safe disposal of hazardous waste contaminated sand soil with toxic heavy metal as lead nitrate (Pb (NO$_3$)$_2$. The curing time and the leaching test through immersion in water are test methods for studying the durability of the stabilized/solidified (S/S) hazardous wastes under the soaking condition that leachate through the compacted samples. These methods simulate the leaching by immersion process; S/S hazardous waste under particular landfill conditions, when the S/S contaminated soil is more permeable than its surrounding materials or when the deterioration of solidified waste has reached a statue that ground water can flow – through the compacted soil via the porosity system of S/S waste matrix. The sandy soil was mixed with mixture binder (10% CKD and 2% SF) and mixed by dissolving three different ratios of Pb(NO$_3$)$_2$ (O.5, 1, and 3%) in tape water and compacted. UCS was determined for 1 and 3 month curing (100% moisture humidity) and for one year soaking in tap water. The concentration of (Pb^{+2} and NO^{-3}) which leachate from the compacted samples to the soaking water was recorded along a time of one year soaking. The UCS was decreased for the compacted samples have high levels of lead nitrate. The leaching of Pb was decreased during the soaking time for the two curing times (30 and 90 days) for the three ratio of Pb(NO$_3$)$_2$.
Keywords: solidification/stabilization, leaching, CKD, SF, heavy metal.

WIT Transactions on Biomedicine and Health, Vol 10, © 2006 WIT Press
www.witpress.com, ISSN 1743-3525 (on-line)
doi:10.2495/ETOX060271

1 Introduction

The type of waste, cost, legislation and technology limit disposal options for industrial waste. Studies have shown that solidification/stabilization (S/S) processes are viable for most metallic waste streams (Vipulanandan [1]), S/S of hydration waste involves mixing the waste with binder material to enhance the physical properties of the waste and to immobilize any contaminants that may be detrimental to the environmental conditions. Several binder systems are currently available and widely used for S/S (Mayers *et al* [2], Jones *et al* [3]). Binding reagents commonly used include Portland cement (OPC), cement kiln dust (CKD) lime and a number of proprietary reagents. Solidification and stabilization S/S immobilization technologies are the most commonly selected treatment options for metal-contaminated sites (Coner *et al* [4]). Solidification involves the formation of a solidified matrix that physically binds the contaminated material. In case of remediation projects, S/S is often the only reasonably available technology for treating the large volumes of heavy metals – contaminated soil, sludge, or sediments resulting from these operations cement or cementing material is uniquely suited for use as a S/S reagent for metal contaminants because it reduces the mobility of inorganic compounds by 1) formation of insoluble hydroxides, carbonates or silicates; 2) substitution of the metal into mineral structure and 3) physical encapsulation (Adaska *et al*. [5] and Bhatty *et al*. [6]). Alkalies in clinker kiln dust accelerated the setting and hydration of cement and quick setting time (QSA) influenced quick setting and increased the compressive strength of cement. The CKD and QSA modified by adding 2% SF accelerated the QSA. The least amount of heavy metals leached and the highest compressive strength was due to a large proportion of the formation of hydrates and the most effective stabilization of hazardous waste containing multiple heavy metals.

Cement stabilization is one of the most widely used waste stabilization methods for metal contaminants (Cheng *et al*. [7], Cocke [8], Li [9], and Poon and Chen [10]). This was due to the high buffering capacity and high pH value, where most metals are insoluble. Stabilization, also referred to as fixation, usually uses chemical reactions to convert the waste to a less mobile form. The general approach for solidification/stabilization treatment processes involves mixing or injecting treatment agents into the contaminated soils. Inorganic binders, such as cement, fly ash or blast furnace slag, and organic binders such bitumen are used to form crystalline glassy or polymeric frameworks around the waste. Abdel-Ghani *et al*. [11], studies the 10% CKD with 2% SF when compacted with sandy soil gives the best result and high resistance to environmental impact. The aim from reused the CKD waste as a binder material to solidified the contaminated sand soil with Pb $(NO_3)_2$ as economic process. Many studies show the effect of the flow rate of water on the solidification processes but this paper studies the effect of brackish water on the solidification process. By studying the durability effect on the compacted contaminated samples and on the effect of the brackish water by using a tap water as soaking media due to seepage or a rise in the groundwater level in landfill constructions as a leaching filled system. The UCS was determined as mechanical properties to study the ability of this stabilized soil to carry external loads and the maximum strength will be calculated to classify this stabilized soil and determine the kind of structure that will be established on it.

2   Methodology and material

2.1 The soil

Sandy soil as sample of a desert in Egypt from 6th October city.

2.2 Binders

The binding material was a mixture of 10% CKD and 2% SF of the total weight of the batch. It was used to solidify a synthetic contaminated sandy soil samples. The chemical compositions of both binders are given in Table 1.

Cement kiln dust (CKD) was obtained from Helwan Portland Cement Company as a waste material. Silica fume (SF) is a by-product of the smelting process for silicon metal and ferrosilicon alloy production. It was taken from the Egyptian company for Ferrosilicon Alloys situated in Edfo, Aswan.

Table 1:   Chemical analysis of the soil and binders.

Symbol	Soil	CKD	SiF
SiO_2 (%)	94.32	14.12	94.64
Al_2O_3 (%)	1.68	4.75	.97
Fe_2O_3 (%)	1.36	2.13	.93
CaO (%)	0.42	55.35	.55
MgO (%)	0.09	2.48	.35
SO_3 (%)	0.47	5.80	0.10
Cl- (%)	.25	3.96	-
Na_2O (%)	0.58	0.58	0.2
K_2O (%)	0.12	3.53	2.5
L.O.I. (%)	1.22	12.60	2.01
PH	7.33	12.2	8.5
TSS	2.74	-	-
OM	--	-	-
TSS	2.74	-	-

2.3 The contaminants

Three different ratios (0.5, 1, 3%) of toxic lead nitrate were used as a synthetic source of pollution with heavy metal, and were separately added to the marked sandy soil samples.

The research aimed to:

(a) use the waste CKD as a factor that chemically binds the polluted soil elements together, protecting the environment from the possible potential leak of the harmful content. CKD is waste material, so it is very economic to use as it serves double purposes: to get rid of it in a useful application instead of searching for an alternative method of disposal, and to limit the negative impact of the polluted sandy soil on the surrounding environment. CKD was deliberately selected because of its high alkalinity which helps to increase the efficiency of solidification of the soil, by chemically reacting with the contents, forming non-soluble salts of heavy metals resulting in immobilization of the toxic material. This action leads to the "imprisoning" of the dangerous content in the solidified soil in the shape of a non-soluble deposit.

(b) use the silica fume, which is also a waste material, with the cement kiln dust to enhance the soil strength and accelerate the setting of the hydration of CKD with the sand soil elements and accelerate the solidification process.

3 Solidified samples preparation

1) The sand soil groups of samples were polluted with 3 different ratios of $Pb(NO_3)_2$ then they were mixed with the binder mixture (10% CKD, 2% SF) with the optimum water content.
2) Samples were compacted employing the "modified effort" according to ASTM D [12].
3) There were typically three groups of samples, each of five sub-groups of samples to be cured for 1 and 3 months in a moisture ratio of 100%, soaked in tap water for one of the selected soaking times (30, 90, 180, 360 days), in enough quantities to afford the leaching pollution mobilized concentration and strength tests. They were kept in covered polyvinyl bottles at room temperature (25-30C°).

4 Testing procedure

The research program was carried out as follows:

- The strength of the curing stabilized soil specimens after soaking was determined using the unconfined compression test accordance ASTMD1633 [13].
- pH values were determined according to BS [14].
- the concentrations of Pb in the soaking water were determined after acidified with concentrated H_2NO_3 for each soaking according to AWWA [15].

5   Results and desiccation

5.1   Unconfined compressive strength (UCS)

UCS tests were initially performed on samples that had undergone two curing conditions at 1 and 3 months as reference values to compare the results that will be obtained for soaked samples in tap water as soaking media for one year was illustrated in table [2].

5.1.1   Effect of soaking time on UCS

The continuous decrease of strength (UCS) of the compacted solidified samples contained 1% and 3% $Pb(NO_3)_2$ after one month curing for the compacted samples, contained 3% $Pb(NO_3)_2$ after 3 months curing and soaking for one year is due to the acidic effect resulting from the presence of 5.4% of SO4 in CKD and the presence of the NO_2 in high percentage in the $Pb(NO_3)_2$, in addition to the presence of CO_2 in the air. All of them, having been dissolved in water, are responsible for the acidification of the soaking water (by the H_2NO_3, H_2SO_4 and H_2CO_3 respectively). These acidic waters in contact with the cementing materials give rise to the formation of calcium nitrates, sulfates, bicarbonates, etc. Some of these salts are highly soluble and easily leached, which causes pores that negatively affect the strength of the solidified samples. However, it has been recently proven that a strong synergy effect is produced by SO_2 and NO_2 when they exist together.

Table 2:   Unconfined compression strength of solidified polluted sandy soil samples.

Time	0.5% Lead Nitrate				1.0 % Lead Nitrate				3.0% Lead Nitrate			
	1	3	6	12	1	3	6	12	1	3	6	12
	UCS Values				UCS Values				UCS Values			
Curing	3416	2949			3198	2797			1143	1099		
Soaking after 1Month curing	1732	2004	2320	3299	1594	1223	1059	900	546	472.3	176	5.4
Decrease of Strength %	49.30	41.33	32.08	3.43	50.16	61.76	66.89	71.86	52.23	58.68	84.60	99.53
Soaking after 3 months curing	2594	2656	3240	4076	2275	1598	2942	3324	862	762	689	548
Decrease in Strength %	12.04	9.94	-9.87	-38.22	18.66	42.87	-5.18	-18.84	21.57	30.66	37.31	50.14

According to the standard guide of the ASTM D4609 6 [16], the UCS of the chemically stabilized soil must be increased by 345(kN/m^2). These values can be attained after one and three months of curing for all solidified samples. And according to U.S.EPA standards [17] the minimum values of UCS are $3.5(kN/m^2)$ for the disposal of solidified hazardous wastes in landfills. UCS values for 0.5% of $Pb(NO_3)_2$ for one year soaking for both curing time also the UCS for 1% and 3% $Pb(NO_3)_2$ after curing or soaking.

5.2 Leaching of Pb by soaking (mobilized concentration)

The initial concentration of Pb for the three different ratios of contamination by Pb $(NO_3)_2$, (0.5, 1 and 3%) were (1220 mg/L, 2240 mg/L and 7320 mg /L) in the contaminated soil/cement samples respectively. The mobilized concentrations of Pb pollution leached from the compacted specimens (cured for 1 and 3 months, soaked in tap water for the scheduled times: 1, 3, 6 and 12 months) are given in Tables 3 and 4. The same results are graphically represented in Figs. 1 to 6. It is obvious from the results that the mobilized (leaching) concentration of lead has continuously decreased with the increase of the soaking time throughout the year's duration of the experiment with the exception of the 3% polluted samples which shown an increase in the mobilized concentration. However, the decrease in concentration of the leaching pollution was more than 99% in all cases compared to the initial concentration. These are extremely positive results. The initial concentration was cut down by more than 99% during the 1st month of soaking irrespective of the initial pollution status (0.5, 1, 3%) Curing the samples for 3 months showed better results (decreasing in the mobilized concentration) than curing them for one month. The compacted samples of 3 months curing emitted less leaching pollution than the samples of 1 month curing did.

Table 3: Mobilized concentration of Pb leached from solidified sandy soil samples polluted with Pb $(NO_3)_2$ different ratios (0.5,1,3)% by weight, soaked for 30, 90, 180 and 365 days, after 1 month curing.

	Efficiency of Solidification ξ%								
	Initial Concentration of Lead								
	% weight = mg/l								
Time of Soaking	0.50%			1%			3%		
	(1220 mg/l)			(2440 mg/l)			(7320 mg/l)		
	Mobilized Concentr mg/l	Wt. of Mob. Concentr mg	ξ%	Mobilized Concentr mg/l	Wt. of Mob. Concentr Mg	ξ%	Mobilized Concentr mg/l	Wt. of Mob. Concentr mg	ξ%
30 Days	6.28	2.6	99.79	12.6	5.2	99.79	97	40.9	99.44
90 Days	3.2	1.35	99.89	21.8	8.99	99.63	115	47.4	99.35
180 Days	1.93	0.8	99.93	15.7	6.48	99.73	153	63.1	99.14
365 Days	0.78	0.32	99.97	6.28	2.59	99.89	187	77.1	98.95

Table 4:   Mobilized concentration of Pb leached from solidified sandy soil samples polluted with Pb $(NO_3)_2$ different ratios (0.5,1,3)% by weight, soaked for 30, 90, 180 and 365 days, after 3months curing.

	Efficiency of Solidification ξ %								
	Initial Concentration of Lead								
	% weight = mg/l								
	0.50%			1%			3%		
Time of Soaking	(1220 mg/l)			(2440 mg/l)			(7320 mg/l)		
	Mobilized Concentr. mg/l	Wt. of Mob. mg Concent	ξ %	Mobilized Concentr. mg/l	Wt. of Mob. mg Concent.	ξ %	Mobilized Concentr. mg/l	Wt. of Mob. mg Concent.	ξ %
30 Days	4.15	2.6	99.79	10.8	4.5	99.82	73	30.1	99.59
90 Days	2.05	1.9	99.84	7.7	3.15	99.87	87	35.9	99.51
180 Days	0.6	0.53	99.96	2.3	0.95	99.96	113	46.6	99.36
365 Days	0.22	0.1	99.99	0.75	0.31	99.99	141	58.2	99.20

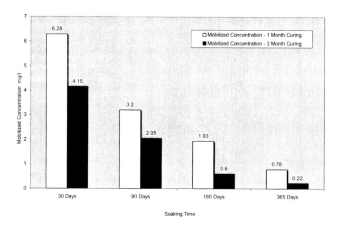

Figure 1:   Mobilized concentration of Pb leached from solidified sandy soil samples polluted with Pb $(NO_3)_2$ 0.5% by weight, soaked for 30, 90, 180 and 365 days, cured for 1 and 3months.

The decrease in concentration of lead might be due the increase in the high pH value of the compacted samples (about 10-11) and the brackish soaking water (about 8.5-9.6) in the early time of soaking but this increased gradually to 10.5 to 11 in up to one year so that the solubility of lead decreased to its lowest solubility and precipitate in the form of insoluble salts (M. D. Largrega *et al.* [18]), the forming of insoluble precipitates in different samples producing higher hydraulic resistance and due to the hydration of a cementitious matrix and the formation of CSH gel which increased with time, so that when the water entered the samples by soaking, the porosity of the compacted samples were consequently reduced.

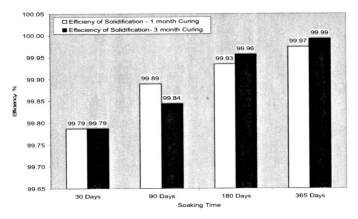

Figure 2:   Efficiency of solidification effect on reduction of leachant pollution from samples polluted with Pb $(NO_3)_2$, 0.5% by weight = 1220 mg, soaked for 30, 90, 180 and 365 days, cured for 1 and 3 months.

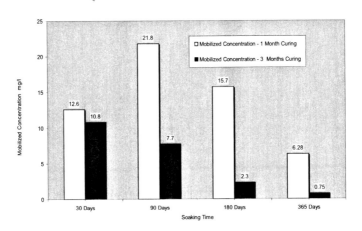

Figure 3:   Mobilized concentration of Pb leached from solidified sandy soil samples polluted with $Pb(NO_3)_2$, 1.0% by weight, soaked for 30, 90, 180 and 365 days, cured for 1 and 3 months.

6   Conclusions

The results proved the feasibility of reducing the concentration of the leaching pollution (mobilized concentration) coming out of the polluted soil that had undergone solidification/stabilization, by curing (for 1 and 3 months), and soaking for up to 1 year.

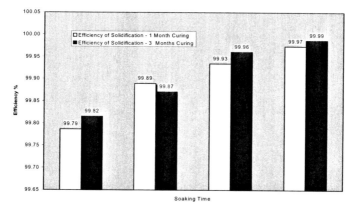

Figure 4:   Efficiency of solidification effect on reduction of leachant pollution by lead after 1 and 3 months curing. Soaking time 30, 90, 180 and 365 days. Values for samples with 1% initial pollution.

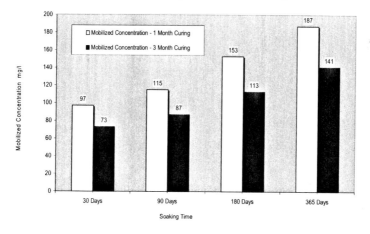

Figure 5:   Mobilized concentration of Pb leached from solidified sandy soil samples polluted with Pb (NO$_3$)$_2$, 3.0% by weight=7320mg. Soaked for 30, 90, 180 and 365 days, cured for 1 and 3 months.

1. The level of pollution (by lead) of the water was reduced to less than the maximum allowable in the applicable Environment Laws. There would be no harmful effect if the water were to be used in agriculture, or if the water leaked into the underground.
2. Lead mobilized concentration was reduced to 0.22 mg/l and 0.75 mg/l for the samples with 0.5% and 1% initial pollution respectively, when cured for 3 months and soaked for 1 year.
3. Lead mobilized concentration was reduced to 0.78 mg/l and 6.28 mg/l for the samples with 0.5% and 1% initial pollution respectively, when cured for 1 month and soaked for 1 year.

4. Lead mobilized concentration was reduced to 187mg/l and 141mg/l for the samples with 3% $Pb(NO_3)_2$ initial pollution, when for 1 month, 3 months respectively and soaked for 1 year.
5. The strength (UCS) of samples with 0.5% and 1% $Pb(NO_3)_2$ initial pollution, was found to have decreased in the first six months of soaking, then increased in the 2nd half of the soaking time (1 year).
6. The strength (UCE) of samples with 3% $Pb(NO_3)_2$ initial pollution experienced a continuous gradual decrease throughout the 1 year of soaking. However, the value of UCS was found to be within the accepted limits, despite the continuous decrease.
7. The cement kiln dust (CKD) and silica fume mixture (10% CKD and 2% SF) can be used as solidified materials for $Pb(NO_3)_2$ contaminated sand soil in small ratios (0.5% and 1%).
8. All compressive strength results can be used in landfill structures.
9. The curing time plays a good role in the solidification/stabilization process.
10. With a high ratio of $Pb(NO_3)_2$, there must be a high ratio of CKD.

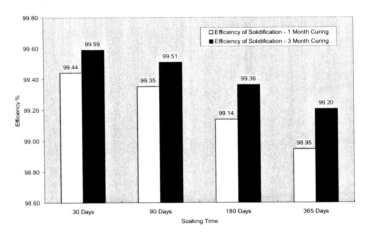

Figure 6: Efficiency of solidification effect on reduction of leachant pollution from samples polluted with Pb $(NO_3)_2$, 3% by weight = 7320 mg, soaked for 30, 90, 180 and 365 days, cured for 1 and 3 months.

References

[1] Vipulanandan, S. Krishnan, XRD analysis and leachability of solidified phenol -Cement mixtures, Cem, Conc. Res. 23(4) (1993) 792-802.
[2] T.E. Myers, A Simple procedure for acceptance testing of freshly prepared solidified waste, in: J.K.Peros Jr., W.J Lacy, R.A. Conway (Eds.), Hazardous and Industrial Solid Waste Testing and Materials, Philadelphia, PA, 1986 .
[3] L. W. Jones, Interference Mechanisms in waste solidification / stabilization processes, EPA-600/s-89/067(NTIS No. PB90-

156209/AS),USEPA. Risk Reduction Engineering Laboratory, Cincinnati, OH,1989 .

[4]   Coner, J.R. Chemical Fixation and solidification of hazardous wastes; Van Nostrand Reinhold: New York, 1990.

[5]   Adaska, W.S. et al. Solidification and stabilization of wastes using Portland cement, 2nd Edition ;EBO71; Portland Cement Association: Skokle, IL,1198

[6]   Bhatty, J.L. et al. stabilization of heavy metals in Portland cement, Silica fume / Portland cement and masonry cement matrices; RP348; Portland Cement Association : Skokle, IL., 1999.

[7]   Cheng, K.Y. (1991). Controlling Mechanisms of Metals Release from Cement –Based waste from in acetic acid solution, PhD. Thesis, University of Cincinnati, Cincinnati.

[8]   Cocke, D.L. (1990) Journal of Hazardous Materials, 24,-231-253.

[9]   Li, X.d,. Poon, C.S., Sun, H., Lo, I.M.C. and Kirk, D.W. (2001) Journal of Hazardous Materials,A82,215-230.

[10]   Poon, C.S., Chen, Z.Q. and Wai, O.W.H. (2001) Journal of Hazardous, B81, 179-192.

[11]   Abdel-Ghani, Kh., I., "Strength Durability of Sandy Soil Stabilized by Preheater Fines", Journal of the Egyptian Geotechnical Society, Cairo, Vol. 13, Part 1, June, 2002.

[12]   ASTM D698-91, "Standard Test Method for Laboratory Compaction Characteristics of Soil Using Standard Effort", 1998.

[13]   ASTM1633 compressive strength of molded soil cement cylinders

[14]   British Standard Institution Methods of test for soils for civil engineering, BS1377, BS.1975

[15]   Standard methods for the examination of water and wastewater .18th edition. Am. Works Assn. (AWWA), Washington, D.C, (1992).

[16]   STM D 4609 "Standard guide for screening chemicals for soil stabilization (2002).

[17]   U.S. EPA.(1996). Test. Methods for Evaluation solid wastes, Physical /chemical methods, SW-846, 3rd Edition. Office of solid waste and emergency response, Washington, D.C.

[18]   Largrega.M.D.Buckingham P.L, and Evans J.C. Hazardous Waste Management, McGraw Hill Inc., USA, 640-696, 1994.

Section 9
Biodegradation and
bioremediation

Toxicity and biodegradation potential of waste water from industry producing fire protection materials

T. Tišler, M. Cotman & J. Vrtovšek
National Institute of Chemistry, Ljubljana, Slovenia

Abstract

Effluents may contain hazardous chemicals, which could have adverse effects on aquatic ecosystems. Important intrinsic properties are toxicity and biodegradability of effluents and they should be assessed in order to protect the aquatic environment. Microorganisms of activated sludge are responsible for biological waste water treatment. The lack of treatment effectiveness could, in case of unexpected toxic influents to waste water treatment plants, lead to a discharge of toxic and undegradable effluents into receiving streams. The assessment of potential adverse effects of influents entering the waste water treatment plant is for this reason necessary. The aim of the study was to predict a suitability of discharging the industrial waste water into waste water treatment plants based on the assessment of waste water toxicity and biodegradability. The waste water sample, originating from the industry producing fire protection materials, was highly polluted with organic substances. At first, the acute toxicity of the investigated sample was determined using activated sludge measuring oxygen consumption rates. It was found that the waste water sample was acutely toxic to microorganisms. Ready biodegradability of waste water was determined in a closed respirometer by measuring oxygen consumption. As diluted samples of the waste water were not readily biodegradable inherent biodegradability of the waste water sample was evaluated using the Zahn-Wellens test. It was found that diluted samples were inherently biodegradable probably due to the fact that higher concentration of microorganisms of activated sludge was added. It can be concluded that the tested waste water could be discharged into the sewerage system on the condition that a dilution of the tested waste water in the municipal waste water treatment plant would be at least 1:20.
Keywords: activated sludge, industrial waste water, inherent biodegradability, oxygen consumption, ready biodegradability, toxicity, Zahn-Wellens test, waste water treatment plant.

WIT Transactions on Biomedicine and Health, Vol 10, © 2006 WIT Press
www.witpress.com, ISSN 1743-3525 (on-line)
doi:10.2495/ETOX060281

1 Introduction

Aquatic ecosystem is vulnerable to pollution as many chemicals are released into the aquatic environment during their production and final use. One of the most important aspects of environmental protection is biodegradability of chemicals as a biodegradable substance is anticipated to cause less ecological problems than a persistent one. Major pollution is contributed to by effluents as they are important sources of chemicals entering aquatic ecosystems. Therefore, efficient treatment systems are needed for reduction of pollution and biological waste water treatment plants (WWTP) using activated sludge are the most important facilities for cleaning effluents. The WWTP system is based on the biodegradation processes of organic substances similar to that in the aquatic environment [1].

Due to the fact that biodegradation is an important process in WWTP for removing organic substances biodegradability of the inflow of waste water should be predicted before entering the WWTP. Many testing strategies have been described for biodegradability assessment of pure chemicals involving standardised tests initiated or updated by ISO and OECD [2]. However an assessment of waste water biodegradability could be a problem due to unknown composition of waste water leading to analytical limitations [3]. Three groups of tests were defined in order to assess biodegradability of pure chemicals: (1) ready biodegradability test (2) potential (inherent) biodegradability test (3) simulation test. Tests for ready biodegradability assessment i.e. carbon dioxide evolution test, manometric respirometry, closed bottle test indicate if a chemical is degradable without problems under natural conditions. These tests basically distinguish between readily biodegradable chemicals and others. In a case of negative result obtained a test for inherent biodegradability is required. Higher concentration of microorganisms is added to samples for providing optimal conditions for degradation and the Zahn-Wellens test is frequently used for this purpose. Finally, a simulation test is used to provide information about behaviour of substance in specific environmental conditions [4–6].

The aim of our study was to predict biodegradability of industrial waste water originating from the industry producing fire protection materials in the WWTP. At first acute toxicity of the waste water sample was determined using activated sludge to eliminate possible inhibition of biodegradation in further experiments due to waste water toxicity. Ready biodegradability of the waste water sample was determined by measuring the oxygen consumption in a closed respirometer. In the next step, the Zahn-Wellens test was used to assess the inherent biodegradability of the waste water sample. A suitability of discharging the investigated waste water into WWTP was predicted on the results obtained.

2 Material and methods

In the waste water sample the pH [7], COD [8], and BOD_5 [9] were determined.

All experiments on toxicity and biodegradability assessment were conducted in a temperature controlled room at $21 \pm 1^\circ C$.

2.1 Acute toxicity of waste water

The acute toxicity test with activated sludge was performed according to the ISO standard [10] using high concentration of microorganisms (1500 mg/L of suspended solids). Microorganisms of activated sludge were obtained from the aeration tank of the laboratory municipal waste water treatment plant and prior testing the sensitivity of microorganims was checked with the reference chemical (3,5-dichlorophenol). Oxygen consumption of the samples was measured with an oxygen electrode (WTW Oximeter, OXI 96) following the biochemical degradation of easy degradable meat extract, peptone, and urea until the oxygen concentration dropped below 1 mg/L. The calculated oxygen consumption rates of different concentrations of waste water were compared to the control without added waste water and the percentages of inhibition were calculated for each waste water concentration. The percentages of inhibition were plotted against corresponding concentrations of waste water on the semi logarithmic paper. The test results were expressed as effective concentrations, which reduced the oxygen consumption by 20%, 50%, and 80% (EC20, EC50, EC80 values) and were determined using linear regression analysis. The EC20 value was taken as a toxicity threshold.

2.2 Ready biodegradability

Aerobic ready biodegradability of the waste water was determined in a closed respirometer (Baromat WTW, BSB – Messgerät, Model 1200) by measuring the oxygen consumption [11]. The final concentration of microorganims was 30 mg/L of suspended solids using the same source of activated sludge as in the toxicity test. The oxygen consumption was measured during 28 days in the following samples; the samples containing 3, 4, 5 v/v % of the waste water (test medium, corresponding volume of waste water, inoculum), a blank control (test medium, inoculum), a control to check the sensitivity of inoculum (test medium, sodium acetate, inoculum), and an abiotic control (test medium, waste water, HgCl$_2$ solution). Biodegradation curves were plotted as the percentages of biodegradation for each concentration of the waste water versus time. A final level of biodegradation, a lag phase and a degradation time were the parameters used for biodegradability assessment.

2.3 Inherent biodegradability

Aerobic inherent biodegradability of the waste water was determined using the Zahn-Wellens test [12]. The same source of microorganisms was used as in the previous experiments with the final concentration of 400 mg/L of suspended solids. The same samples with continuous homogenisation and aeration were tested as in the ready biodegradation experiment. The elimination of organic matter in the samples was followed by dissolved organic carbon (DOC) measurements (TOC Analyser - 5000A, Shimadzu) during 28 days. The evaluation of biodegradability was carried out in the same manner as in the ready biodegradability assessment.

3   Results and discussion

The pH of the sample was 10.3, the waste water was polluted with organic matter; the COD and BOD_5 were 3659 mg/L and 901 mg/L, respectively.

3.1 Acute toxicity to activated sludge

A waste water toxicity was determined using the acute toxicity test with activated sludge (Table 1) to eliminate possible inhibition of biodegradation in further experiments.

Table 1:   Inhibition of oxygen consumption in different concentrations of waste water.

Concentration of waste water (v/v %)	Oxygen consumption rates (mg/L.h)	Inhibition (%)
0	0.493	/
1	0.506	0
3	0.468	5
7	0.173	65
10	0.064	87
20	0.015	97
40	0.005	100
70	0.005	100
Abiotic - 40	0	/

The EC values were calculated from the obtained results (Table 1):

30 min EC20 = 3.7 v/v %
30 min EC50 = 5.9 v/v %
30 min EC80 = 8.9 v/v %

The waste water sample was highly toxic to microorganisms of activated sludge. The sample of 5.9 v/v % of waste water inhibited the oxygen consumption by 50% and a toxicity threshold was found in the sample containing 3.7 v/v % of waste water.

3.2 Ready biodegradability

In regard to the obtained toxicity, the concentrations of waste water sample used in biodegradability tests were between 3 and 5 v/v% of the waste water sample. The results obtained in the ready biodegradability test are given in Fig. 1.

   In all tested samples a lag phase (2-4 days) was observed; the highest and the lowest final levels of biodegradation were found in the samples containing 3 v/v % (47%) and 5 v/v % (31%) of the waste water, respectively. The non-toxic sample containing 3 v/v % of the waste water was not readily biodegradable according to the recommendations for the ready biodegradability classification of pure chemicals as the "pass level" of biodegradation in the O_2 and CO_2 tests was not achieved [6].

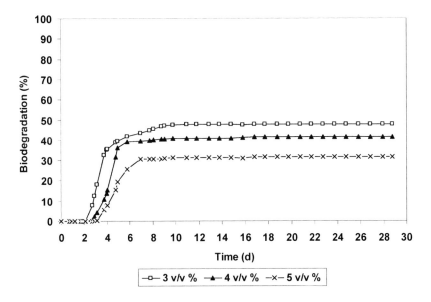

Figure 1:   Biodegradation curves of the samples containing 3, 4, and 5 v/v % of waste water – closed respirometer.

3.3   Inherent biodegradation

The results obtained in a test for inherent biodegradability assessment are given in Fig. 2.

In the Zahn - Wellens test biodegradation started immediately without a lag phase and the biodegradation was completed in 12 days. The same final levels of biodegradation (about 80%) were determined in the samples containing 3 and 5 v/v % of the waste water (Fig. 2). We concluded that both samples were inherently biodegradable as biodegradation was higher than 70% [2]. Better biodegradability of the tested sample was probably due to higher concentration of microorganisms added in the Zahn – Wellens test.

The final levels of biodegradation obtained in the O_2 test (31% – 47%) and the Zahn-Wellens test (80%) were higher than the BOD_5/COD ratio (25%). The reasons are probably more favourable conditions in biodegradability tests such as duration of experiments and amount of added microorganisms.

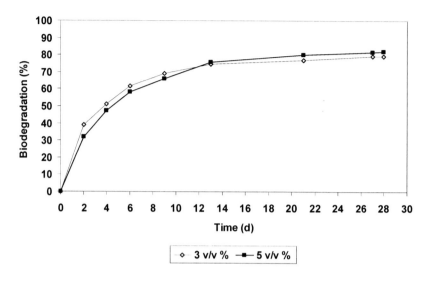

Figure 2: Biodegradation curves of the samples containing 3 and 5 v/v % of waste water – Zahn – Wellens test.

4 Conclusions

The waste water sample, obtained from the industry producing fire protection materials, was acutely toxic to activated sludge; a toxicity threshold was found even at 3.7 v/v %. All tested samples of the waste water were not readily biodegradable. However it was found that the investigated sample were biodegradable in the Zahn - Wellens test in the samples containing 3 and 5 v/v % of the waste water. The experimental conditions in the Zahn-Wellens test are closer to a WWTP system than an oxygen consumption measurements in a closed respirometer. Based on the results obtained we can conclude that the investigated waste water could be treated in the municipal waste water treatment plant on condition that the inflow of this waste water will not exceed 5 v/v % of total volume of waste waters. It means that a ratio of the investigated waste water and total inflow in the WWTP should never be less than 1:20.

References

[1] Pagga, U., Testing biodegradability with standardized methods. *Chemosphere*, **35(12)**, pp. 2953-2972, 1997.
[2] Painter, H.A., Detailed review paper on biodegradability testing, OECD Guidelines for the testing of chemicals, OECD Paris, 1995.
[3] Nyholm, N., Biodegradability characterisation of mixtures of chemical contaminants in waste water – The utility of biotests. *Wat Sci Tech*, **33(6)**, pp. 195-206, 1996.

[4]   Žgajnar-Gotvajn, A., Zagorc-Končan, J., Whole effluent and single substances approach: A tool for hazardous waste water management. *Wat Sci Tech*, **37(8)**, pp. 219-227, 1998.
[5]   Lapertot, M.E., Pulgarin, C., Biodegradability assessment of several priority hazardous substances: Choice, application and relevance regarding toxicity and bacterial activity. *Chemosphere*, in press.
[6]   Struijs, J., van den Berg, R., Standardized biodegradability tests: Extrapolation to aerobic environments. *Wat Res*, **29(1)**, pp. 255-262, 1995.
[7]   International Organisation for Standardization, International Standard ISO 10523. Determination of pH. Geneve, 1996.
[8]   International Organisation for Standardization, International Standard ISO 6060. Water Quality – Determination of chemical oxygen demand. Geneve, 1996.
[9]   International Organisation for Standardization, International Standard ISO 5815 - 1. Water Quality – Determination of biochemical oxygen demand after n days (BODn), Dilution and seeding method with allylthiourea addition. Geneva, 2003.
[10]   International Organisation for Standardization, International Standard ISO 8192: Water Quality - Test for inhibition of oxygen consumption by activated sludge. Geneve, 1986.
[11]   International Organisation for Standardization, International Standard ISO 9408: Water Quality – Evaluation of ultimate aerobic biodegradability of organic compounds in aqueous medium by determination of oxygen demand in a closed respirometer. Geneve, 1999.
[12]   International Organisation for Standardization, International Standard ISO 9888: Water Quality - Evaluation of the aerobic biodegradability of organic compounds in an aqueous medium - Static test (Zahn-Wellens method). Geneve, 1999.

Isolation and characterization of endophytic bacteria of coffee plants and their potential in caffeine degradation

F. V. Nunes & I. S. de Melo
Laboratory of Environmental Microbiology,
Embrapa Environment Jaguariúna, SP, Brazil

Abstract

The plant kingdom is colonized by a diverse array of endophytic bacteria, which form a non-pathogenic relationship with their hosts. Endophytic bacteria may confer benefits to the plant, and the benefits may be reciprocal, resulting in an enhanced symbiotic system for specific plant characteristics. The objective of this study was to isolate endophytic bacteria from coffee (*Coffea arabica* and *C. robusta*) and evaluate their potential in degrading caffeine. The isolates were identified by partial 16S rDNA sequence analysis and by FAME and the analysis of biodegradation of caffeine was carried out by HPLC. As a result, 252 bacterial strains were isolated, with most of them belonging to the species *Bacillus lentimorbus, B. megaterium, B. subtilis, B. cereus, Pseudomonas putida, P. chlororaphis, Pantoea ananatis, P. agglomerans, Stenotrophomonas malthophilia, Kluyvera cryocrescens, Kocuria kristinae* etc. Approximately 20% of the bacterial strains showed the ability to grow in the presence of high concentrations (5.000 mg.L^{-1}) of caffeine and two *Pseudomonas putida* strains completely degraded the alkaloid, showing the potential of endophytic bacteria in decaffeination processes.
Keywords: endophytic microorganisms, caffeine, coffee, biodegradation.

1 Introduction

Endophytes are microorganisms that live within the tissues of healthy plants where they cause symptomless infections. Their significance to the host is still unclear; pathogens as well as non-pathogens can live endophytically. Endophytic bacteria can contribute to the health, growth and development of plants. Plant growth promotion by endophytic bacteria may results from direct effects such as the production of phytohormones or by providing the host plant with fixed nitrogen or the solubilization of soil phosphorus and iron (Glick [1]; Shishido et al. [2]; Kinkel et al. [3]; Sturz et al. [4]). Some endophytic microorganisms have the ability to grow and degrade many xenobiotics. Several studies were carried out to investigate the use of caffeine, as a source of energy for microorganisms. The fungi *Penicillium* and *Aspergillus* and the bacteria *Pseudomonas* are the most frequent caffeine – degrading genera. (Asano et al. [5]; Kurtzman and Shwimmer [6]). The objective of this study was to identify endophytic bacteria from coffee and evaluate their potential to degrade caffeine.

2   Material and methods

2.1 Isolation of endophytic bacteria

Leaves, stems and roots of healthy plants of *Coffea arabica* and *C. robusta* were surface sterilized with ethanol 70% per 1 min., sodium hipoclorite 2,5% per 20 min.; and then in ethanol 70% per 30 sec., following three successive washes in distilled sterilized water. The fragments (1 cm) were transferred to tryptic soy-agar medium. The cultures were incubated at 28°C for seven days and the colonies were purified and stored in glycerol at –80°C.

2.2 Strain identification

Isolated strains were screened for the selection of caffeine-resistant strains. Each strain was identified by analysis of fatty acids methyl-esters (FAME) using the Microbial Identification System developed by Microbial ID (MIDI, Neward, DE) and by 16S rDNA sequence analysis.

2.3 Biodegradation of caffeine

The dissipation of caffeine by endophytic bacteria was monitored in mineral medium (Czapeck) supplemented with methylxanthine (5 g.L^{-1}). The analysis of caffeine was carried out on a Shimadzu 10 AS-UP HPLC with photodiode array detector and inertsil 5 ODS-3 (5 µm, 150x4,6 mm) column. The mobile phase was acetic acid/ acetronitrile (90:10, v/v), with a flow rate of 1,0 mL min^{-1}, by an isocratic system.

3   Results and discussion

A total of 252 endophytic bacteria were isolated from coffee, with most of them assigned to the genera *Bacillus*, *Pseudomonas*, *Pantoea*, *Enterobacter*, *Stenotrophomonas*, *Kluyvera* and *Kocuria*. Approximately 20% of the bacterial strains showed the ability to grow in the presence of high concentrations (5.000 mg L^{-1}) of caffeine. Fig. 1 and 2 illustrate the growth of different strains in Czapeck medium supplemented with 5.000 mg L^{-1} of 1,3,7-trimethylxanthine. It was observed that two *Pseudomonas putida* strains, O1G and 13R presented the ability to grow in high concentration of caffeine. It is also shown that *Bacillus lentimorbus* (54G), *Enterobacter* sp. (111G), *Alcaligenes xylosoxydans* (06F) and *B. megaterium* (45R and 51R) grew in medium supplemented with caffeine (5.000 µg.mL^{-1}).

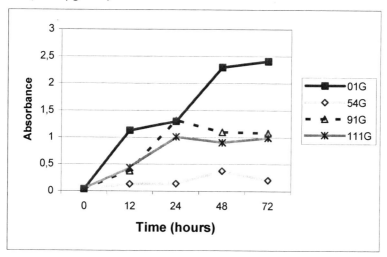

Figure 1:   Growth of endophytic bacteria, isolated from coffee stems, supplement with caffeine (5.000 mg L^{-1}): *Pseudomonas putida* (01G), *Bacillus lentimorbus* (54G), *Escherichia coli* (91G) and *Enterobacter* sp. (111G).

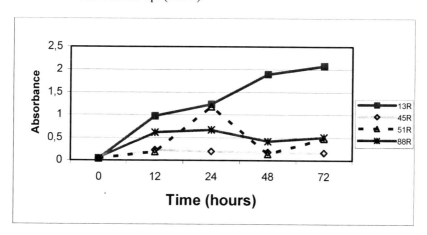

Figure 2:   Growth of endophytic bacteria, isolated from coffee roots in liquid medium supplemented with caffeine (5.000 mg L^{-1}): *Pseudomonas putida* (13R), *Bacillus megaterium* (45R), *Bacillus megaterium* (51R) and *Bacillus* sp. (88R).

Twenty endophytic bacterial strains grew in liquid medium containing concentrations as high as 0.05M caffeine. Growth was accompanied by complete removal of caffeine from culture medium. Two *Pseudomonas putida* strains completely degraded the alkaloid (table 1), showing the potential of endophytic bacteria in decaffeination processes. The other bacterial strains evaluated did not metabolized the alkaloid. It is verified that caffeine did not inhibited the normal growth of these bacteria. Caffeine is considered toxic for many microorganisms, however, some microorganisms have the ability to grow in the presence of caffeine and the capacity to degrade this alkaloid.

Table 1: Quantification of caffeine degradation by endophytic bacteria isolated from leaves, stems and roots of *Coffea arabica* and *Coffea robusta*, carried out by HPLC. The initial concentration of caffeine was 5.000 mg L^{-1}.

Treatments	CAFFEINE (g 100 mL^{-1})	Treatments	CAFFEINE (g 100 mL^{-1})
Control	0,5809	147G- *Acinetobacter* sp.	0,6571
13R - *Pseudomonas putida*	**0,0000**	54G-*Bacillus lentimorbus*	0,5961
01G - *Pseudomonas putida*	**0,0000**	11G-*	0,6076
06F- *Alcaligenes xylosoxydans*	0,6418	16F- *Bacillus cereus*	0,6215
12F- *Pandoraea* sp.	0,6216	43G-*Pseudomonas* sp.	0,6321
114R- *P. putida*	0,6278	167G-*Stenotrophomonas maltophilia*	0,6259
45R- *Bacillus megaterium*	0,6078	100G- *Stenotrophomonas maltophilia*	0,6542
7F- *Pantoea agglomerans*	0,5997	152G-*	0.6168
51R- *Bacillus megaterium*	0,6370	129G- *Ochrobactrum* sp.	0,6472
88R-*Bacillus* sp.	0,6184	91G- *Escherichia coli*	0,6049
07G- *Pseudomonas chlororaphis*	0,6357		

*- Bacteria not identified.

One of these strains of *Pseudomonas putida*, strains 13R, presented high pectinolytic activity (pectin lyase) 96,92 nMols of unsaturated product per mL^{-1}. Plant cell walls consist mainly of cellulose, whereas midle lamella, that connects the cells, consists mainly of pectin. Although pectinases might play an important role in plant-microbial interactions and intercellular colonization of roots they have not yet been studied intensively in the endophytic bacteria.

References

[1] Glick, B.R. The enhancement of plant growth by free-living bacteria. *Canadian Journal of Microbiology*, 41: 109-117. (1995).

[2] Shishido, M., Breuil, C., & Chanway, C.P. Endophytic colonization of spruce by plant growth-promoting rhizobacteria. *FEMS Microbiol. Ecol.* 29:191-196. 1999.

[3] Kinkel, L.L., Wilson, M., & Lindow, S.E. Plant species and plant incubation conditions influence variability in epiphytic bacterial population size. *Microb Ecol* 39: 1– 11. 2000.

[4] Sturz, A.V., Christie, B.R. & Nowak, J. Bacterial endophytes: potential role in developing sustainable systems of crop production. *Crit. Rev. Plant Sci.*, v. 19, p. 1-30, 2000.

[5] Asano, Y., Komeda. T. & Yamada. H. Microbial production of theobromine from caffeine. *Bioscience Biotechnology Biochemistry*, v.57, p.1286-1289, 1993.

[6] Kurtzman, R.H. & Shwimmer, S. Caffeine removal from growth media by microorganisms. *Experientia*, 127:481-482, 1971.

Contamination of agricultural soil by arsenic containing irrigation water in Bangladesh: overview of status and a proposal for novel biological remediation

S. M. A. Islam[1], K. Fukushi[2] & K. Yamamoto[3]
[1]Department of Civil Engineering,
Dhaka University of Engineering and Technology, Bangladesh
[2]Integrated Research System for Sustainability Science,
The University of Tokyo, Japan
[3]Environmental Science Center, The University of Tokyo, Japan

Abstract

Arsenic toxicity in the groundwater of Bangladesh poses a serious threat to public health. More than 35 million people are consuming arsenic-polluted groundwater there. So far, a lot of effort has been made to find safe drinking water but no suitable solution has yet been established, while continued cropping with arsenic contaminated irrigation water increases the extent of contamination in agricultural land. The objective of this paper is to review information on the arsenic concentration in the agricultural soil of Bangladesh irrigated with arsenic-contaminated water. In addition, this study overviews the literature, which focus on the biological remediation of arsenic from soil. Arsenic as high as 10 kg/ha per year is cycled through irrigated water and deposited on surface soil of Bangladesh, which results in its cumulative accumulation. It was observed to be as high as 83 mgAs/kg arsenic in the topsoil of irrigated agricultural land whereas in non-irrigated land arsenic was reported as 3-9 mgAs/kg. Considering the increasing trend of arsenic concentration in soil, it is necessary to establish some procedure for arsenic cleanup from soil. Biological cleanup could be a feasible solution for a developing country like Bangladesh. The feasibility of biological removal of arsenic from soil has been established in one study.
Keywords: arsenic concentration, agricultural land soil, rice plant, biological removal of arsenic, biological gasification of arsenic, soil column.

WIT Transactions on Biomedicine and Health, Vol 10, © 2006 WIT Press
www.witpress.com, ISSN 1743-3525 (on-line)
doi:10.2495/ETOX060301

1   Introduction

Arsenic is a ubiquitous element and is assumed to be the 20th most abundant element in the biosphere [1, 2]. It is mostly naturally occurring and is mainly distributed in the environment as a consequence of weathering of rocks, volcanic activity, evaporation of water, anthropogenic input and biological activity [3]. Being a metalloid arsenic can present in soil, water, air and all living matter in any of the form of solid, liquid and gas. Arsenic, primarily in the inorganic form, is present in the earth's crust at an average of 2-5 mg/kg [4]. However, arsenic contamination has become a common problem in many parts of the world. Arsenic leaching has occurred from mine tailings in Australia, Canada, Japan, Mexico, Thailand, the United Kingdom and the United States. Arsenic contamination in natural aquifers has occurred in Argentina, Bangladesh, Cambodia, Chile, China, Ghana, Hungary, India, Mexico, Nepal, New Zealand, the Philippines, Taiwan, the United States and Vietnam [5].

Arsenic pollution has occurred most severely in Bangladesh and India (West Bengal). It is estimated that more than 35 million people are consuming arsenic-polluted ground water alone in Bangladesh where underground water is used mainly for drinking and cooking [5, 17]. So far, lot of effort has been devoted to find safe drinking water there but no suitable measure has been established yet. In addition to the drinking water problem, continued irrigation with arsenic-contaminated water increases the extent of arsenic contamination in agricultural land soil in Bangladesh [6–9]. Objective of this paper is to review information on the arsenic concentration in soil of Bangladesh, where arsenic-contaminated water is used for irrigation and also to provide information on subsequent effect on plant growing on contaminated soil. In addition, this study overviews the literatures, which focus on the biological remediation of arsenic from soil and proposes a novel biological remediation process for soil cleanup.

2   Arsenic pollution in Bangladesh: overview of status

2.1   Heath effect of arsenic and its major exposure pathway to human in affected area

Like other creatures, arsenic is toxic to human. The mode of toxicity depends on the chemical form of arsenic and concentration and duration of exposure. Inorganic forms of arsenic like arsenate and arsenite, are usually more toxic than those of organic forms. Arsenate, a molecular analog of phosphate, inhibits oxidative phosphorylation and inactivates life's main energy-generation system. Arsenite is even more broadly toxic as it binds to sulfhydryl groups and impairs the function of many proteins [10]. Acute exposure of arsenic is lethal to body depending on exposure time and concentration. Chronic arsenic poisoning can cause many serious health effects such as melanosis (hyperpigmentation or dark spots and hypopigmentation or white spots), hyperkeratosis (hardened skin), restrictive lung disease, peripheral vascular disease (blackfoot disease), gangrene, diabetes mellitus, hypertension, and ischaemic heart disease [5, 12]. It is recognized that chronic consumption of arsenic, even at a low level, can cause carcinogenesis [2]. Malnutrition and poor socio-conomic conditions intensify the hazards of arsenic toxicity [13].

Human beings can be exposed to arsenic in many different ways: mainly by ingestion of contaminated water and food and by inhalation of metal-containing dust. As in arsenic polluted areas, arsenic-contaminated groundwater is used for drinking, cooking and irrigation, direct ingestion of drinking water, as well as cooking and soil–crop–food via food chain transfer may be the major exposure pathways of arsenic [15].

2.2 Arsenic concentration in groundwater of Bangladesh

From arsenic surveys conducted by different organizations in 60 districts out of 64 administrative districts of Bangladesh, it is observed that many tube wells of shallow depth (less than 100m) exceed the arsenic concentration level of 0.05 mg/l (Bangladesh Standard for arsenic in drinking water) in almost all 60 districts [12, 16]. In a study of arsenic survey conducted by BGS/DPHE [16] in 60 districts observed that about 61% of samples exceed 0.01 mg/l (WHO Guideline for arsenic concentration in drinking water), about 45% samples exceed 0.05 mg/l, and 2% exceed 1 mg/l of arsenic concentration in shallow tubewell among 9089 samples [13, 16]. The latest statistics on the arsenic contamination in groundwater of Bangladesh are presented in table 1 by summarizing different literature [12, 13, 16].

Table 1: Latest statistics on arsenic concentration in groundwater of Bangladesh [12, 13, 16].

Total area in Bangladesh (km^2)	147570
Total population (million)	128
Total number of administrative districts	64
Total number of districts surveyed	60
Total number of districts where arsenic concentration exceeds 0.01 mg/l	52
Total number of districts where arsenic concentration exceeds 0.05 mg/l	41
Total area where arsenic concentration exceeds 0.05 mg/l (km^2)	89186
Total population where arsenic concentration exceeds 0.05 mg/l (million)	85
Median value of arsenic concentration observed in tested samples (mg/l)	0.0108
Maximum arsenic concentration observed in tested samples (mg/l)	1.67

2.3 Arsenic concentration in soil of Bangladesh

Due to the scarcity of water in dry season, groundwater is very necessary for irrigation especially in the northern part of Bangladesh. It is more likely that the last 30 years of irrigation have led to diffuse contamination of land throughout the districts depending on arsenic-contaminated groundwater [15]. A preliminary estimate of arsenic deposition on irrigated agricultural land has been conducted by DPHE/BGS [16] shows that a large amount of arsenic as high as 10 kg/ha per year is cycled through irrigated water and deposited on surface soil of Bangladesh. A large portion of this arsenic remains in soil due to its affinity for iron, manganese, aluminum and other minerals in soil and not likely to be washed out by flood or rainwater, which results in a cumulative accumulation of arsenic in soil [8].

Some studies reported arsenic concentration in uncontaminated land of Bangladesh, which varies from 3-9 mgAs/kg [6, 7]. On the other hand, elevated arsenic concentrations were observed in many studies in agricultural land soil irrigated with arsenic-contaminated water, which is in some cases found about 10-20 times higher than arsenic concentration in non-irrigated land. Ullah [6] reported arsenic concentration in top agricultural land soil (up to 0-30 cm depth) up to 83 mgAs/kg, Islam et al [18] up to 80.9 mgAs/kg, Alam and Sattar [7] up to 57 mgAs/kg of soil for samples collected from different districts of Bangladesh. The reported arsenic concentration for agricultural land soil in different district of Bangladesh is summarized in table 2.

It is obvious from the result that high arsenic concentration in source tube well supplying the irrigation water is mainly responsible for accumulation of high arsenic content in soil. However, no significant correlation between water-As and soil-As was observed by Islam et al [18] and Huq et al [14] as shown in fig.1. In contrast, some sort of positive relation was observed by some other studies [5, 19]. However, the arsenic accumulation in agricultural soil not only depends on the source well concentration but also depends on soil properties, especially the clay content (permeability), organic content (biological transformation), etc.

Arsenic concentration in agricultural soil is usually found higher for those source tube wells, which have been operated for longer period of time. Regression of soil arsenic levels with tube-well age was found significant [19]. Agricultural soil arsenic concentration varies with depth of soil. Ali et al [8] obtained high arsenic concentration in topsoil (0-15 cm) with a decreasing profile along depth. Soil arsenic concentration in agricultural field also varies along with crop season. Higher arsenic concentration is usually observed at the end of irrigation period than the beginning [8, 9]. Lowest concentration is usually obtained after flood. It can be inferred that the accumulated arsenic during irrigation period is partially washed by rain and flood water [8].

2.4 Arsenic concentration in rice, vegetables and other plants

It can be accounted that higher arsenic concentration in irrigation water and soil results in higher concentration of arsenic in crops. Arsenic accumulation was observed in roots, stem and leaves of rice plants [5, 8, 19–22]. Table 3 summarises the arsenic concentration level in rice grains observed in different districts of Bangladesh by some studies. The average arsenic concentration observed in rice grains in different studies is lower than the food hygiene concentration limit of 1 mg/kg [5]. The different varieties of rice tested for accumulation of arsenic by different studies are khatobada, paijam, swarna, gocha, kalia, swarna and different types of BR. Among them BR11 from Naogaon was found to have high arsenic concentration of 1.84 mg/kg [19]. Arsenic concentration observed in much higher range in roots and straw than leaves and rice grains [5, 8]. The high arsenic concentrations in straw may have

Table 2: Arsenic concentration in agricultural soils irrigated with arsenic-contaminated water in Bangladesh [14, 18–19].

Location (Districts)	Water-arsenic (mg/kg)	Soil Depth (cm)	Soil-arsenic (mg/kg)	Ref	Location	Water-arsenic (mg/kg)	Soil Depth (cm)	Soil-arsenic (mg/kg)	Ref.
Sharsha	0.041	0-15	13.67	[14]	Comilla	NR	0-15	11.00	[19]
Sirajidikhan	0.544	15-30	10.66	[14]	Laksham	NR	0-15	7.40	[19]
Alamdanga	0.021	0-15	16.65	[14]	Gazipur	NR	0-15	14.60	[19]
Meherpur	0.163	0-15	33.91	[14]	Jamalpur	NR	0-15	16.50	[19]
Laksham	0.261	15-30	42.61	[14]	Manikgonj	NR	0-15	13.50	[19]
Chandina	0.16	15-30	19.27	[14]	Mymensingh	NR	0-15	25.40	[19]
Sonargaon	0.682	0-15	38.93	[14]	Tangail	NR	0-15	24.00	[19]
Bancharampur	0.092	0-15	17.15	[14]	Chuadanga	NR	0-15	33.30	[19]
Nagarkanda	0.077	0-15	81.25	[14]	Jessore	NR	0-15	23.30	[19]
B. Baria	0.23	0-30	9.80	[18]	Kushtia	NR	0-15	42.50	[19]
Comilla	0.56	0-30	40.40	[18]	Meherpur	NR	0-15	36.60	[19]
Faridpur	0.11	0-30	7.30	[18]	Rajbari	NR	0-15	23.90	[19]
Munshigonj	0.15	0-50	3.90	[18]	Bogra	NR	0-15	15.50	[19]
Munshigonj	0.21	0-50	80.90	[18]	Dinajpur	NR	0-15	11.70	[19]
Manikgonj	0.17	0-50	46.50	[18]	Kurigram	NR	0-15	9.60	[19]
Manikgonj	0.36	0-50	55.60	[18]	Nawabgonj	NR	0-15	20.90	[19]
Manikgonj	0.23	0-50	60.10	[18]	Naogaon	NR	0-15	26.70	[19]
Bagalkati	NR	0-15	24.50	[19]	Panchagarh	NR	0-15	8.10	[19]
Barisal	NR	0-15	26.10	[19]	Pabna	NR	0-15	14.40	[19]
Bhola	NR	0-15	16.30	[19]	Rajshahi	NR	0-15	7.80	[19]
Chandpur	NR	0-15	18.40	[19]	Rangpur	NR	0-15	11.50	[19]
Comilla	NR	0-15	21.60	[19]	Thakurgaon	NR	0-15	12.40	[19]

*NR- Not Reported

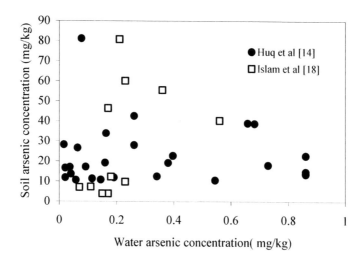

Figure 1:   Correlation between arsenic concentration in agricultural soils and source tube wells supplying the irrigation water obtained in different studies.

adverse health effects on the cattle and thus can increase chance of arsenic exposure in humans via the plant–animal–human pathway [20].

Vegetables and other plants growing on arsenic contaminated soil exhibit arsenic accumulation [5]. Table 4 summarizes the arsenic concentration level in vegetables and other plant varieties in some districts of Bangladesh reported by different studies. Among the vegetables korola (*Momordica charantia*), gourd, potol (*Trichosanthesdioeca*), okra/vindi/dharosh (*Hibiscus esculentus*), data stem (*Amaranthus lividus*), etc. were not found to contain high amounts of arsenic [5]. It was observed that plant uptake of arsenic is highly variable among plant species and is also influenced by soil characteristics, soil fertility, and concentration and chemical forms of arsenic in soil [8].

3   Perspective of biological removal of arsenic from soil: a novel approach for permanent clean up

Biological transformation of arsenic plays an important role in the distribution and cycling of arsenic in arsenic-rich environment. Microbes are able to metabolize inorganic arsenic to less toxic both volatile and nonvolatile organic species such as monomethylarsine (MMA), dimethylarsine (DMA), and trimethylarsine (TMA) and non-volatile species such as methylarsonic acid, dimethylarsinic acid, and trimethylarsenic oxide. Arsenic biomethylation can be considered as a detoxification of arsenic in the environment because the toxicity of organic arsenic is much less than that of inorganic arsenic [11]. Production of volatile methylated species has further importance in the removal of arsenic from environmental samples (soil/sediment) and biosolids [24]. In soil, biological removal of arsenic can be achieved by 1) reduction of arsenate to mobile arsenite and subsequent collection and treatment of water, 2) biosorption and bioaccumulation 3) biogasification. The last one is advantageous as it is a phase change of arsenic and can be involved in the removal of arsenic from soil as it is basically a natural process.

Table 3:   Arsenic concentration level in rice plants observed in different districts of Bangladesh reported in different literature.

Location (District)	Area of Survey	Arsenic conc. range (mg/kg)	Mean arsenic conc. (mg/kg)	No. of samp. (n)	Ref
Rajshahi	District	0.03-0.34	0.10	24	[21]
Rajshahi	Market	0.03-0.28	0.11	17	[21]
Pabna	Cooked food	0.11-0.36	0.24	4	[14]
Chadpur, Jamalpur	Random	0.04-0.27	0.14	10	[5]
Jessore	Village	0.11-0.94	0.28	21	[21]
Nawabganj	Boro, District	0.01-0.99	0.33	8	[23]
Sonargaon	Field	0.05-1.23	0.46	12	[8]
Srinagar	Random	0.09-1.84	0.5	13	[19]
Dinajpur	Field	-	0.54	9	[8]
Gopalganj	Boro, district	0.01-1.18	0.57	6	[23]
Rajbari	Boro, district	0.05-2.05	0.76	8	[23]
Faridpur	Boro, District	0.21-1.5	0.95	7	[23]

Table 4:   Arsenic concentration level in rice and other plant verities in Bangladesh reported in different studies.

Name of verity	Arsenic conc. (mg/kg)	No. of samp.	Ref.
Arum leaves (*Colocasia antiquorum*)	0.09-3.99	9	[5]
Potatoes (*Solanum tuberosum*)	0.07-1.39	5	[5]
Kalmi sak (*Ipomoea reptans*)	0.1-1.53	6	[5]
Ghotkol (*Typhonium trilobatum*)	0.45	NR	[15]
Taro/Kachu lati (*Colocasia esculenta*)	0.44	NR	[15]
Snake gourd/Chichinga (*Trichosanthes anguina*)	0.49	NR	[15]
Brinjal (*Solanum melongena*)	0.20	NR	[15]
Lau sak/Bottle ground leaf (*Lagenaria siceraria*)	0.30	NR	[15]
Green papaya (*Carica papaya*)	0.45	NR	[15]

*NR- Not Reported

Though biological gasification of arsenic was evidenced earlier, Islam *et al* [9, 18] was first started working on biological gasification of arsenic as an engineering process for arsenic removal from soil/solids. Islam *et al* [9, 18] observed microbial volatilization of arsenic as MMA, TMA and arsine from different soil samples collected from agricultural soil of Bangladesh cultured in batch under anaerobic conditions at 37^0C using substrate as acetate and formate. The observed results positively indicate the possibility of using biogasification of arsenic in bioremediation from soil.

In order to observe arsenic gasification in original soil for application of arsenic bioremediation, Islam [24] further conducted soil column test where native soil microbial growth was enhanced by supplying substrate favorable for methanogenic growth. In that study microbial gasification was observed from existing arsenic concentration in the soil, no external arsenic was added. Five soil columns were run with triplicate as sample and two as control. The columns were filled with soil of around 300 grams under anaerobic condition and were incubated at room temperature ($25\pm3°C$). The steady state rate of gasification was observed after 90 days of operation as 0.4 μgAs/d, which is very low. However, the results established the feasibility of biological removal of arsenic from soil while it requires longer time for total clean up. Islam [24] further achieved bioaugmentation in the column by adding microorganisms grown in cow dung-seeded reactor, which were previously acclimated to arsenic. It was observed that gasification rate of arsenic was rapidly increased after addition of new microrganisms even up to 8 times. This observation undoubtedly established the possibility of arsenic clean up from the contaminated soil of Bangladesh within a reasonable time.

4 Conclusion

This study summarized the present status of agricultural soil arsenic concentrations in Bangladesh, which are irrigated with arsenic-contaminated water. There is no regulatory guideline for soil or plant arsenic concentration in Bangladesh. Some developed countries provide such guideline, for example, the regulatory limit of arsenic for agricultural soil is set for environmental health investigation in Australia is 20 mgAs/kg. This suggests that in case of exceeding this level requires soil clean up [25]. Considering increasing concentration in agricultural soil and its subsequent effect on plants, establishment of such a guideline is necessary and as a result some soil may require cleanup of arsenic. Biological cleanup could be fruitful solution for a developing country like Bangladesh. The feasibility of biological removal of arsenic from soil has already been established by Islam [24]. The application of biological removal as a novel process can be introduced after investigation of the fate of volatile arsenic compound in the environment, which is hitherto unknown.

References

[1] Woolson, E. A., Generation of alkylarsines from soil. *Weed Science*, **25**, pp. 412-416, 1977.

[2] Mandal, K.M. & Suzuki, K.T., Arsenic round the world: a review. *Talanta* **58**, pp. 241–235, 2002

[3] Cullen, W.R. & Reimer, K. J., Arsenic speciation in the environment. *Chemical Reviews*, **89**, pp. 713–764, 1989.

[4] Tamaki, S. & Frankenberger, W.T., Environmental Biochemistry of Arsenic. *Reviews of Environmental Contamination and Toxicology*, **124**, pp. 79-110, 1992.

[5] Das, H. K., Mitra, A. K., Sengupta, P. K., Hossain, A., Islam, F. & Rabbani, G. H. Arsenic concentrations in rice, vegetables, and fish in Bangladesh: a preliminary study. *Environment International*, **30**, pp. 383-387, 2004.

[6] Ullah, S.M., Arsenic Contamination of Groundwater and Irrigated Soils in Bangladesh, *Proc. of the International Conference on Arsenic Pollution on Groundwater in Bangladesh: Causes, Effects and Remedies*, Dhaka, Bangladesh, 1998.

[7] Alam, M.B. & Satter M.A., Assessment of Arsenic Contamination in Soils and Waters in Some areas of Bangladesh. *Water Science Technology*, **42** (7-8), pp. 185-192, 2000.

[8] Ali M.A., Badruzzaman A.B.M., Jalil M.A., Hossain M.D., Ahmed M.F., Masud A.A., Kamruzzaman M. & Rahman M.A., Arsenic in Plant-Soil Environment in Bangladesh. *Fate of Arsenic in the Environment*, eds. M. F. Ahmed, M. A. Ali & Z. Adeel, BUET-UNU, Dhaka, pp 85-112, 2003 .

[9] Islam, S.M.A, Fukushi, K. & Yamamoto, K., Severity of arsenic concentration in soil and arsenic-rich sludge of Bangladesh and potential of their biological removal: a novel approach for tropical region. *Proc. of the Second International Symposium on Southeast Asian Water Environment*, Hanoi, Vietnam, 2004.

[10] Oremland, R.S. & Stolz, J.F., The ecology of arsenic. *Science*, **300**, pp 939-944, 2003

[11] Kaise, T., Yamauchi, H., Horiguchi, Y., Tani, T., Watanabe, S., Hirayama, T. & Fukui, S., A comparative study on acute toxicity of methylarsonic acid, dimethylarsinic acid and trimethylarsine oxide in mice. *Applied Organometallic Chemistry*, **3**, pp. 273–277, 1989.

[12] Rahman, M.H., Rahman, M.M., Watanabe, C. & Yamamoto., K., (2002). Arsenic contamination of groundwater in Bangladesh and its remedial measures, *Arsenic Contamination in Groundwater -Technical and Policy Dimensions. Proc. on the UNU-NIES International Workshop*, Tokyo. United Nations University (UNU) and Japan National Institute for Environmental Studies (NIES), Tokyo.

[13] Hossain, M.F., Arsenic contamination in Bangladesh-An overview. *Agriculture Ecosystem & Environment*, **in press,** 2005.

[14] Huq, S.M.I., Rahman, A., Sultana, N. & Naidu R., Extent and Severity of Arsenic Contamination in Soil of Bangladesh. *Fate of Arsenic in the Environment,* eds. M. F. Ahmed, M. A. Ali & Z. Adeel, BUET-UNU, , Dhaka, pp 69-84, 2003.

[15] Alam, M.G.M., Snow, E.T. & Tanaka, A., Arsenic and heavy metal contamination of vegetables grown in Samta village, Bangladesh. *The Science of the Total Environment,* **308**, pp. 83-96, 2003.

[16] BGS/DPHE, Groundwater studies for arsenic contamination in Bangladesh, *Final Report, Summary,* Department of Public Health and Engineering, Government of Bangladesh, DFID, British Geological Survey, Dhaka, Bangladesh, 2000.

[17] Rabbani, G.H., Chowdhury, A.K., Shaha, S.K. & Nasir, M. Mass arsenic poisoning of ground water in Bangladesh. *Proc. of the Global Health Council Annual Conference,* Washington DC, 2002.

[18] Islam, S.M.A, Fukushi, K. and Yamamoto, K., Bioreduction and biomethylation of arsenic in soil of Bangladesh: novel process for permanent removal of arsenic. *Proc. of the 1st IWA-ASPIRE Conference & Exhibition,* Singapore, 2005.

[19] Meharg, A.A. & Rahman, M.M., Arsenic contamination of Bangladesh paddy field soils: implications for rice contribution to arsenic consumption. *Environmental Science & Technology,* **37**, pp. 229-234, 2003.

[20] Abedin, M.J., Cresser, M.S., Meharg, A.A., Feldmann, J. & Cotter-Howells, J., Arsenic Accumulation and Metabolism in Rice. *Environmental Science Technology,* **36(5)**, pp. 962-968, 2002.

[21] Hironaka, H. & Ahmed, S.A., Arsenic Concentration of Rice in Bangladesh. *Fate of Arsenic in the Environment,* eds. M. F. Ahmed, M. A. Ali & Z. Adeel, BUET-UNU, Dhaka, pp 69-84, 2003

[22] Williams, P.N., Price, A.H., Raab, A., Hossain, S.A., Feldmann, J., & Meharg, A.A., Variation in arsenic speciation and concentration in paddy rice related to dietary exposure. *Environmental Science & Technology,* **39**, pp. 5531-5540, 2005.

[23] Islam, M. R., Jahiruddin, M. & Islam, S., Assessment of arsenic in the water-soil-plant systems in gangetic flood plains of Bangladesh. *Asian Journal of Plant Science,* **3**, pp. 489-493, 2004.

[24] Islam S.M.A., (2005) Gasification of arsenic from contaminated solids by anaerobic microorganisms. PhD Dissertation, The University of Tokyo, Tokyo, Japan.

[25] O'Neil, P., Arsenic. *Heavy Metals in Soils,* ed. B.J. Alloway, John Wiley & Sons, New York, 1990.

Synthesis and biodegradation of three poly(alkylene succinate)s: mathematical modelling of the esterification reaction and the enzymatic hydrolysis

D. S. Achilias & D. N. Bikiaris
Laboratory of Organic Chemical Technology, Department of Chemistry, Aristotle University of Thessaloniki, Thessaloniki, Greece

Abstract

Synthesis of three biodegradable aliphatic polyesters, namely poly(ethylene succinate), poly(propylene succinate) and poly(butylene succinate), is presented using the appropriate diols and succinic acid. A theoretical mathematical model for the esterification reaction is developed and applied successfully in the simulation of all experimental data. Different glycols do not influence the average degree of polymerization of the oligomers produced much, even though they slightly affect esterification rates in the order BG>PG>EG. In contrast, these values are affected by the amount of catalyst, with a larger catalyst molar ratio giving a polymer with a larger than average molecular weight. Biodegradability studies of the polyesters included enzymatic hydrolysis for several days using *Rhizopus delemar* lipase at pH 7.2 and 30°C. The biodegradation rates of the polymers decreased following the order PPSu>PESu≥PBSu and it was attributed to the lower crystallinity of PPSu compared to other polyesters, rather than to differences in chemical structure. Finally, a simple theoretical kinetic model was developed and Michaelis-Menten parameters were estimated.
Keywords: biodegradable polymers, enzymatic hydrolysis, aliphatic polyesters, mathematical modelling.

1 Introduction

Continuous environmental surcharge due to domestic and industrial waste consists of one of the major problems of growth nowadays. Plastic waste materials represent a high percentage of domestic waste and their amount continues to increase, despite some increasing attempts to reduce, reuse, recycle and recover. This is mainly due to their wide application in the manufacture of packaging for the food industry as well as in other goods of daily life. Although most plastic wastes are not toxic and do not create a direct hazard to the environment, they are seen as noxious materials due to their substantial fraction by volume in the waste stream and their high resistance to the atmospheric and biological agents. Therefore, the development and application of biodegradable polymers, instead of those traditionally used, has received a great attention from the viewpoint of environmental protection and resource recycle. Among synthetic polymers, aliphatic polyesters combine the features of biodegradability, biocompatibility and physical or chemical properties, comparable with some of the extensively used polymers like polyethylene (LDPE), polypropylene (PP), etc.

In the present investigation, three aliphatic polyesters, namely poly(ethylene succinate), poly(propylene succinate) and poly(butylene succinate) were prepared from succinic acid and ethylene, propylene and butylene glycol, respectively. The advantage of these polyesters is that despite their low melting point they have very high thermal stability, which is familiar to aromatic polyesters like poly(ethylene terephthalate) and poly(butylene terephthalate). The kinetics of the esterification reaction of these polyesters are examined and a detailed mathematical model is developed illustrating the effect of the monomer type together with the catalyst used on the time required to complete the reaction. Furthermore, a comparative biodegradability study is carried out and a theoretical enzymatic hydrolysis kinetic model is introduced.

2 Experimental

2.1 Materials

Succinic acid (purum 99 %), ethylene glycol (purum 99 %) and butylene glycol (purum 99 %) were purchased from Aldrich Chemical Co. 1,3-Propanediol (CAS Number: 504-63-2, Purity: > 99,7 %) was kindly supplied by Du Pont de Nemours Co. Tetrabutoxy-titanium (TBT) used as catalyst was analytical grade and purchased from Aldrich Chemical Co. Polyphosphoric acid (PPA) used as heat stabilizer was supplied from Fluka. *Rhizopus delemar* lipase was purchased from BioChemika. All other materials used were of analytical grade.

2.2 Synthesis of polyesters

Synthesis of aliphatic polyesters was performed following the two-stage melt polycondensation method (esterification and polycondensation) in a glass batch reactor (Karayannidis et al. [1]). In brief, the proper amount of succinic acid (0.55 mol) and appropriate glycols (ethylene glycol, 1,3-propanediol and butylene glycol) in a molar ratio 1/1.1 and the catalyst TBT (3×10^{-4} mol TBT/mol SA) were charged into the reaction tube of the polycondensation apparatus. The reaction mixture was heated at 190°C under argon atmosphere and stirring at a constant speed (500 rpm) was applied. This first step (esterification) is considered to complete after the collection of almost the theoretical amount of H_2O, which was removed from the reaction mixture by distillation and collected in a graduate cylinder. In the case of PPSu two additional catalyst concentrations, namely 1.5×10^{-4} and 6×10^{-4} mol TBT/mol SA were used in order to study the effect of catalyst amount on esterification and polycondensation reactions.

In the second step of polycondensation, PPA was added (5×10^{-4} mol PPA/mol SA) and a vacuum (5.0 Pa) was applied slowly over a period time of about 30 min. The temperature was increased to 230°C while stirring speed was slowly increased to 720 rpm. The polycondensation reaction time was always kept constant at 1h and afterwards the polyesters were maintained at room temperature in order to be cooled.

2.3 Biodegradation studies (enzymatic hydrolysis)

Polyesters in the form of films with 5 x 5 cm in size and approximately 2 mm thickness, prepared in a hydraulic press, were placed in petries containing phosphate buffer solution (pH 7.2) with 1 mg/mL *Rhizopus delemar* lipase. The petries were then incubated at 30±1°C in an oven for several days while the media were replaced every 3 days. After a specific period of incubation, the films were removed from the petri, washed with distilled water and weighted until constant weight. The degree of biodegradation was estimated from the mass loss and molecular weight reduction as measured by GPC. In all polyester films intrinsic viscosity, end groups analysis and thermal analysis were performed.

3   Mathematical modeling of the esterification reaction for the poly(alkylene succinate) synthesis

3.1 Reaction scheme

In this paper the polymer segment approach, which is in the framework of functional group approaches was used (Kang et al. [2]) following our previous article on modeling of the poly(propylene terephthalate) esterification kinetics (Karayannidis et al. [1]). According to this approach, the polymerization reaction is regarded as a reaction between two functional groups. The molecular structure of the components considered in the reaction scheme, is presented in Table 1. The esterification reaction of succinic acid (SA) with three glycols is considered here (i.e. EG ethylene glycol, PG propylene glycol and BG butylene glycol). Five different oligomeric segments are used: tSA, tG, bSA, bG and bDG (the term t and b refer to the terminal functional group and bound monomeric repeating unit).

Based on these functional groups, the following mechanism is assumed to represent the kinetics of the esterification process [1].

$$SA + G \underset{k'_1}{\overset{k_1}{\rightleftharpoons}} tSA + tG + W \tag{1}$$

$$tSA + G \underset{k'_2}{\overset{k_2}{\rightleftharpoons}} bSA + tG + W \tag{2}$$

$$SA + tG \underset{k'_3}{\overset{k_3}{\rightleftharpoons}} tSA + bG + W \tag{3}$$

$$tSA + tG \underset{k'_4}{\overset{k_4}{\rightleftharpoons}} bSA + bG + W \tag{4}$$

$$tG + tG \underset{k'_5}{\overset{k_5}{\rightleftharpoons}} bG + G \tag{5}$$

$$tG + tG \overset{k_6}{\longrightarrow} bDG + W \tag{6}$$

Table 1:   Molecular structure of components considered.

Symbol	Description	Molecular Structure
SA	Succinic acid	$HOOC\text{-}CH_2\text{-}CH_2\text{-}COOH$
G	ethylene glycol	$HO\text{-}CH_2\text{-}CH_2\text{-}OH$
	propylene glycol	$HO\text{-}CH_2\text{-}CH_2\text{-}CH_2\text{-}OH$
	butylene glycol	$HO\text{-}CH_2\text{-}CH_2\text{-}CH_2\text{-}CH_2\text{-}OH$
W	Water	H_2O
tSA	SA end group	$HOOC\text{-}CH_2\text{-}CH_2\text{-}CO\text{-}$
tG	EG end group	$HO\text{-}CH_2\text{-}CH_2\text{-}O\text{-}$
	PG end group	$HO\text{-}CH_2\text{-}CH_2\text{-}CH_2\text{-}O\text{-}$
	BG end group	$HO\text{-}CH_2\text{-}CH_2\text{-}CH_2\text{-}CH_2\text{-}O\text{-}$
bSA	SA repeating unit	$\text{-}OC\text{-}CH_2\text{-}CH_2\text{-}CO\text{-}$
bG	EG repeating unit	$\text{-}O\text{-}CH_2\text{-}CH_2\text{-}O\text{-}$
	PG repeating unit	$\text{-}O\text{-}CH_2\text{-}CH_2\text{-}CH_2\text{-}O\text{-}$
	BG repeating unit	$\text{-}O\text{-}CH_2\text{-}CH_2\text{-}CH_2\text{-}CH_2\text{-}O\text{-}$
bDG	Diethylene glycol repeat. unit	$\text{-}O\text{-}CH_2CH_2\text{-}O\text{-}CH_2CH_2\text{-}O\text{-}$
	Dipropylene glycol rept. unit	$\text{-}O\text{-}CH_2CH_2CH_2\text{-}O\text{-}CH_2CH_2CH_2\text{-}O\text{-}$
	Dibutylene glycol rept. unit	$\text{-}O\text{-}CH_2CH_2CH_2CH_2\text{-}O\text{-}CH_2CH_2CH_2CH_2\text{-}O\text{-}$

Reactions 1-4 represent the typical esterification reactions, while reaction 5 is the polycondensation reaction, occurring mainly in the second step of polyester formation. Finally, reaction 6 is a side reaction resulting in di-glycol repeating units, with ether linkages in the oligomeric chain. k_i (i = 1, 6) and k'_i (i = 1, 5) represent the kinetic rate constants of the 6 elementary reactions (L mol^{-1} min^{-1}).

3.2 Development of the mathematical model

In order to develop a mathematical model for the esterification reaction the following assumptions are made: all kinetic rate constants are independent of polymer chain length; all the water produced during the reaction is instantaneously vaporized and removed; all glycol vaporized is totally returned to the reactor. Based then on the reaction mechanism, the reaction rates can be expressed in terms of the different functional groups present in the reactor and the corresponding rate constants [2].

$$R_1 = \{4\,k_1\,(SA)\,(G) - (k_1/K_1)\,(tSA)\,(W)\}/\,V^2 \tag{7}$$

$$R_2 = \{2\,k_2\,(tSA)\,(G) - 2\,(k_2/K_2)\,(bSA)\,(W)\}/\,V^2 \tag{8}$$

$$R_3 = \{2\,k_3\,(SA)\,(tG) - (k_3/K_3)\,(tSA)\,(W)\}/\,V^2 \tag{9}$$

$$R_4 = \{k_4\,(tSA)\,(tG) - 2\,(k_4/K_4)\,(bSA)\,(W)\}/\,V^2 \tag{10}$$

$$R_5 = \{k_5\,(tG)\,(tG) - 4\,(k_5/K_5)\,(bG)\,(G)\}/\,V^2 \tag{11}$$

$$R_6 = \{k_6\,(tG)\,(tG)\}/\,V^2 \tag{12}$$

Furthermore the material mole balance equations for a semi-batch reactor can be written as:

$$\frac{1}{V}\frac{d(SA)}{dt} = -R_1 - R_3 \tag{13}$$

$$\frac{1}{V}\frac{d(G)}{dt} = -R_1 - R_2 + R_5 \tag{14}$$

$$\frac{d(W)}{dt} = V\left(R_1 + R_2 + R_3 + R_4 + R_6\right) - F_w = 0 \tag{15}$$

$$\frac{1}{V}\frac{d(tSA)}{dt} = R_1 - R_2 + R_3 - R_4 \tag{16}$$

$$\frac{1}{V}\frac{d(tG)}{dt} = R_1 + R_2 - R_3 - R_4 - 2R_5 - 2R_6 \tag{17}$$

$$\frac{1}{V}\frac{d(bSA)}{dt} = R_2 + R_4 \tag{18}$$

$$\frac{1}{V}\frac{d(bG)}{dt} = R_3 + R_4 + R_5 \tag{19}$$

$$\frac{1}{V}\frac{d(bDG)}{dt} = R_6 \tag{20}$$

In eqn (15), F_W is used to represent the flow rate of the water vaporized and removed from the reactor. The total moles of water removed up to time t, N_W, are:

$$N_W = \int_0^t F_W \cdot dt \tag{21}$$

Finally, conversion according to the water produced and removed can be calculated according to the following equation:

$$X_W = \frac{N_W\,MW_W}{19.8} \tag{22}$$

Using the polymer segment approach the number average degree of polymerization (NADP) of oligomers produced can be expressed as [2]:

$$\overline{DP_n} = \frac{(tSA) + (bSA) + (tG) + (bG) + (bDG)}{(tSA) + (tG)} \tag{23}$$

The system of differential eqns (13-20) was integrated by applying the Runge-Kutta 4th order method with varying step size. In order to have the mole number of every component in the reaction mixture as a function of time the kinetic rate constants have to be evaluated. According to the reaction mechanism, 11 rate constants (k_i, i = 1-6 and k'_i, i = 1-5) should be determined. Since there are not any experimental data in the literature on the polymerization of SA with either EG, PG or BG, all these parameters were estimated according to the following assumptions. In PET and poly(propylene terephthalate) polymerization it was assumed that the reactivity of the acid end group on (terephthalic acid) TPA is equivalent to the reactivity on oligomer chain (tTPA), whereas the reactivity of hydroxyl end group on EG was twice of the reactivity on half-esterified EG (tEG) [1,2]. Since in the polymerization of either PG or BG the end groups are identical, we used the same assumption, i.e. $k_1 = k_2 = 2k_3 = 2k_4$. In order to simplify further the calculations it was assumed that $k_5 = 0$, i.e. that the poly-condensation reaction is not carried out in a great extent during the esterification step. The equilibrium constants K_i, $i = 1\text{-}5$ need not to be evaluated since according to the assumptions made the reverse reaction rates are zero [$(W)=0$ in the liquid phase]. From the above analysis the number of parameters that need to be evaluated is only two, namely k_1 and k_6.

4 Results and discussion

4.1 Esterification reaction

The values of the kinetic parameters were calculated for every different system studied from fitting to the experimental data and they are reported in Table 2. The effect of the catalyst molar ratio on the rate of the PPSu esterification reaction is presented in Figure 1. It is obvious that the theoretical model simulates the experimental data very well at all catalyst amounts. As it was also reported for PBSu [3], an increase in the catalyst concentration leads to increased reaction rates. Furthermore, it was observed that when plotting k_1 versus the catalyst concentration in a double logarithmic plot a very good straight line was obtained and the equation thus calculated was: $k_1 = 10^{-0.91}(\text{mol TBT/mol SA})^{0.5}$. Very interesting to note, that for PBSu an almost equal catalyst dependence (i.e. 0.51) was observed in the literature [3].

Table 2:   Numerical values of the kinetic rate constants.

Polymer	Symbol	Catalyst amount ($\times 10^4$ mol TBT/mol SA)	$k_1 = k_2$ ($\times 10^3$ L/mol· min)	k_6 ($\times 10^3$ L/mol· min)
Poly(propylene succinate)	PPSu(H)	1.5	1.5	1.0
Poly(propylene succinate)	PPSu(T)	3.0	2.2	0.8
Poly(propylene succinate)	PPSu(D)	6.0	3.0	0.7
Poly(ethylene succinate)	PESu	3.0	1.8	0.6
Poly(butylene succinate)	PBSu	3.0	2.7	1.1

The effect of the type of glycol used (i.e. EG, PG or BG) on the esterification reaction is examined next. Results on the water conversion as a function of time appear in Figure 2. It is seen that use of BG leads to slightly higher reaction rates than PG, which in turn is also slightly faster compared to EG. Again the theoretical simulation model fits the experimental data very well.

From theoretical results, it was found that different glycols do not influence much the number average degree of polymerization of the oligomers produced during the esterification step. In contrast, these values were affected much from the amount of catalyst used.

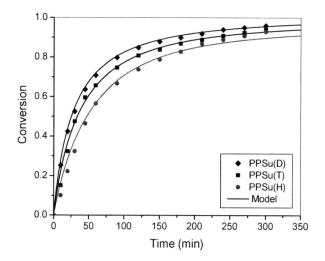

Figure 1:   Conversion (X_w) versus time for the esterification reaction of PPSu at 190°C with different amounts of added catalyst. Experimental data (discrete points) and theoretical model simulation results.

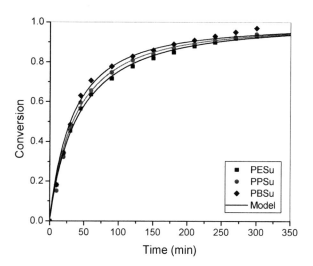

Figure 2:   Conversion (X_w) versus time for the esterification reaction of PESu, PPSu and PBSu at 190°C and 3×10^{-4} mol TBT/mol SA. Experimental data (discrete points) and simulation results

4.2 Biodegradation studies (kinetics of the enzymatic hydrolysis)

The kinetics of enzymatic hydrolysis has been studied in literature for several polyesters [4,5]. The reaction temperature mainly affects hydrolysis rate and the enzyme used. For aliphatic polyesters, as the temperature increases, approximating melting temperature, enzymatic hydrolysis is accelerated [4]. Thus, quite satisfactory hydrolysis rates are obtained when the temperature is 10-20°C lower than the polymers melting point. The most commonly used temperature for enzymatic hydrolysis studies is that of 37°C. In this study, it was decided to study hydrolysis at 30°C, since PPSu melts at a relatively low temperature (T_m=44°C) and at higher temperatures it would be significantly softened.

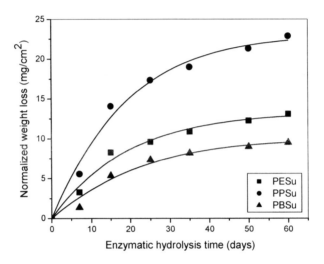

Figure 3: Weight loss of aliphatic polyester films during enzymatic hydrolysis. Curves through the experimental points represent the best fit of eqn (30).

As it is well known enzymatic hydrolysis is a heterogeneous process. Enzymes are attached on the surface of an insoluble substrate and hydrolysis takes place via surface erosion. In general, the internal parts of polyesters specimens are not attacked until extended holes are created onto surface allowing the enzymes to penetrate and attack to the main body. Thus, hydrolysis rates are decreased after consumption of the amorphous material of the surface. Afterwards, a layer of crystalline domains remains, where only slow degradation may occur. This inhibits penetration of water into the body of the film resulting in decreased rates of hydrolysis. In Figure 3 it was observed that PPSu exhibits the highest weight loss values, while PBSu showed the lowest ones. Since in our study the prepared polyesters have almost identical molecular weights, crystallinity seems to be the predominant factor that controls the biodegradation rates. It is therefore the higher degree of crystallinity of the PBSu samples, as was found by DSC and WAXD, [6] which resulted in lower degrees of biodegradation comparing to that of PESu or PPSu. PPSu, with the lower crystallinity, hydrolyses faster comparing to the other polyesters.

In order to quantify these results a simple theoretical kinetic model is developed to predict the time evolution of the polymer weight loss. It should be pointed out that the derivation of a detailed mathematical model is rather complicated, since mass transfer phenomena (between the solid-polymer phase and the enzyme) occur in parallel to the chemical reaction. This model is formally based on the kinetic mechanism of enzymatic hydrolysis according to the Michaelis-Menten scheme:

$$E + S \underset{k_{-1}}{\overset{k_1}{\rightleftarrows}} ES \tag{24}$$

$$ES \xrightarrow{k_2} P + E \tag{25}$$

where E, S and P represent the enzyme, substrate-polymer and hydrolysis reaction products, respectively and ES is the enzyme/substrate complex.

According to this kinetic scheme the mass balance of the substrate – polymer can be expressed from:

$$\frac{dC_S}{dt} = -k_1 C_E C_S + k_{-1} C_{ES} \tag{26}$$

However, since reaction (24) involves a solid surface it is better to transform eqn (26) in surface area terms:

$$\frac{d(A(1-\theta))}{dt} = -k_1 C_E A(1-\theta) + k_{-1} A\theta \quad \Rightarrow \quad \frac{d(\theta)}{dt} = k_1 C_E - (k_1 C_E + k_{-1})\theta \tag{27}$$

where A is the substrate surface area, θ is the fraction of substrate occupied by the ES complex, $(1-\theta)$ is the "free" surface fraction and C_E is the enzyme concentration. The same terms have been also used in literature [5].

Assuming C_E to be constant eqn (27), can be easily integrated to give:

$$\theta = \frac{k_1 C_E}{k_1 C_E + k_{-1}} \left\{ 1 - \exp\left[-\left(k_1 C_E + k_{-1}\right)t\right] \right\} \tag{28}$$

The experimentally measured normalized weight loss in mg/cm^2, M_{S0} - M_S can be correlated to θ according to the following equation:

$$M_{S0} - M_S = \frac{N_{S0}}{A_0} - \frac{N_S}{A} \tag{29}$$

where the symbol N is used to denote the substrate mass and the subscript 0 refers to initial conditions (i.e. time equal to zero).

Assuming that the substrate surface area, A, is proportional to the free surface fraction powered to a constant value, α, according to $A = A_0(1-\theta)^\alpha$, eqn (29) is expressed as:

$$M_{S0} - M_S = \frac{N_{S0}}{A_0} \left[1 - (1-\theta)^{1-\alpha} \right] \tag{30}$$

If a is set equal to zero a constant surface area during the reaction is assumed, furthermore, if α is set equal to 1 a constant normalized weight loss is obtained. In this study the arithmetic mean value was considered, i.e. $\alpha=0.5$.

Eqns (30) and (28) can then be used to fit the experimental data of the normalized substrate weight loss as a function of time for the three individual polymers used. In these equations three different fitting parameters were identified, i.e. the kinetic rate constants k_1, k_{-1} and the initial substrate mass N_{S0}/A_0, which was kept as a fitting parameter in order to have an estimation of the initial effective polymer mass, which actually is biodegradable. A non-linear curve-fitting algorithm was used based on the Levenberg–Marquardt method and the best fitting values for the parameters are presented in Table 3.

Table 3: Kinetic and physical parameters determined by fitting eqn (30) to the experimental data of enzymatic hydrolysis of aliphatic polyesters.

Polyester	k_1 (mL/mg×days)	k_{-1} (days^{-1})	N_{S0}/A_0 (mg/cm^2)
PESu	0.021	0.036	66
PPSu	0.032	0.028	72
PBSu	0.053	0.009	16

As it can be seen going from PESu to PBSu an increase in the right direction kinetic rate constant was obtained with the reverse effect observed in the left direction rate constant. This means that as the spacing between ester groups increases the polymer is more susceptible to enzymatic attack. Furthermore, the best fitting value of the initial normalized substrate mass, N_{S0}/A_0, taking part in the enzymatic hydrolysis is larger for PPSu followed by PESu and PBSu. This result is in direct accordance with the polyester crystallinity data. As it was reported [6], PPSu exhibits the less degree of crystallinity followed by PESu and PBSu. Hence the assumption that the effective material mass, taking place in the degradation, is that of the amorphous polymer is confirmed.

Acknowledgement

This work was funded by the E.K.T. / E.Π.E.A.E.K. II in the framework of the research program PYTHAGORAS II, Metro 2.6.

References

[1] Karayannidis, G.P., Roupakias, C.P., Bikiaris, D.N. & Achilias, D.S., *Polymer,* **44**, 931, 2003.
[2] Kang, C.K., Lee, B.C. & Ihm, D.W., *Journal of Applied Polymer Science*, **60**, 2007, 1996.
[3] Park, S.S., Jun, H.W. & Im, S.S., Polymer Engineering & Science, 38(6), 905, 1998.
[4] Marten, E., Müller, R.-J. & Deckwer, W.-D., *Polymer Degradation & Stability*, **80**, 485, 2003.
[5] Scandola, M., Focarete, M.L. & Frisoni, G., *Macromolecules*, **31**, 3846, 1998.
[6] Bikiaris, D.N., Papageorgiou, G.Z. & Achilias, D.S., *Polymer Degradation & Stability*, **80**, 485, 2003.

Basic evaluation of sorting technologies for CCA treated wood waste.

K. Yasuda[1], M. Tanaka[1] & Y. Deguchi[2]
*[1]Department of Environmental Science and Engineering,
Okayama University, Japan
[2]Advanced Technology Research Center, Technical Headquarters,
Mitsubishi Heavy Industries Ltd, Japan*

Abstract

Two sorting technologies including an X-ray fluorescence technique and a Laser-induced breakdown spectroscopy (LIBS) technique were investigated for separating chromate copper arsenate (CCA) treated wood from other wood types in the wood waste stream. X-ray fluorescence was tested in the laboratory using a portably available X-ray fluorescence spectrometer. Operational parameters for continuous sorting using LIBS technology were established. These parameters concluded that chromate was the most sensitive metal for analysis, analysis time was less than 1 second per wood sample. LIBS technology shows considerable promise for continuous separation of large quantities of CCA treated wood from other wood types in the field using an on-line sorting system.
Keywords: CCA treated wood, continuous sorting, color reacted with chemical stain, X-ray fluorescence(XRF), Laser-induced breakdown spectroscopy (LIBS).

1 Introduction

Chromated copper arsenate (CCA) is a wood treatment preservative containing copper, chromium and arsenic. The purpose of the chemical is to protect wood from biological deterioration. CCA treated wood comes in a range of retention levels, from 4 kgm^{-3} to 40 kg m^{-3}. Lower retention levels are used for above ground applications and in freshwater contact, whereas higher retention levels are used for wood that is in contact with the ground or saltwater. The types of products treated with CCA include plywood, lumber, timbers, fence posts, utility poles and others (AWPI [1]).

WIT Transactions on Biomedicine and Health, Vol 10, © 2006 WIT Press
www.witpress.com, ISSN 1743-3525 (on-line)
doi:10.2495/ETOX060321

The use of CCA chemical in Japan grew from 1970s until 1997. During this time frame the total volume of treated products also significantly increased. In 1970, the total volume of CCA treated products was 0.8 million m^3. By 1997, 4.4 million m^3 of CCA treated wood were produced. Since CCA treated wood has a life span of about 25 years, it is projected that the wood waste stream will mirror the production statistics of CCA treated wood and amount of CCA treated wood being disposed will multiply (Solo-Gabriele and Townsend [4]).

Since CCA contains metals, there are concerns with the ultimate fate of chromium, copper and arsenic once the CCA treated wood is disposed. In Japan, much of the CCA treated wood waste is frequently mixed with untreated wood at construction and demolition (C&D) debris recycling facilities. When CCA treated wood is burned with C&D wood, the ash created contains high amounts of chromium, copper and arsenic, posing an environmental problem.

One solution to avoid the contamination associated with CCA treated wood in recovered C&D wood to develop technologies for separating the CCA treated wood from the rest of the wood waste. The rest of the wood waste can then be burned in cogeneration plants or recycled as mulch without the added metals from CCA, whereas CCA treated wood can be recycled or disposed in a different manner (Felton and De Groot [3]). The chemical stains followed by manual sorting is one option. PAN(1-(2-pyridylazo)-2-naphthol) indicator performed the best in the field, mainly due to its faster reaction time. PAN, however, was subject to interference reactions with paints, adhesive, nails and other metal fasteners. It is known that PAN might react with metals other than those found in CCA. Chemical stains are suitable for sorting small quantities of wood waste and as a method for manual separation (Blassino et al. [2]). This study addresses two alternative methods of continuous sorting CCA treated wood from other wood waste, use of X-ray fluorescence (XRF) and Laser-induced breakdown spectroscopy (LIBS).

2   Sample description and preparation

Two kinds of samples were utilised in XRF and LIBS analyses: CCA treated wood and wood chips from construction demolition (C&D) recycling facilities. The column in house during demolition was used as the CCA treated wood sample.

The wood chips samples were taken from two C&D recycling facilities located throughout Okayama-city, Japan.

3   X-ray fluorescence analyser

X-ray fluorescence (XRF) is based on the ability of an atom to emit its characteristic energy in form of an X-ray when an external source of X-ray is applied. XRF analysers register this incoming energy and compare it to the known energies of a given element.

The XRF detector used for measurement in this study (energy dispersive X-ray fluorescence analyzer, OURSTEX 100FA) is routinely used by the recycling facility. The detector is a portable type and convenience to measure in the field.

3.1 Analyses conducted

Fig. 1 shows the result of the analysis as the CCA treated wood. As noted from Fig. 1, three elements of chromium, copper and arsenic were confirmed.

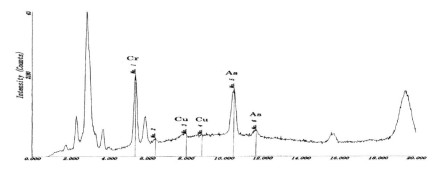

Figure 1:   Measurement result of the CCA treated wood by XRF technology.

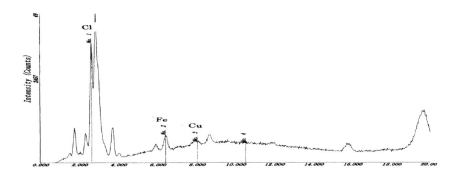

Figure 2:   Measurement result of wood chip color reacted with the chemical stain (PAN) by XRF technology(1).

Fig. 2 to Fig. 4 show the measurement results of wood chips color reacted with the chemical stain (PAN) by XRF technology.

As seen in those Figures, the peak of copper was confirmed for every wood chip, but there were no peaks of chromium and arsenic. It is known that PAN might react with metals other than those found in CCA. It was concluded that those wood chips color reacted with PAN were not CCA treated wood.

The sorting of CCA treated wood by XRF is more accurate than that by chemical stain.

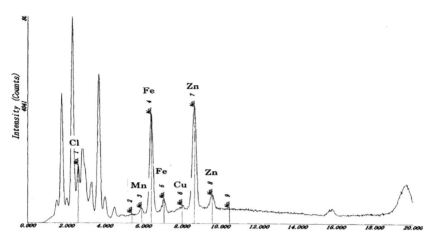

Figure 3: Measurement result of wood chip color reacted with the chemical stain (PAN) by XRF technology (2).

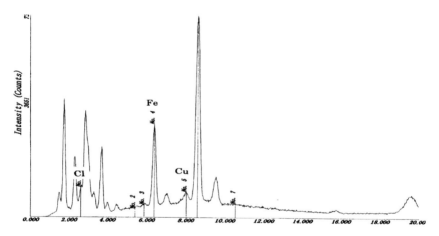

Figure 4: Measurement result of wood chip color reacted with the chemical stain (PAN) by XRF technology (3).

4 Laser-induced breakdown spectroscopy analyser

Laser-induced breakdown spectroscopy (LIBS) is based on the ability of plasma emission spectroscopy to measure the concentrations of metals (Cr, Cu, As etc.) by making plasma ion when laser beam is focussed to solid materials (ex. Wood chips). LIBS also has the ability of real time and on-line measurement of CCA treated wood. Fig. 5 shows the result of chromium measurement in wood chips.

As seen in Fig. 5, chromium in wood chip could be identified with clarity by LIBS technology. From Fig. 6 to Fig. 7 show the measurement results of copper and arsenic in wood chips by LIBS.

Measurement metal : chromium (Cr)

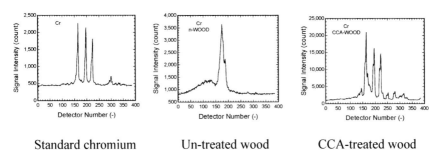

Standard chromium   Un-treated wood   CCA-treated wood

Figure 5:   Analysis of wood chips by LIBS technology (1).

Measurement metal : copper (Cu)

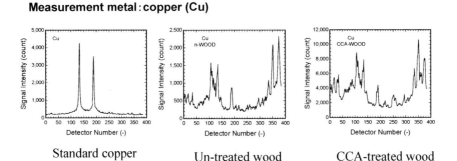

Standard copper   Un-treated wood   CCA-treated wood

Figure 6:   Analysis of wood chips by LIBS technology (2).

As shown in Fig. 6 and Fig. 7, Arsenic in wood chip could be also identified by LIBS, but not be selected so clear as chromium. In the case of copper, it was concluded that the identification of copper in wood chip was very difficult because copper was used many kinds of wood as a preservative in Japan (Ukishima et al. [5]).

Measurement metal: arsenic (As)

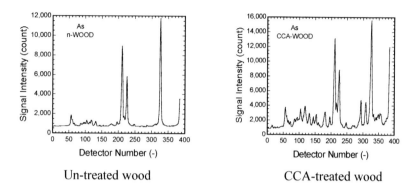

Un-treated wood   CCA-treated wood

Figure 7:   Analysis of wood chips by LIBS technology (3).

As seen in Table 1, chromium was found in the highest concentration in demolished CCA treated wood.

Table 1:   The range of metal concentrations contained in CCA treated wood (demolished wood).

	Range of metal concentrations in demolished CCA treated wood (mg/kg-wood)	
	Min.	Max.
Cr	212	5,980
Cu	79	2,750
As	62	2,670

Also there are no other preservative contain chromium but CCA. Therefore, chromium was considered the best indicator for presence of CCA in wood samples.

Fig. 8 shows the result of quantitative analysis in the case of chromium by LIBS.

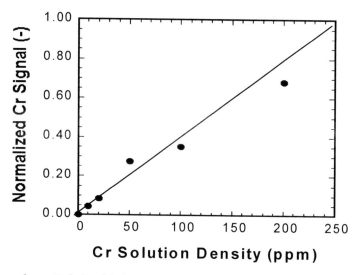

Figure 8: Figure 8:   Relationship between chromium content and signal intensity.

As shown in Fig. 8, it was confirmed the linearity of quantitative analysis between 0ppm and 200ppm of the concentration. Therefore, it was proved the ability of wide-ranging retention levels to detect CCA treated wood.

5   Continuous separation of CCA treated wood

It needs continuous sorting for separating large quantities of CCA treated wood from other wood types in the field. There are two sorting technologies including an XRF and a LIBS. Table 2 shows comparison between XRF and LIBS of the function. As seen in Table 2, LIBS technology shows considerable promise for continuous separation of large quantities of CCA treated wood from other wood in the field using an on-line sorting system.

Table 2:   Comparison of XRF and LIBS.

Method	Window for the detection	Measurement point	Detection limit	Measurement time	Distance attenuation
XRF	About 5 mm	Fixing	>50 ppm	>20 second	Profound effect
LIBS	About 1 mm	Scanning	ppb~ppm	Real time	Lightly effect

Fig. 9 shows the pattern diagrams for continuous separation of large quantities of CCA treated wood from other shredded wood chips in the field using an on-line sorting system.

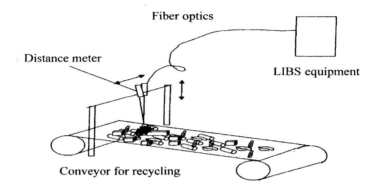

Figure 9:   Continuous sorting of CCA-treated wood using LIBS technology.

LIBS equipment is fixed in the air-conditioned room. LIBS is connected through fiber optics to the conveyor for recycling in the field. Laser beam from LIBS is focussed to shredded wood chips through scanning fiber optics. CCA treated wood is continuously separated from other wood chips when chromium is detected.

6 Conclusions

Overall, the use of technologies for distinguishing CCA treated wood from other wood types should be encouraged in recycling operations where the quality of the recycled wood impacts its potential for re-use. Two sorting technologies including an XRF technique and a LIBS technique were investigated for separating CCA treated wood from other wood types in the wood waste stream. XRF technology is suitable for sorting small quantities of wood waste and as a method for checking wood quality. LIBS technology, which is characterised by higher capital costs, has the potential for sorting large quantities of wood waste using an on-line system. Future efforts will focus on testing LIBS technology in the field using an on-line sorting system for separating CCA treated wood from other wood types.

References

[1] American Wood Preservers Institute (1996): *Wood Preserving Industry Production Substantial Report*. Fairfax, VA, U.S.A.: Elsevier Science Publishing Company, Inc.

[2] Blassino, M., Solo-Gabriele, H.M. and Townsend, T. (2002): Pilot Scale Evaluation of Sorting Technologies for CCA Treated Wood Waste. *Waste Management & Research* **20**, 290-301.

[3] Felton, C., De Groot, R.C. (1996): The Recycling Potential of Preservative Treated Wood. *Forest Products Journal*. **46** (7/8), 37-46.

[4] Solo-Gabriele, H.M., Townsend, T (1999): Disposal Practices and Management Alternatives for CCA-treated Wood Waste. *Waste Management & Research* **17**, 378-389.

[5] Ukishima, Y., Furuya, Y., Maenami, K. and Ajioka, Y. (2001): Study on Recycling Scraped Wood with which Medicine Processing of Antiseptics, Antifungal Agents and Insecticides were Performed. *Bulletin of Shizuoka Institute of Environ. and Hyg.* **44**, 67-74.

Kinetics of nitrobenzene biotransformation

A. V. N. Swamy[1] & Y. Anjaneyulu[2]
[1]*Department of Chemical Engineering, JNTU College of Engineering, Anantapur, India*
[2]*JNTU, Kukatpally, Hyderabad, India*

Abstract

Nitrobenzene and substituted aromatics like nitro-chlorobenzene, aniline, and nitro-toluene containing chlorine, amino and methyl groups are recalcitrant in nature. These molecules are both anthropogenic and xenobiotic. Nitro benzene and nitro aromatic compounds are environmental pollutants discharged through wastewaters from nitro aromatic manufacturing plants. Nitrobenzene and other aromatics are toxic to several forms of aquatic life. However, biological transformation of nitrobenzene to non-toxic entities exists in specialized microbes, which have enzymes of aromatic catabolic pathways. Transformation of nitrobenzene is also inhibited by the presence of other toxic materials such as cyanides and sulphides that are present in industrial waste when nitrobenzene is the dominant carbonaceous material. Kinetics of the biotransformation of nitrobenzene using pure cultures isolated in the laboratory, mixture of consortium of the pooled cultures and enriched activated sludge biomass in pure substrates, mixed substrates and actual nitrobenzene plant waste has been estimated under varying input concentrations. Experimental results from these studies have been subjected to analysis by mathematical models using Monod's and Haldane's equations to test their validity in interpreting the data on inhibitory substances under stable as well as unstable state operations of wastewater treatment plants. Comparative evaluation of the kinetic parameters reveals that Monod's model can be employed for the estimation of kinetic constant μ only while Haldane's model has to be used for the calculation of μ_{max} and K_i.
Keywords: kinetics, nitrobenzene, nitrochlorobenzene, and biotransformation.

1 Introduction

Nitrobenzene is a substituted aromatic molecule containing nitro group. The origin of nitrobenzene to the atmosphere is due to anthropogenic activity. Anthropogenic sources are industrial wastes derived from chemical manufacturing processes such as nitro aromatic chemicals production units. Mostly the industrial wastes are originated from the wash waters of final product purification. The summary of microbial transformations of some substituted aromatic compounds together with isolated microorganisms is given in a research paper (Gibson [1]). Anaerobic degradation studies on benzene nucleus were carried out (Taylor *et al.* [2]) by using facultative and anaerobic microorganisms

Activated sludge process (ASP) is one of the most widely accepted biological systems for the treatment of nitro aromatic compounds. The biodegradation of aromatic pollutants under aerobic conditions was studied (Arcangeli and Arvin [3]). Existing theory suggests that operational difficulties associated with inhibitory compounds present in the waste is due to process dynamics. The significance of inhibitory mechanisms, associated with operational problems in the waste treatment, is to avoid metastable and unstable regimes. Bacteria capable of utilizing nitro aromatic compounds are found in soil and water environment. The wastewater from nitro aromatic production units predominantly contains nitrobenzene, nitrochlorobenzene, nitro toluene, and nitro cresols. Microorganisms attack these aromatics ring through a degradative pathway. The kinetics of microbial degradation of toluene were studied (Jirgensen *et al.* [4]). Aromatic compounds are rich in carbon content and once the rings are cleaved by enzymes the products (organic acids) enter the energy cycle. During the studies of biodegradation of nitro aromatic compounds it was observed (Hu and Sheih [5]) that nitro groups substituted by hydroxyl groups, with the elimination of the group as nitrite ion. Kinetics of the enzymatic reactions is hence significant in understanding the biotransformation of these molecules. Techniques for obtaining bio kinetic parameter values were studied (Grady [6]).

Table 1:   Physico chemical characteristics of nitrobenzene wastewater.

S.No	Parameter	Value
1	PH	4.5-5.6
2	Color	Reddish brown
3	C.O.D.	1800-2200
4	Total solids	100-120
5	B.O.D.	650-800
6	Nitrobenzene	80-150
7	Suspended solids	50-80
8	Ammonical nitrogen	120-200
9	Oil and Grease	10-25

All values expressed in mg/L except pH and colour.

2 Theory

Evaluation of bio kinetic constants is significant for understanding the capacities of microorganisms for the operation of biological reactors. Kinetic studies on bacterial growth have given rise to considerable amount of literature. Particularly with regard to application of Monod's equation in respect of batch results.

$$\mu = \mu_{max}\left(\frac{S}{K_s + S}\right) \tag{1}$$

In calculation of kinetic constants, μ, μ_{max}, and K are related to reactor functions. These constants have been derived from the original Michaelis-Menton's enzyme equation:

$$v_0 = V_{max}\left(\frac{S}{K_m + S}\right) \tag{2}$$

Monod's model has extensively been used for the estimation of biokinetic constants for bacterial growth on non-inhibitory substrates and for inhibitory substrates.

Nitrobenzene exhibits substrate inhibition and the fact that metastable conditions existed in the ASP lead to a review in the application of Monod's model to inhibitory growth models; resulting in the selection of Haldane's equation:

$$\mu = \mu_{max}\left[\frac{S}{S + K_i + \left(S^2 / K_i\right)}\right] \tag{3}$$

where K_i is the inhibitor constant.

3 Materials and method

Pure cultures of *Pseudomonas aeruginosa* and *Pseudomonas putida*, were used in monoculture experiments. The mixed culture consisting of Pseudomonas *aeruginosa* and *Pseudomonas putida* has been used as mixed cultures in the growth responsible studies. Activated sludge biomass from a sewage treatment plant was used as the seed for plant waste studies. The seeds were then transferred into the growth medium with respective carbon sources. Specific growth rate.μ has been measured from the exponential growth phase and used in the model equations.

The wastewater from NB plant was collected as a composite sample, transported in airtight containers, and characterized. The waste after proper dilution was sterilized with a membrane filter before it was used in the growth experiments. Compositions of the NB plant waste are given in Table 1.

3.1 Analytical procedures

Nitrobenzene was estimated by gas chromatography (Varian3700). Waste water was analyzed for other parameters according to standard methods APHA, AWWA, WPCF [7].Growth of biomass was measured by turbid metric method (OD at 610nm) which was then calibrated against dry weight. The cultures were incubated at a constant temperature of $30^{\circ}C$.

3.2 Experimental observations

The synthetic waste was constituted based on proximate composition and supplemented with other essential nutrients. The specific growth rate μ derived from various initial substrate concentrations has been plotted against initial constant concentrations to derive the kinetic constants using the kinetic models. Fig. 1(a)-(d) represents the reciprocal plot deriving the kinetic constants by Monod's model while Fig. 2(a)-(d) represents the typical line of best fit by Haldane's model.

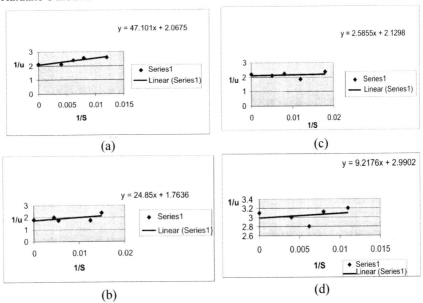

Figure 1:   (a) Kinetic constants Monod model K, u_{max}, *P. aeruginosa*. (b) kinetic constants Monod model k, u_{max} *P. putida*, (c) kinetic constants Monod model K, u_{max}, mixed culture synthetic waste and (d) kinetic constant Monod model K, μ_{max}, mixed culture, plant waste.

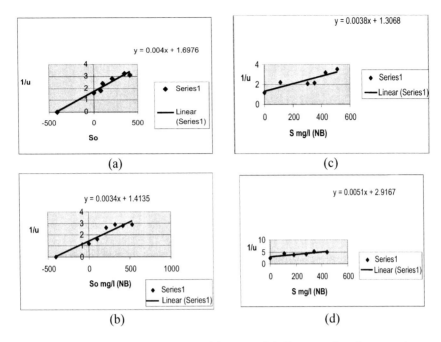

Figure 2:   (a) Kinetic constant Haldane model K, μ_{max} for *P. aeruginosa,*
(b) kinetic constant Haldane model K, μ_{max} for *P. putida,* (c) kinetic
constants Haldane model K, μ_{max}, mixed culture, synthetic waste
and (d) kinetic constant Haldane Model K, μ_{max}, mixed culture
plant waste.

Table 2:   Biokinetic constants for NB biodegradation of the values with
Monod's and Haldane's models.

S.No	Parameter	Monod's value		Haldane's value	
		$\mu_{max}(_h^{-1})$	$K_s(\,mgl^{-1})$	$\mu_{max}(_h^{-1})$	$K_i(mgl^{-1})$
1	*Pseudomonas aeruginosa*	0.52	31.2	0.62	480
2	*Pseudomonas putida*	0.66	19.5	0.83	490
3	Mixed culture with Synthetic NB	0.55	12.06	0.76	440
4	Mixed culture with plant waste	0.68	103.52	0.62	380
5	Sewage plant sludge with plant waste	0.42	32.25	0.45	370

These expressions clearly indicate that μ_{max} is a reaction rate constant achieved by a culture when it is grown under no limiting growth conditions. The value should then be the one that is obtained as the highest value. Thus μ_{max} calculated by Haldane's model should be considered the true value.

Growth data from a continuously operated completely mixed activated sludge (CMAS) system has also been subjected to treatment using Monod's and Haldane's model equations. Experimental results obtained from batch and continuous culture experimentation has been subject to validation of mathematical models suggested by various investigators. The apprehensions and difficulties experienced in comparing the values obtained with synthetic media in nitrobenzene biodegradation and in actual waste water systems have been examined in the present study. One of the major difficulties experienced by previous investigators is the lack of techniques and tools in conducting pure culture experiments. Isolation, enrichment and monoculture experimentation of laboratory stored cultures have been possible in the author's laboratory to study the kinetics of biodegradation of nitrobenzene in absence and presence of secondary inhibitors. These studies have been found necessary to understand the maximum degradation rate (the maximum substrate utilization rate) of the substrate so that in the event of the absence of all environmental pressure how far the organisms can express themselves. The μ_{max} for *Pseudomonas aeruginosa* of 0.62 hr^{-1} could not be achieved when the culture is mixed with other cultures and cultivated in a defined medium along with other strains of bacteria. The μ_{max} is further reduced when the mixed cultures are grown in actual wastewater (μ_{max} =0.76 $hr^{-1)}$, indicating that there are several factors which influence the growth of microorganisms in a given environment. Competition for the common substrate is one of the possible reasons for a lower growth rate while noncompetitive inhibition secondary inhibitors are another. Laboratory-stored cultures are highly induced having multiple copies of plasmid genomes, which can express themselves, so that high efficiency can be achieved with such specialized seeds. Treatment of kinetic data with various model equations shows that Monod's model has limitations when used with inhibitory compounds. When the substrate induced cultures are grown at high initial substrate concentrations they tend to show lower growth rates (observations by Hill and Robinson, [8]; Gaudy *et.al,* [9]). This means that the organisms are under the inhibitory effect of the substrate. It is possible to achieve the exponential growth very fast and reproducibility of the experimentation is possible when the cultures are grown with initial substrate concentration is below the K_s. for experimentation on any kinetic studies with inhibitory constituents, it is necessary to obtain the K_s of the system, in the first instance and then with this as the critical concentration further experiments have to be conducted to obtain value for Ks. Biodegradation can considered a feasible and reasonable system to eliminate, NB, NT from the effluents (Swamy *et al* [10]).

4 Conclusions

Analysis of data on continuously operated reactor systems shows that under feed starve conditions μ_{max} is much lower than the batch fed systems, indicating high detention time has to be given in the reactor. However, for the treatment of nitrobenzene, conventional ASPs with a HRT of 12 to 24 hrs is optimum. K_i of the continuously operated system is low: a situation that was postulated by Monod. The death rate of the organisms is very low which is ideal for detoxification of the toxic substrate. Engineering significance of kinetic constants is that, when hydraulic shock is effected beyond μ^* there is every possibility of biomass being washed out. Similarly, when organic shock is made beyond K_i value the rate of degradation of the substrate is retarded, resulting in substrate build up which also reduces the treatment efficiency and functioning of the treatment plant.

Acknowledgement

Authors would like to thank M/s Hindustan Organic Chemicals Ltd., Rasayani, Dist.Raigad., Maharashtra, India, for generously funding this research work.

References

[1]   Gibson, D.T. (editor), "Microbial Degradation of Organic Compounds," Marcel Decker, Inc., New York, 1986.

[2]   Taylor, B. F., W.L. Campbell and T. Chinoy, "Anaerobic Degradation of the Benzene Nucleus by a Facultative Anaerobic Microorganisms." *J. Bacteriol.*, 102(2), pp.430-437, 1970,

[3]   Arcangeli, J.P. and Arvin, E., Kinetics of toluene degradation in a bio film reactor under denitrifying conditions. *Wat. Sci. Tech.*, 29(10-11), 343-400, 1993.

[4]   Jirgensen, C., Flyvbjerg J., Arvin E. and Jensen. B.K., Stoichiometry and kinetics of microbial toluene degradation under denitrifying conditions, *Biodegradation* 6, 147-156, 1995

[5]   Hu, L.Z. and W.K. Sheih., , Anoxic bio film degradation of monocyclic aromatic compounds. *Biotech. Bioengg.* 30:1077-1083, 1987

[6]   Grady. C.P.L., Jr., Biodegradation of Toxic Organics: Status and Potential. *J. Environ. Eng.*, 116, 805, 1990

[7]   APHA, Standard methods for the examination of water and wastewater. American Public Health Assoc., New York, N.Y., 1985.

[8]   Hill G.A. and Robinson C.W, , Substrate inhibition kinetics. Phenol degradation by P.putida. *Biotechnol. Bioengg* 17, 1599-1615, 1975.

[9]   Gaudy A.F., Rozich A.R and Gaudy E.T., Activated sludge process models for treatment of toxic and non-toxic wastes. *Environ. Sci. Technol.* 123-137, 1986,

[10]   Swamy A.V.N and K.S.Rao, Biodegradation of Nitrobenzene and nitrotoluene by acclimated bacteria, National Journal of Life Sciences, 2(1), 2005.

Section 10
Biotests

Microbiotests in aquatic toxicology: the way forward

C. Blaise[1] & J.-F. Férard[2]
[1]Environnement Canada, Montréal, Canada
[2]Université de Metz, France

Abstract

The industrial revolution has driven the need for ecotoxicology and shaped its evolution. Indeed, the increased use and transformation of (non)renewable resources for over a century to benefit mankind have had a downside and created a plethora of contaminants harmful to the receiving environments. With time, we have gone from an age of darkness in the 1950s (i.e., diagnostic ignorance in terms of recognizing and dealing with contamination) to one of enlightenment as the 21st century unfolds (i.e., use of tools and strategies to identify and correct environmental pollutions). Effects measurements, reflected by toxicity testing conducted at different levels of biological organization, have proven especially useful to achieve proper hazard/risk assessments of contaminants. Knowing why toxicity testing has been conducted over the past decades to protect and conserve freshwater environments is also essential to grasp the importance and breadth of this field. For this purpose, we have recently reviewed a substantial number of articles describing numerous bioanalytical endeavours undertaken to comprehend toxic effects associated with the discharge of xenobiotics to aquatic environments. Scrutiny of publications identified in our literature search has enabled us to uncover the various ways in which laboratory toxicity tests have been applied, many of which are small-scale in nature. In essence, freshwater toxicity testing has significantly focussed on liquid (complex environmental samples, chemical and biological contaminants) and solid media (sediments) assessment. For both media, miscellaneous studies/initiatives linked to toxicity testing applications have again promoted the development, validation, refinement and use of toxicity testing procedures. Bioassays are clearly an essential component of environmental management programs and several small-scale tests (microbiotests) can be employed to generate cost-effective toxicity data that assist decision-making.
Keywords: microbiotests, aquatic toxicology, effects measurements, freshwater toxicity testing, contaminants, evolution of ecotoxicology.

 WIT Transactions on Biomedicine and Health, Vol 10, © 2006 WIT Press
www.witpress.com, ISSN 1743-3525 (on-line)
doi:10.2495/ETOX060341

1   Introduction

Innovations in technology periodically spark changes in human society and drastically transform our way of doing things. The most recent industrial revolution, which began in England as early as the 18th century, spread to most of Europe and elsewhere during the 19th and 20th centuries, and continues onwards to other developing parts of the world to this day, is unquestionably the most significant of modern times. Extensive mechanization, urban concentration of labour and large-scale production of goods and materials from increased use (and transformation) of (non)renewable resources characterize this indefectible quest to enhance the quality of life of *Homo sapiens*. While human progress is a worthwhile endeavour, unabated industrial growth over time has produced, as its downside, numerous pollution problems for receiving environments of our biosphere.

2   Ecotoxicological evolution

With respect to freshwater aquatic environments – the focus of this paper-growing chemical contamination generated by industrial processes eventually led, as we shall see, to the creation of a new discipline of the environmental sciences termed "ecotoxicology".

Figure 1 illustrates, by way of decadal snapshots, how ecotoxicological evolution occurred starting from the middle of the 20th century. Awareness of pollution problems finally began to sink in by the 1950s when degraded environments became obvious and decision-makers (depicted by the friendly dinosaur in Fig. 1a) wished to reverse the trend and hoped for a return to more pristine conditions. Because know-how, tools and strategies were simply not available to assess impacted areas, however, curative responses were either limited or ineffective. The 1950s and before could thus be referred to as "the age of darkness" in terms of environmental action.

In developed countries in the 1960s, toxicity testing conducted with fish confirmed that large volumes of wastewaters discharged from various industries and municipalities, as well as specific contaminants (e.g., cadmium, mercury, pesticides), were indeed toxic, very often causing rapid lethal effects on exposed organisms [1]. This decade, in particular, highlighted the intrinsic need for effects-based measurements to properly assess contaminants and certainly merits being remembered as "the beginning of enlightenment" in terms of enhancing basic knowledge on the hazards posed by aquatic contaminants (Fig. 1b).

Owing in part to bioanalytical data amassed in the 1960s demonstrating the adverse effects of point source discharges and their potential (or real) impact on aquatic biota and habitat in some cases, environmental agencies/departments were created in several countries (e.g., U.S. Environmental Protection Agency, Environment Canada, French Environment Department) at the start of the 1970s. Chemical characterization of effluents and regulatory compliance based on both chemical and toxicological parameters soon proved helpful in eradicating acute

Figure 1:   Ecotoxicological evolution: contaminant pressures trigger the need for effects-based measurements.

lethal effects from such wastewaters. As a result, fish were once again able to survive in the vicinity of effluent outfalls, but their yet unaddressed chronic (insidious) effects remained a serious concern, as suggested by the three-eyed fish in Fig. 1c. The 1970s can also be remembered as those marking the official birth of the field of ecotoxicology, by and large the study of environmental pollutants and their effects on biota. Essentially a fusion of the words "ecology" and toxicology", the appellation "ecotoxicology", makes its debut in the scientific literature. First reported in France [2], it is soon adopted by several others who propose definitions to specify its scope and breadth [3–5].

Sustained biotesting carried out throughout the 1970s and beyond showed that toxic effects could often be trophic level-specific (e.g., herbicides on photosynthesis of micro-algae) and consequently stimulated the development of acute/chronic toxicity testing at different levels of biological organization. In the 1980s, several small-scale tests (microbiotests) employing bacteria, algae and micro-invertebrates, for example, were used in "battery approaches" in an attempt to circumscribe the full toxic potential of chemicals and environmental samples [6]. This decade clearly popularized microbiotesting and holistic thinking in that integrated biological/chemical strategies were able to identify the species most at risk linked to a particular contaminant or pollution event (Fig. 1d). By protecting the most sensitive life form, all others are indeed secure.

The need for applying cost-effective toxicity tests capable of high throughput became urgent in the 1990s owing to the increasing number of environmental samples requiring assessment (Fig. 1e). Because of the attractive features intrinsic to several microbiotests (e.g., low-volume requirements, miniaturisation and automation potential), microplate-based tests such as the SOS Chromotest were indeed capable of achieving remarkable performances [7].

At the dawn of the 21st century where issues of sustainable development and biodiversity are inextricably interdependent, ecotoxicology, with its tools and strategies which include microbiotesting of xenobiotics, will have a positive role to play in the conservation and protection of aquatic systems. Effects measurements applied internationally to estimate contaminant hazard of (non)point sources of pollution, coupled with effective technology, should provide the cognitive function necessary to drive subsequent curative and preventive actions to ensure clean water for all living creatures including mankind (Fig. 1f). This goal will only be achievable pending concerted efforts of knowledgeable decision-makers on a global scale.

3 The "cart before the oxen" syndrome

Attempting to infer effects on biota from chemistry alone is simply a no-no nowadays, but it was once prevalent during a time period (possibly up to 1975) when toxicity tests, still to be developed or applied, had not yet come to the fore. An example of the "cart before the oxen syndrome" in this respect depicting moot management of environmental funds is given by way of the narrative in Table 1. Not so obvious then for environmental managers was to think that thorough chemical scans of complex effluents, which are seldom exhaustive, would suffice to report their hazardous status and identify corrective clean-up actions. Knowledge today dictates that a chemical-based approach alone says little on bioavailability and possible interactions of the cocktail of pollutants that can be present in effluent mixtures. Nor does it give information on effects (acute/chronic toxicity, genotoxicity, etc.) or on the trophic level(s) that can be targeted by specific wastewaters.

Table 1:   The cart-before-the-oxen-syndrome (see text for details).

An environmental manager, fond of chemical parameters, was mandated to determine the environmental hazard of a particular industrial effluent. He thus called upon a private laboratory to undertake an exhaustive scan of priority pollutants suspected of being present in the effluent. After spending 25K for this study and obtaining chemical data on 50 substances in return, his superior urged him to conclude categorically on the potential impact this wastewater discharge was having on receiving water biota and on the clean-up actions that should be taken. He suddenly realized that he could not infer very much!

- Where did he go wrong?
- What should he have envisaged instead?

This "data rich/information poor" situation can only be offset by contemporary recognition that a combined biological/chemical strategy should be implemented. Indeed, the demonstrated presence of adverse effects by first conducting biological testing will then justify the generation of chemical analysis in an attempt to link specific contaminants to observed effects via validated approaches calling upon toxicity identification evaluations, or TIEs, and toxicity reduction evaluations, or TREs [8]. To be effective, ecotoxicology now thrives on multi-disciplinary partnerships struck with toxicology, chemistry and ecology (and other related fields of science) such that the cart-before-the-oxen-syndrome is rapidly disappearing.

4   Microbiotests

Small-scale toxicity testing, an offshoot of aquatic toxicology, blossomed in the 1980s as development and application of microbiotests began to proliferate owing to an increasing demand for cost-effective tests conducted at various levels of biological organization. Simply defined as the "exposure of a unicellular or small multi-cellular organism to a liquid/solid sample to measure a specific effect" [9], microbiotests are clearly at the forefront in the daily war being waged against contaminants owing to their effective toxicity screening potential. Coupled with other effects-based measurements that include biomarkers and *in situ* approaches (e.g., biotic indices), ecotoxicological tools can convincingly combine with chemical analysis to identify culprit pollutants responsible for biotic impacts in the laboratory and in the field (Fig. 2). Several reliable and relevant small-scale tests exist to appraise both liquid and solid media, although a larger number of methods are presently available to measure toxicity in the former over the latter (Table 2). While representative of applicable micro-scale assays, this list is nevertheless far from exhaustive.

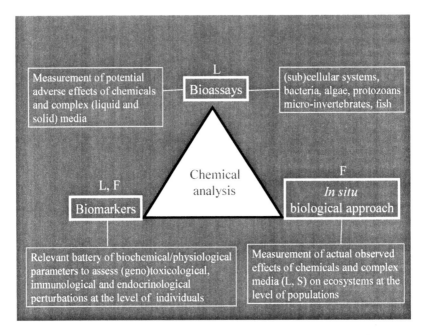

Figure 2:   Complementary tools in ecotoxicology for laboratory- (L) or field-(F) based studies and their essential link with chemical analysis to establish cause-effect relationships.

5   Richness of toxicity testing applications

Published articles in the scientific literature (both primary and grey) spanning over three decades (1970s and beyond) have contributed a wealth of information linked to diverse aspects of toxicity testing. In surveying several databases, we have recently reviewed a substantial number of articles describing numerous bioanalytical endeavours undertaken to comprehend toxic effects associated with the discharge of xenobiotics to aquatic environments [10]. This search enabled us to uncover the various ways in which laboratory toxicity tests have been applied, many of which are small-scale in nature. In essence, freshwater toxicity testing has significantly focussed on liquid (complex environmental samples, chemical and biological contaminants) and solid media (sediments) assessment. For both media, miscellaneous studies/initiatives linked to toxicity testing applications have again promoted the development, validation, refinement and use of toxicity testing procedures (Fig. 3).

Table 2:   Examples of typical (micro)biotests available for freshwater toxicity investigations.

Liquid media assessment
Algae: Flask growth inhibition assay [11]; Microplate growth inhibition assay [12]; Flow cytometric techniques [13]
Aquatic macrophytes: *Lemna minor* chronic assay [14]
Bacteria: Microtox acute assay [15]; SOS Chromotest [7]; MetPlate [16]
Fish cells: Trout primary hepatocyte cytotoxicity assay [17]; Trout gill cell line cytotoxicity assay [18]
Micro-invertebrates: *T. platyurus* acute assay (http://www.microbiotests.be); *Hydra* population reproduction assay [19]; *D. magna* acute/chronic assays [20]
Protozoa: *S. ambiguum* acute assay [21]; *T. thermophila* chronic assay (http://www.microbiotests.be)
Solid media assessment
Algae: Algal solid phase assay [22]; LuminoTox solid phase assay [23]
Bacteria: Microtox acute solid phase assay [24]; SOS Chromotest solid phase assay [25]
Micro-invertebrates: *C. riparius* survival and growth assay [26] *H. azteca* survival and growth assay [27]; *H. incongruens* survival and growth assay (http://www.microbiotests.be)

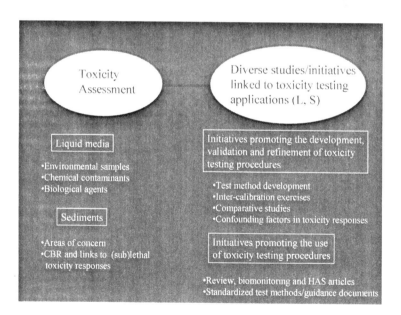

Figure 3:   Overview of contemporary toxicity testing: adapted from Blaise and Férard, 2005 [10]. L (liquid media); S (solid media); CRB (critical body residues); HAS (hazard assessment schemes).

6 Conclusions

Adequate assessment of pollutants calls for an effects-based approach complemented by chemical analysis, now an essential cog of the multi-disciplinary field of ecotoxicology. Within this combined biological/chemical strategy, toxicity testing can provide key information to guide decisions that will ensure effective protection and conservation of freshwater biota from adverse effects of harmful chemicals. As diagnostic tools of ecotoxicity, bioassays have an important role to play in this respect.

In addition and also owing to their attractive characteristics, several microbiotests are able to confer much needed cost-effectiveness and throughput for appraisal of contaminants. Recent books reporting on the benefits of small-scale testing applications reflect the marked influence of microbiotests and suggest that their popularity and use will grow in the future [28–30].

Acknowledgments

The authors are indebted to their respective managements (Environment Canada and University of Metz) for supporting this work presented during the "Environmental Toxicology 2006" conference held in Mykonos, Greece, from September 11-13, 2006. The remarkable drawings incorporated in Fig. 1 are those of Patrick Bermingham (Montreal, Quebec, Canada).

References

[1]   Blaise, C., Sergy, G., Wells, P., Bermigham, N. & van Coillie, R., Biological Testing - Development, Application and Trends in Canadian Environmental protection Laboratories. *Toxicity Assessment* **3**, pp. 385-406. 1988.

[2]   Jouany, J. M. Nuisances et Ecologie. *Actualités Pharmaceutiques* **69**, pp. 11-22, 1971.

[3]   Butler, G.C. (ed.). *Principles of Ecotoxicology, SCOPE series, Volume 12*, J. Wiley and Sons: New York, 372 pages. 1978.

[4]   Ramade, F. (ed.). *Ecotoxicologie. Collection d'écologie, no. 9*, Masson éditeur: Paris, pp. 228, 1979.

[5]   Moriarty, F. (ed.). *Ecotoxicology, the study of pollutants in ecosystems*. Academic Press: London/New York, pp. 233, 1983.

[6]   Dutka, B. Priority setting of hazards in waters and sediments by proposed ranking scheme and battery of tests approach. *Zeitschift für Angewandte Zool.* **75**, pp. 303-316, 1988.

[7]   White, P., Rasmussen, J. & Blaise, C. A semi-automated, microplate version of the SOS Chromotest for the analysis of complex environmental extracts. *Mutation Res.* **360**, pp. 51-74, 1996.

[8]   Novak, L. & Holtze, K. Overview of toxicity reduction and identification evaluations for use with small-scale tests. *Small-scale Freshwater Toxicity*

Investigations, Volume 2, eds. C. Blaise & J.F. Férard, Springer Publishers Dordrecht, The Netherlands, pp. 169-213. 2005.

[9] Blaise, C., Wells, P. & Lee, K., Microscale testing in aquatic toxicology: Introduction, historical perspective, and context. *Microscale testing in Aquatic Toxicology Advances, Techniques and Practice*, eds. P. Wells, K. Lee & C. Blaise, CRC Lewis Publishers: Boca Raton, Florida, pp. 1-9, 1998.

[10] Blaise, C. & Férard, J.F., Overview of contemporary toxicity testing. *Small-scale Freshwater Toxicity Investigations, Volumes 1 and 2*, eds. C. Blaise & J.F. Férard, Springer Publishers: Dordrecht, The Netherlands, pp. 1-68, 2005

[11] Staveley, J.P. & Smrchek, J.C. Algal toxicity test. *Small-scale Freshwater Toxicity Investigations, Volume 1*, eds. C. Blaise and J.F. Férard, Springer Publishers: Dordrecht, The Netherlands, pp. 181-202, 2005.

[12] Blaise, C. & Vasseur, P., Algal Microplate Toxicity test. *Small-scale Freshwater Toxicity Investigations, Volume 1*, eds. C. Blaise and J.F. Férard: Springer Publishers: Dordrecht, The Netherlands, pp. 137-179, 2005.

[13] Stauber, J., Franklin, N. & Adams, M. Microalgal toxicity tests using flow cytometry. *Small-scale Freshwater Toxicity Investigations, Volume 1*, eds. C. Blaise & J.F. Férard, Springer Publishers Dordrecht, The Netherlands, pp. 203-241, 2005.

[14] Moody, M. & Miller, J. *Lemna minor* growth inhibition test. *Small-scale Freshwater Toxicity Investigations, Volume 1*, eds. C. Blaise & J.F. Férard, Springer Publishers: Dordrecht, The Netherlands, pp. 271-298, 2005.

[15] Johnson, B.T. Microtox acute toxicity test. *Small-scale Freshwater Toxicity Investigations, Volume 1*, eds. C. Blaise & J.F. Férard, Springer Publishers: Dordrecht, The Netherlands pp. 69-105, 2005.

[16] Bitton, G., Ward, M. & Dagan, R. Determination of the heavy metal binding capacity (HMBC) of environmental samples. *Small-scale Freshwater Toxicity Investigations, Volume 2*, eds. C. Blaise and J.F. Férard, Springer Publishers: Dordrecht, The Netherlands, pp. 215-231. 2005.

[17] Gagné, F. Acute toxicity assessment of liquid samples with primary cultures of rainbow trout hepatocytes. *Small-scale Freshwater Toxicity Investigations, Volume 1*, eds. C. Blaise & J.F. Férard, Springer Publishers Dordrecht, The Netherlands, pp. 453-472, 2005.

[18] Dayeh, V.R., Schirmer, K, Lee, L.E. & Bols, N. Rainbow trout gill cell line microplate cytotoxicity test. *Small-scale Freshwater Toxicity Investigations, Volume 1*, eds. C. Blaise and J.F. Férard, Springer Publishers: Dordrecht, The Netherlands, pp. 473-503. 2005.

[19] Holdway, D. Hydra population reproduction toxicity test method. *Small-scale Freshwater Toxicity Investigations, Volume 1*, eds. C. Blaise & J.F. Férard, Springer Publishers Dordrecht, The Netherlands, pp. 395-411, 2005.

[20] Jonczyk, E. & Gilron, G. Acute and chronic toxicity testing with *Daphnia* sp. *Small-scale Freshwater Toxicity Investigations, Volume 1*, eds. C. Blaise & J.F. Férard, Springer Publishers: Dordrecht, The Netherlands, pp. 337-393, 2005.

[21] Nalecz-Jawecki, G. *Spirostomum ambiguum* acute toxicity test. *Small-scale Freshwater Toxicity Investigations, Volume 1*, eds. C. Blaise & J.F. Férard, Springer Publishers Dordrecht, The Netherlands, pp. 299-322. 2005.

[22] Blaise, C. & Ménard, L. A micro-algal solid phase test to assess the toxic potential of freshwater sediments. *Water Qual. Res. J. Canada* **33**, pp. 133-151, 1998.

[23] Dellamatrice, P. Monteiro, R., Blaise, C., Slabbert, J.L., Gagné F. & Alleau. S. Toxicity Assessment of reference and natural freshwater sediments with the Luminotox assay. *Environ. Toxicol.* **21**(4), 2006 (in press).

[24] Doe, K., Jackman, P., Scroggins, R., McLeay, D. & Wohlgeshaffen, G. Solid-phase test for sediment toxicity using the luminescent bacterium, *Vibrio fischeri. Small-scale Freshwater Toxicity Investigations, Volume 1*, eds. C. Blaise & J.F. Férard, Springer Publishers: Dordrecht, The Netherlands, pp. 107-136. 2005.

[25] Dutka, B., Teichgräber, K. & Lifshitz, R. A modified SOS-Chromotest procedure to test for genotoxicity and cytotoxicity in sediments directly without extraction. *Chemosphere* **31**, pp. 3273-3289, 1995.

[26] Péry, A., Mons, R. & Garric, J. *Chironomus riparius* solid-phase assay. *Small-scale Freshwater Toxicity Investigations, Volume 1*, eds. C. Blaise & J.F. Férard, Springer Publishers Dordrecht, The Netherlands, pp. 437-451. 2005.

[27] Borgmann, U., Norwood, W.P. & Nowierski, M., Amphipod (*Hyalella azteca*) solid-phase toxicity test using high water sediment ratios. *Small-scale Freshwater Toxicity Investigations, Volume 1*, eds. C. Blaise & J.F. Férard, Springer Publishers: Dordrecht, The Netherlands, pp. 413-436. 2005.

[28] Wells, P., K. Lee & C. Blaise (eds.). *Microscale testing in Aquatic Toxicology Advances, Techniques and Practice*, CRC Lewis Publishers: Boca Raton, Florida, pp. 679, 1998.

[29] Persoone, G. Janssen, C. & De Coen, W. (eds.). *New microbiotests for routine toxicity screening and biomonitoring*, Kluwer Academic/Plenum Publishers: New York, pp. 550, 2000.

[30] Blaise, C. & Férard, J.F. (eds). *Small-scale Freshwater Toxicity Investigations, Volume 1 (Toxicity test methods, 551 pages) and Volume 2 (Hazard assessment schemes, 422 pages)*, Springer Publishers, Dordrecht, The Netherlands, 2005.

Combining three *in vitro* assays for detecting early signs of UVB cytotoxicity in cultured human skin fibroblasts

C. Khalil
School of Safety Science,
Chemical Safety and Applied Toxicology Laboratories,
University of New South Wales, Sydney, Australia

Abstract

The aim of this study was to determine the most sensitive approach for detecting the early signs of UVB-induced cellular damage using human skin fibroblasts. UVB-induced cell damage was assessed immediately and 24 h post irradiation using 3 *in vitro* colorimetric assays: neutral red (NR); 5-(3-carboxy-methoxyphenyl)-2-(4.5-dimethylthiazolyl)-3-(4-sulphonyl) MTS; and (iii) LDH enzyme release. A good correlation was observed immediately post exposure between the MTS and NR in measuring damage levels, but was lost 24 h post exposure. This loss of correlation was the result of delayed expression of lysosomal damage and led to investigating cell membrane damage using LDH cell leakage assay. LDH levels observed immediately post UVB irradiation indicated significant LDH release at exposure doses of 2.2 and 2.8 J/cm^2, while LDH release reported 24 h post exposure was recorded for doses as low as 0.70 J/cm^2. The data reported in this paper indicated that cell viability and damage were significantly affected in a dose dependent manner as a result of exposure doses. The assays used displayed different sensitivities in detecting damage with the earliest signs of cellular UVB damage best-measured 24 h following exposure using the LDH assay. Furthermore UVB contributed to denaturing the cellular LDH released during irradiation. Therefore the use of LDH cytotoxicity based assays with UVB exposure must be considered with extreme care.
Keywords: cells, viability, MTS, neutral red, UVB, LDH, cytotoxicity.

WIT Transactions on Biomedicine and Health, Vol 10, © 2006 WIT Press
www.witpress.com, ISSN 1743-3525 (on-line)
doi:10.2495/ETOX060351

1 Introduction

The current degradation of the ozone layer by CFC's and other man made chemicals triggered major concerns about the harmful effects of environmental exposure to solar radiation and their impacts on climate, environmental processes and life on earth (Kledsen and Scheutz [1]). Among the spectrum of UV radiation emitted by the sun UVB is thought to be the major cause of skin damage and long-term health risks. The effects of UVB (290-320 nm) radiation on human skin can be seen as acute (sunburn and inflammation) (Grassen and van Loveren [2]) or chronic (photoaging and skin cancer) ([2]; Tebbe et al. [3]; Deliconstantinos et al. [4]). The acute effects of UVB are thought to be mediated by cytokine production leading to cell death (early apoptosis) (Petit-Frere et al. [5]; Norris et al. [6]; Gniadecki et al. [7]; Corsini et al. [8]).

The human skin is composed of a variety of cells including skin fibroblasts. These cells by virtue of their location, numbers in the dermis, and ease of growth in culture could be used as possible indicators of cellular damage caused by UVB irradiation.

This study explored a number of approaches for detecting the early signs of UVB damage through the use of an artificial UVB source coupled with primary cell cultures of human skin fibroblast to mimic exposure pattern observed *in vivo*. This approach was adopted to investigate UVB exposure induced damage (mitochondrial, lysosomal and membrane damage) and to understand the different level of damage expression within cell organelles.

This non-invasive *in vitro* study and the data generated represented a suitable alternative for using whole animal models for assessment of toxicity of UVB exposure and assisted in reducing uncertainties in data extrapolation from animals to humans due to the use of human skin fibroblasts cells. Furthermore, the techniques used are rapid and reproducible as they generated toxicity profiles within hours of running the assays (Bakand et al. [9]).

The *in vitro* assays selected are modified cytotoxicity tests (MTS, NR and LDH) developed by our lab and used extensively in measuring different biological end points ([9]; Lestari et al. [10]).

The main purpose of this study was to identify the most sensitive assay for measuring early signs of UVB exposure cytotoxicity immediately and 24 h post UVB exposure in addition to understanding the limitations of the *in vitro* assays used.

2 Materials and methods

2.1 Cell cultures

Primary fibroblast cell cultures were derived from human skin biopsies [Children's Hospital Westmead (Australia)] and maintained in short term cell culture. Cells were subcultured as adherent cells in 75 cm^2 tissue culture flasks with 0.2 μm vented seals (Falcon). The culture media consisted of colour free

Dulbeco's modified eagle medium (DMEM): RPMI 1640 (1:1) purchased from Sigma Chemicals, supplemented with 5% foetal calf serum (Trace Bioscience), 3% Sigma antibiotics [penicillin (100 U/ml), streptomycin (0.1 mg/ml) and L-glutamine (2 mM)]. The cell lines were cultured at 37°C at sub-confluence in a humidified incubator set to a mixture of 5% CO_2/95% air. Cell viability was over 95% as measured by tryptan blue dye exclusion.

Confluent cells in log phase of growth were released from the bottom of the culture flask using Trypsin EDTA (Gibco, USA), and then washed three times with cell culture medium. This was followed by a cell count/cell viability assessment before the cells were seeded on 24 well plates and incubated overnight to allow cells to reattach to the bottom of plates before UVB irradiation.

The seeding density of cells to the 24 well plates was previously determined and cells were seeded at a density of 500,000 cells/ml based on the linearity range studies (cell concentration versus absorbance) previously conducted [9].

2.2 UVB irradiation

For UVB irradiation medium was replaced with Hank's Balanced Salt solution (HBBS Gibco, USA), the coverlids of the 24 well microtiter plate removed and cells exposed to UVB irradiation (3.92×10^{-4} W/cm^2) from a 6 lamps (FS40212) supplied by Wayne Electronics (Somersby, Australia). The output of the lamps was measured by an IL-1700 research radiometer (International Light, Newbury Port, MA).

Cells were irradiated with UVB doses ranging from 0.0078 to 5.6 J/cm^2. The procedure followed in the assay immediately post exposure consisted of removal of exposure medium (HBSS) following UV exposure and subjecting the cells to the cytotoxicity assays. The 24 hour exposure protocol consisted of replacing the UVB exposure medium (HBSS) post exposure with culture medium, incubating the cells for 24 h in a CO_2 incubator then subjecting the cells to the cytotoxicity assays.

2.3 Cytotoxicity assays

The MTS assay (Promega, USA) was selected for measuring the number of active cells in the culture (based on the lactated dehydrogenase activity in the mitochondria). The MTS assay measuring the conversion of a soluble tetrazolium salt to a formazan product by viable cells. The assay consisted of an MTS solution prepared by mixing a solution of MTS (42 mg MTS powder in 21 ml of DPBS pH 6.0-6.5) with a PMS solution (0.92 mg/ml PMS in DPBS) to the cells to be tested in a ratio of 1:5. The MTS was then incubated with the cells for a period of 2 h at 37°C in the dark. After 2 h, the cellular supernatant was removed for measurement. The amount of reduced Formazan was assessed by measuring the optical density at 492 nm using a Labsystem Multiskan MS plate reader. Data was plotted as a dose response curve exposure versus absorbance reading.

The NR assay (Sigma, USA) is a cell survival assessment technique that measures the uptake of the neutral red dye by viable cells. The procedure consisted in removal of exposure medium following UV exposure (under sterile conditions), this was followed by the addition of 0.33% neutral red solution (Sigma Chemicals, USA) in an amount equal to 10% of the initial culture media volume, followed by an incubation period of 2 h, then cells are fixed (NR Assay Fixative (1% (wt/v) $CaCl_2$: 0.5% (v/v) Formalin)) before addition of an assay solubilization solution (1% (v/v) Acetic acid: 50% (v/v) Ethanol). The amount of dye incorporated in the cells measured using the Multiskan plate reader using a 540 nm reading wavelength with background absorbance reading measured at 690 nm and subtracted from the 540 nm measurement.

The LDH assay (Promega Corporation) measured the amount of lactate dehydrogenase released by the cells upon UV insult. Released LDH in cultures was measured with a 30 minutes coupled enzymatic assay which resulted in the conversion of the tetrazolium salt into a red colored formazan product. The protocol supplied by Promega (Technical Bulletin 163) was followed without any alterations. 50 μl of exposure media was transferred to a mutliwell plate. 12 ml of the assay buffer was mixed with the substrate mix and 50 μl of this solution added to each well of the plate. Cells were incubated for 30 minutes in the dark. This incubation period was followed by the addition of 50 μl of the stop solution (0.1M HCL) to each well. The plates were spectrophotometrically processed within 1 h at 492 nm.

2.4 Statistical analysis

The dose response curves reported were plotted from the experimental data and the background absorbance subtracted from the presented graphical pots. All the data reported was expressed as the mean ± SD of 4 replicated wells. Statistical procedures and graphical analysis were performed using Microsoft Excel software.

3 Results

3.1 Cytotoxicity assessment of UVB-exposed on human skin fibroblasts with the MTS and NR assays immediately post exposure

The experimental investigations used skin fibroblasts cultures at a density of 50,000 cells/100 μl exposed to UVB doses (0 - 5.6 J/cm^2) with damage levels assessed by two cytotoxicity assays MTS and NR are summarized in Figure 1.

Correlation analysis between the two cytotoxicity assays (Figure 2) showed that there was a direct agreement between the MTS and NR assays in measuring damage extent (R^2 = 0.70) immediately post exposure although they were measuring damage levels at different cellular compartments.

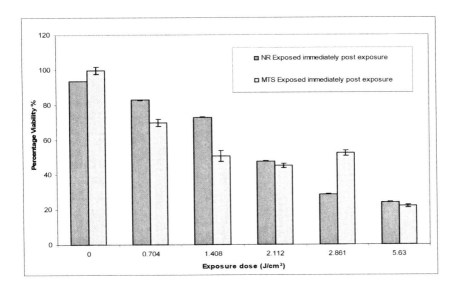

Figure 1:   Relationship between UVB dose and cell viability as measured by the MTS and NR assays immediately post exposure. Values are mean ± SD of at least 4 experimental replicates.

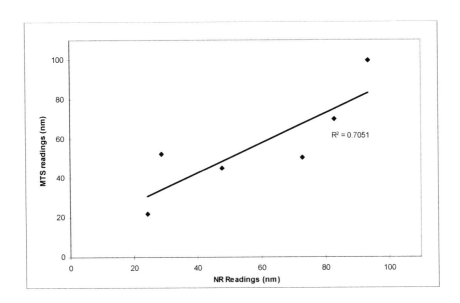

Figure 2:   Correlation between MTS and neutral red assay in evaluation of UVB-induced toxicity following skin fibroblasts exposure.

3.2 Cytotoxicity assessment of UVB-exposed on human skin fibroblasts with the MTS and NR assays 24 h post exposure

Primary skin fibroblasts (46 x 10^3 cells/100 μl) were exposed to UVB and cultured for 24 h post exposure before cytotoxicity assays. The exposure of skin fibroblasts cells to UVB doses and their impacts on cellular viability 24 h post-exposure are outlined in Figure 3.

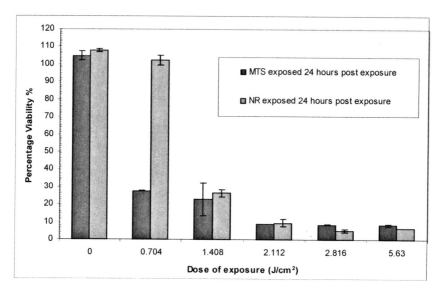

Figure 3: Relationship between UVB dose and cell viability as measured by the MTS and NR assays 24 h post exposure. Cell media was removed and replaced with 400 μL HBSS for UVB exposure.

Cytotoxicity results using the MTS assay illustrated in Figure 3 showed the sensitivity of the MTS test in detecting the effects of UVB radiation on cellular processes. By contrast to the MTS assay, the NR assay did not exhibit the same level of sensitivity in detecting early cell membrane damage under the same exposure conditions especially at low exposure doses. This discrepancy at exposure dose of 0.704 J/cm^2 between the 2 assays contributed to the loss of correlation 24 h post exposure. Therefore, no correlation could be established between the MTS and neutral red cytotoxicity test for UVB exposure of skin fibroblasts 24 h post exposure.

3.3 Measurement of LDH activity in UVB exposed skin fibroblasts

The discrepancy in measuring damage levels observed in Figure 3 lead to further experimentation to investigate the extent of cell membrane damage upon UVB exposure. These experiments consisted of exposing skin fibroblasts cells to variable UVB doses and determining cell membrane integrity immediately and

24 h post-exposure by assessing LDH leakage from damaged cells. In the experiments LDH release assay was performed on both the HBSS from exposed cells and the media of incubation (the background readings were considered and subtracted from all values). The data was reported in Figure 4.

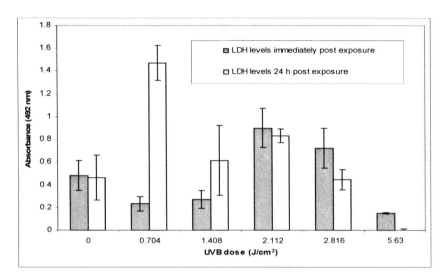

Figure 4:   Skin fibroblasts LDH release immediately and 24h post UVB dose exposure. LDH levels in HBSS (exposure media) following UVB irradiation of skin fibroblasts.

The readings reported at 0.704 J/cm^2 UVB dose, especially the lower than expected result for the immediate measurement of LDH release led to exposing cell fibroblasts cultures to low doses and trying to further clarify this result discrepancy. Cells were exposed UVB doses ranging from 7.8 to 250 mJ/cm^2 and release of LDH immediately and 24 h post exposure compared in Figure 5.

4   Discussion

Ozone depletion is of major concern to the future of life on the planet due to its important role in filtering ultraviolet radiation (Kalbin et al. [11]). This paper attempts to investigate cellular damage triggered by a range of UVB radiation *in vitro* using human skin fibroblasts as the target cells. The assessment of cellular damage was achieved by 3 colorimetric assays selected for their reliability and reproducibility [9, 10].

The cytotoxicity of UVB exposure on human skin fibroblast was studied using the MTS, NT and LDH assays. This study compared and correlated the results of those assays immediately and 24 h post exposure.

The results in Figure 1 showed a dose response relationship between exposure and cellular mitochondrial dehydrogenase activity immediately post UVB exposure as measured by the MTS assay. The decrease in the mitochondrial activity was dose dependent and directly associated with increased UVB dose. The exposure of human skin fibroblast to a dose of 2.11 J/cm^2 of UVB reduced the mitochondrial activity to 50%, while at 5.6 J/cm^2 of UVB radiation activity was reduced to 20%. The relationship between UVB exposure and cellular viability immediately post exposure was further investigated using the NR assay. The damage caused by UVB was assessed by determining the reduction in NR dye uptake by the lysosomes as cells with damaged cellular membranes cannot retain the dye. NR decreased uptake by exposed cells could be directly related to UVB exposure (Figure 1). The data in Figure 1 showed a decrease in the uptake of NR dye with increased UVB dose. The exposure of the cellular cultures to UVB dose of 2.1 J/cm^2 decreased the viability to less than 50%. The existence of a potential correlation between membrane damage and reduced mitochondrial activity was investigated in Figure 2. The R^2 value (0.70) for this correlation showed a good agreement between the two assays in measuring damage at different levels (mitochondrial and lysosomal damage) immediately post exposure within the exposed cells.

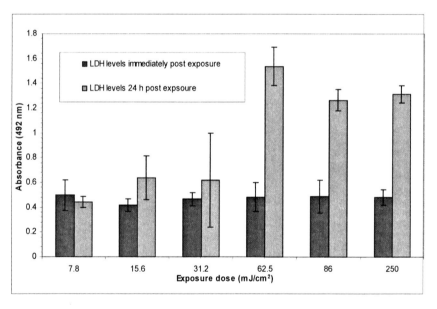

Figure 5:   Upregulation of LDH extracellular release upon low dose exposures immediately and 24 h post exposure.

The cytotoxicity of skin fibroblast UVB exposure was also investigated 24 h post exposure and reported in Figure 3. Comparison of the observed damage 24 h post exposure to a UVB dose of 0.70 J/cm^2 clearly indicated the higher sensitivity of the MTS assay (compared to NR assay) in measuring early damage. The NR data showed no significant difference between the viability of controls and exposed cells (0.7 J/cm^2).

The correlation between the MTS and NR assays absorbance readings reported immediately post exposure was absent 24 h post exposure. This loss of correlation could be mainly attributed to the greater sensitivity of the MTS assay (by comparison to NR) at a dose of 0.70 J/cm^2. No significant difference (p>0.01) could be reported at a dose of 0.70 J/cm^2 between exposed and control cells as measured by NR assay. This indicated a delayed expression of lysosomal damage upon UVB exposure by skin fibroblasts 24 h post exposure, although published research in this regard (Debacq-Chainiaux et al. [12]) established that doses higher than 0.5 J/cm^2 were needed to trigger a significant reduction in fibroblast cell viability.

The delayed expression of lysosomal damage reported 24 h post UVB exposure (0.7 J/cm^2) led to further examination of lysosomal toxicity by determining whether the low lysosomal damage observed in the NR data could be compared to overall cell membrane UVB exposure damage as measured by the LDH leakage assay (Figure 4).

UVB exposure (0.7 to 5.6 J/cm^2) of skin fibroblasts triggered significant LDH release immediately and 24 h post exposures. UVB exposure triggered an LDH peak observed at 2.1 J/cm^2 immediately post exposure. This was followed by a decrease in the LDH levels to below control cell levels for a dose of 5.6 J/cm^2 (Figure 4). The examination of LDH release for doses lower than 2.11 J/cm^2 indicated a significant reduction in the amount of LDH detected by comparison to the control levels. The extracellular LDH release (24 h post exposure) was also reported in Figure 4. UVB was found to cause a peak in LDH activity at a UVB dose of 0.7 J/cm^2 24 h post exposure. This peak was followed by a decrease in the extracellular LDH detected.

These deviations in LDH levels from the expected results (significant upregulation of LDH leakage for doses of 0.70 and 1.4 J/cm^2 immediately post UVB exposure) led to studying lower UVB doses (7.8-250 mJ/cm^2) to establish whether LDH release was affected. It was observed that there was no releases of LDH immediately post exposure while a significant dose dependent LDH release was observed 24 h post exposure. The LDH extracellular release 24 h post exposure peaked at UVB dose of 62.5 mJ/cm^2 before reaching a plateau due to reaching the upper detection limit of the Multiskan reader used. Furthermore, no decrease in LDH release to levels below control levels was recorded. This increase in LDH release upon UVB exposure agrees with published literature on UVB exposed murine fibroblast (Kimura et al. [14]) where it was reported that LDH activity increased in cells treated with UVB up to 72 h post exposure.

This increase in LDH release was followed by a decrease for doses higher than 2.5 J/cm^2 (Figure 4). The observed decrease in LDH release was possibly the result of LDH deactivation by UVB rays. The LDH levels detected 24 h post UVB exposure to doses ranging from 0.7 to 5.6 J/cm^2 appeared to peak at a dose of 0.7 J/cm^2 with a gradual decrease. This was predictable since the cellular viability was significantly reduced 24 h post UVB exposure as reported by the MTS and NR data (Figure 3).

Based on these results it was hypothesized that the observed reduction of LDH levels observed for doses ranging from 1.7-0.4 J/cm^2 could be the result of

LDH de-activation by UVB rays. This hypothesis was validated by subjecting a serial LDH dilutions (0.066 u/ml, 0.033 u/ml, 0.0165...) to variable UVB doses (0 - 5.6 J/cm²). It was observed that high UVB doses combined with low LDH concentration significantly reduced LDH activity as shown in Figure 6. The data reported in Figure 6 agreed with other research (Artiukhova et al. [13]) that reported similar decrease in LDH activity in UV exposed LDH.

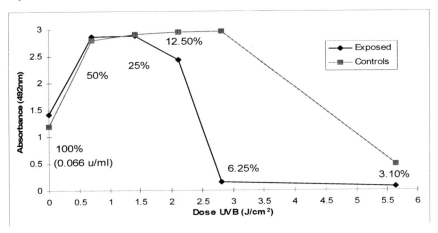

Figure 6: Effect of UVB doses on the enzymatic activity of serial dilutions of an LDH stock solution (0.066u/ml) subjected to UVB to determine its effects on LDH activity. Low levels of LDH (12.5 and 6.25%) were affected by UVB irradiation higher than 2.5 J/cm² with lower enzymatic activity recorded.

It is concluded from the data presented in this paper that the three *in vitro* methods used were suitable for testing the impact of UVB on cellular functions. We found that the three assays complement each other as they were measuring different end-points. The detection of the early damage levels triggered by UV exposure was best achieved 24 h post exposure using the LDH assay. The MTS and NR were also valuable tools for damage assessment immediately and 24 h post although they displayed lower sensitivity than the LDH assay.

References

[1]   Kledsen, P. & Scheutz, C., Short and long term release of Fluorocarbons from disposal of polyurethane foam waste. *Environmental Science Technology.* **37**, pp 5071-5079, 2003.

[2]   Garssen, J., van Loveren, H., Effects of ultraviolet exposure on the immune system (review), *Critical Reviews in Immunology* **21 (4)**, pp 359-397, 2001.

[3]   Tebbe, B., Wu, S., Geilen, C.C., Eberle, J., Kodelja, V., Orfanos, C.E., L-ascorbic acid inhibits UVA-induced lipid peroxidation and secretion of

IL-1α and IL-6 in cultured human keratinocytes in vitro, *Journal of Investigative Dermatology* **108**, pp 302-306, 1997.

[4]   Deliconstantinos, G., Villiotu, V., Stavrides, J.C. Nitric oxide and peroxynitrite released by ultraviolet B-irradiated human endothelial cells are possibly involved in skin erythema and inflammation. *Experimental Physiology* **81**, pp 1021-1033, 1996.

[5]   Petit-Frere, C., Capulas, E., Lyon D.S., Norbury, C.J., Lowe, J.E., Clingen, P.H., Riballo, E., Green, M.H., Arlett, C.F., Apoptosis and cytokine release induced by ionizing or ultraviolet B radiation in primary and immortalized human keratinocytes. *Carcinogenesis.* **21(6)**, pp 1087-1095, 2000.

[6]   Norris, D.A., Whang, K., David-Bajar, K., Bennion, S.D., The influence of ultraviolet light on immunological cytotoxicity in the skin. *Photochemistry Photobiology.* **65(4)**, pp 636-646, 1997.

[7]   Gniadecki, R., Hansen, M., Wulf, H.C., Two pathways for induction of Apoptosis by ultraviolet radiation in cultured human keratinocytes. *Journal of Investigative Dermatology* **109**, pp 163-169, 1997.

[8]   Corsini, E., Bruccoleri, A., Marinovich, M., and Galli, M.L., In Vitro mechanism(s) of ultraviolet induced tumor necrosis factor-α release in human keratinocytes cell line. *Photodermatology Photoimmunology Photomedicine.* **11**, pp 112-118, 1995.

[9]   Bakand, S., Winder, C., Khalil, C., and Hayes A., A novel in vitro exposure technique for toxicity of selected volatile organic compounds. *Journal of Environmental Monioring.* **7**, pp 1-5, 2005.

[10]   Lestari, F., Hayes, A., Green, A.R., and Markovic B., In vitro cytotoxicity of selected chemicals commonly produced during fire combustion using human cell lines *Toxicology in Vitro* **19**, pp 653-663, 2005.

[11]   Kalbin, G., Shaoshan, L., Olsman, H., Petterson, M., Engwall, M., Strid, A., Effects of UVB in biological systems: equipments for wavelength dependence determination. *Journal of Biochemical and Biophysical Methods* **65**, pp 1-12, 2005.

[12]   Debacq-Chainiaux, F., Borlon, C., Pascal, T., Royer, V., Eliaers, F., Ninane, N., Carrard, G., Friguet, B., de Longueville, F., Boffe, S., Remacle, J., Toussaint, O., Repeated exposure of human skin fibroblasts to UVB at subcytotoxic level triggers premature senescence through the TGF-beta1 signaling pathway. *Journal of Cell Science.* **118(4)**, pp 743-58, 2005.

[13]   Artiukhova, V.G., Nakvasina, M.A., Lysenko, IuA., Active forms of oxygen and the degree of UV modification of the structural and functional properties of lactate dehydrogenase. *Radiatsionnaia Biologiia, Radioecologiia.* **37(3)**, pp 453-460, 1997.

[14]   Kimura, H., Minakami, H., Otsuki, K., Shoji, A.: Cu-Zn superoxide dismutase inhibits lactate dehydrogenase release and protects against cell death in murine fibroblasts pretreated with ultraviolet radiation. *Cell Biology International.* **24(7)**, pp 459-65, 2000.

Author Index

Disposal of Hazardous Waste in Underground Mines

Edited by: V. POPOV, Wessex Institute of Technology, UK, R. PUSCH, Geodevelopment AB, Sweden

This book contains the results of a three-year research programme by a joint team of experts from four different EU countries. The main focus of this research was on investigating the possibility of using abandoned underground mines for the disposal of hazardous chemical waste with negligible pollution of the environment.

The contributors address many aspects that are common to underground disposal of nuclear waste, such as: the properties and behaviour of waste-isolating clay materials and practical ways of preparing and applying them, development of tools/software to assess the stability, performance, transport of contaminants inside and outside the repository, and risks associated with different repository concepts considering the long-term safety of the biosphere. Information is also included on the selection of site location, design and construction of repositories, predicting degrees of contamination of groundwater in the surroundings, estimation of isolating capacity of reference repositories, cost estimation of this approach in comparison with some other approaches, and many other relevant issues.

Invaluable to researchers and engineers working in the field of hazardous (chemical) waste disposal, this title will also significantly aid experts dealing with nuclear waste.

Series: The Sustainable World Vol 11
ISBN: 1-85312-750-7 2006 288pp
£95.00/US$170.00/€142.50

Environmental Exposure and Health

Edited by: M. M. ARAL, Georgia Institute of Technology, USA, C. A. BREBBIA, Wessex Institute of Technology, UK, M. L. MASLIA, ATSDR/CDC, USA, T. SINKS, NCEH, USA

Current environmental management policies aim to achieve sustainability while improving the health, safety and prosperity of the population. This is an interdisciplinary activity that requires close cooperation between different sciences.

Featuring contributions from health specialists, social and physical scientists and engineers this volume evaluates current issues in exposure and epidemiology and highlights future directions and needs.

Originally presented at the First International Conference on Environmental Exposure and Health, the papers included cover areas such as: METHODOLOGICAL TOPICS - Methods Of Linking Epidemiology, Exposure and Health Risk; Multipathway Exposure Analysis and Epidemiology; Statistical and Numerical Methods. SITE RELATED TOPICS - Work Place and Industrial Exposure; Soil Dust and Particulate Exposure; Water Distribution Systems, Exposure and Epidemiology; Air Pollution Exposure and Epidemiology. DATA COLLECTION TOPICS - Use of Remote Sensing and GIS; Data Mining and Applications in Epidemiology. SPECIAL TOPICS - Exposure Specific to the Developing World; Epidemiology of Mixed Chemical and Microbial Exposure; Effects of Rapid Transportation in Epidemiology; Interaction of Social and Environmental Issues and Health Risk.

WIT Transactions on Ecology and the Environment, Volume 85
ISBN: 1-84564-029-2 2005 528pp
£185.00/US$325.00/€277.50

Environmental Health Risk III

Editors: C.A. BREBBIA, Wessex Institute of Technology, UK, V. POPOV, Wessex Institute of Technology and D. FAYZIEVA, Academy of Sciences, Uzbekistan

As problems caused by environmental exploitation increase, related health issues are also becoming a major worldwide concern.

Containing papers presented at the Third International Conference on the Impact of Environmental Factors on Health, this volume includes contributions from a variety of different disciplines and countries. The papers featured are divided into the following sections: Water Quality Issues; Air Pollution; Radiation Fields; Accident and Man-Made Risks; Risk Analysis; Emergency Response; Food Contamination; Electromagnetic Fields; Noise Pollution; Housing and Health; Occupational Health; Chemical Risk Assessment; Remediation; Social and Economic Issues; Education; and Ecotoxicology Studies.

WIT Transactions on Biomedicine and Health, Volume 9
**ISBN: 1-84564-026-8 2005 528pp
£185.00/US$325.00/€277.50**

Environmental Health Risk II

Editors: C.A. BREBBIA, Wessex Institute of Technology, UK and D. FAYZIEVA, Academy of Sciences, Uzbekistan

The proceedings of the second international conference on this topic, this book contains papers under headings such as: Water Quality Issues; Air Pollution; Accident and Man-Made Risks; Risk Analysis; Analysis of Urban Road Transportation Systems in Emergency Conditions.

WIT Transactions on Biomedicine and Health, Volume 7
**ISBN: 1-85312-983-6 2003 260pp
£89.00/US$142.00/€133.50**

Environmental Health in Central Asia

The Present and Future

Editor: D. FAYZIEVA, Academy of Sciences, Uzbekistan

This book provides information on how environmental conditions in Central Asia have been affected by anthropogenic activity. It also reviews research carried out during the last decades on the impact of the environment on the health of the region's people.

Partial Contents: Air Quality and Population Health in Central Asia; Hydrosphere and Health of Population in the Aral Sea Basin; Influence of Environmental Factors on Development of Non-Communicable Diseases; Environment and Infectious Diseases; Environment and Children's Health in Central Asia.

Series: Advances in Ecological Sciences, Vol 17
**ISBN: 1-85312-945-3 2004 284pp
£84.00/US$134.00/€126.00**